Principles of Brain Functioning

Springer
*Berlin
Heidelberg
New York
Barcelona
Budapest
Hong Kong
London
Milano
Paris
Santa Clara
Singapur
Tokyo*

Springer Series in Synergetics

Editor: Hermann Haken

An ever increasing number of scientific disciplines deal with complex systems. These are systems that are composed of many parts which interact with one another in a more or less complicated manner. One of the most striking features of many such systems is their ability to spontaneously form spatial or temporal structures. A great variety of these structures are found, in both the inanimate and the living world. In the inanimate world of physics and chemistry, examples include the growth of crystals, coherent oscillations of laser light, and the spiral structures formed in fluids and chemical reactions. In biology we encounter the growth of plants and animals (morphogenesis) and the evolution of species. In medicine we observe, for instance, the electromagnetic activity of the brain with its pronounced spatio-temporal structures. Psychology deals with characteristic features of human behavior ranging from simple pattern recognition tasks to complex patterns of social behavior. Examples from sociology include the formation of public opinion and cooperation or competition between social groups.

In recent decades, it has become increasingly evident that all these seemingly quite different kinds of structure formation have a number of important features in common. The task of studying analogies as well as differences between structure formation in these different fields has proved to be an ambitious but highly rewarding endeavor. The Springer Series in Synergetics provides a forum for interdisciplinary research and discussions on this fascinating new scientific challenge. It deals with both experimental and theoretical aspects. The scientific community and the interested layman are becoming ever more conscious of concepts such as self-organization, instabilities, deterministic chaos, nonlinearity, dynamical systems, stochastic processes, and complexity. All of these concepts are facets of a field that tackles complex systems, namely synergetics. Students, research workers, university teachers, and interested laymen can find the details and latest developments in the Springer Series in Synergetics, which publishes textbooks, monographs and, occasionally, proceedings. As witnessed by the previously published volumes, this series has always been at the forefront of modern research in the above mentioned fields. It includes textbooks on all aspects of this rapidly growing field, books which provide a sound basis for the study of complex systems.

A selection of volumes in the Springer Series in Synergetics:

Hermann Haken

Principles of Brain Functioning

A Synergetic Approach to Brain Activity, Behavior and Cognition

With 220 Figures

 Springer

Professor Dr. Dr. h.c.mult. Hermann Haken

Institut für Theoretische Physik und Synergetik der Universität Stuttgart
D-70550 Stuttgart, Germany and
Center for Complex Systems, Florida Atlantic University
Boca Raton, FL 33431, USA

Series Editor:

Professor Dr. Dr. h.c.mult. Hermann Haken

Institut für Theoretische Physik und Synergetik der Universität Stuttgart
D-70550 Stuttgart, Germany and
Center for Complex Systems, Florida Atlantic University
Boca Raton, FL 33431, USA

Library of Congress Cataloging-in-Publication Data
Haken, H.
    Principles of brain functioning : a synergetic approach to brain
activity, behavior, and cognition / Hermann Haken.
        p.   cm. -- (Springer series in synergetics ; 67)
    Includes bibliographical references.
    ISBN-13: 978-3-642-79572-5   e-ISBN-13: 978-3-642-79570-1
    DOI: 10.1007/ 978-3-642-79570-1
    1. Brain--Mathematical models. 2. System theory.   I. Title.
II. Series: Springer series in synergetics ; v. 67.
    [DNLM: 1. Brain--physiology. 2. Models, Theoretical.
3. Movement--physiology. 4. Thinking--physiology. 5. Visual
Perception--physiology. 6. Electroencephalography.   WL 300 H155p
1996]
QP376.H25   1996
612.8'2--dc20   95-41276

© Springer-Verlag Berlin Heidelberg 1996
Softcover reprint of the hardcover 1st edition 1996
The use of general descriptive names, registered names, trademarks, etc. in this publication does not imply,
even in the absence of a specific statement, that such names are exempt from the relevant protective laws and
regulations and therefore free for general use.

Typesetting: Camera ready copy from the author using a Springer TEX macro package
SPIN 10493001   55/3144 - 5 4 3 2 1 0 - Printed on acid-free paper

To my friends
Waltraut and Achim Finke
with deep gratitude

Preface

It is increasingly being recognized that the experimental and theoretical study of the complex system *brain* requires the cooperation of many disciplines, including biology, medicine, physics, chemistry, mathematics, computer science, linguistics, and others. In this way brain research has become a truly interdisciplinary endeavor. Indeed, the most important progress is quite often made when different disciplines cooperate. Thus it becomes necessary for scientists to look across the fence surrounding their disciplines. The present book is written precisely in this spirit. It addresses graduate students, professors and scientists in a variety of fields, such as biology, medicine and physics. Beyond its mathematical representation the book gives ample space to verbal and pictorial descriptions of the main and, as I believe, fundamental new insights, so that it will be of interest to a general readership, too.

I use this opportunity to thank my former students, some of whom are my present co-workers, for their cooperation over many years. Among them I wish to mention in particular M. Bestehorn, L. Borland, H. Bunz, A. Daffertshofer, T. Ditzinger, E. Fischer, A. Fuchs, R. Haas, R. Hönlinger, V. Jirsa, M. Neufeld, M. Ossig, D. Reimann, M. Schanz, G. Schöner, P. Tass, C. Uhl. My particular thanks go to R. Friedrich and A. Wunderlin for their constant help in many respects. Stimulating discussions with a number of colleagues from a variety of fields are also highly appreciated. At the risk of omitting some important names, I wish to mention P. Kruse and M. Stadler on the connections between Gestalt theory and synergetics, E. Başar, T. Bullock, W. Freeman, D. Lehmann, H. Petsche, and G. Pfurtscheller on various aspects of EEG measurements and their interpretation, H. Körndle and K.H. Leist for analysis of movements, especially on the pedalo device. My thanks go further to P. Beek and W. Beek, O. Meijer, and L. Peper for discussions on movement coordination and the role of order parameters, P. Vanger for his cooperation on the recognition of facial expressions, and to J. Portugali on his concept of interrelational networks. My special thanks go to my friend and colleague Scott Kelso, who, in particular through his ingeniously devised and accurate experiments, decisively contributed to the proof that the concepts of synergetics play a fundamental role in movement coordination and related fields. I think it is fair to say that a paradigm shift has occurred here;

namely, instead of studying stable states, we are now studying transitions close to instability points, transitions that occur in our brain.

This book could have never been completed without the tireless help of my secretary Mrs. I. Möller, who typed several versions of this manuscript including all its formulas and performed the miracle of combining great speed with utmost accuracy. In addition, Mrs. Möller, together with my co-worker R. Haas, brought the manuscript into its final form, ready for printing. A. Daffertshofer and further members of my institute carefully read the manuscript. He, as well as R. Haas, V. Jirsa, D. Reimann and C. Uhl made valuable suggestions. The figures were prepared by A. Daffertshofer and M. Neufeld. I thank all of them for their great help.

Last but not least I thank the members of the Springer-Verlag for the traditionally excellent cooperation, in particular Prof. W. Beiglböck, Dr. A. Lahee, and Ms. Petra Treiber.

Stuttgart, October 1995   *Hermann Haken*

Table of Contents

---

* Marks chapters or sections that are mathematically somewhat more involved

Prologue

The human brain is both the most complex system and the most enigmatic organ we know of in our world. In view of its enormous complexity, a fundamental problem is this: What are the relevant questions we can ask about brain activity? These questions are closely connected with the methods of analysis we have at hand. The most prominent method is that of decomposing a complex system. In the case of the brain we then find individual parts, such as neurons and glia cells. When we decompose these cells further, we find, for instance, membranes, receptors, organelles and, at a still more fundamental level, biomolecules. Within these cells and among them, numerous chemical, electrical and electro-chemical processes are going on. But here the puzzle begins: In order to steer movements, to recognize patterns, or to make decisions – to mention but a few examples – myriads of neurons must cooperate in a highly regulated, well-ordered fashion. But to put this question in anthropomorphic terms, who or what steers the behavior of the neurons? Everybody who has thought about this problem will recognize that here we are touching upon the mystery of the mind-body problem. The answer that I shall expostulate in this book is based on synergetics, an interdisciplinary field of research that I initiated in 1969 and that has found numerous applications in physics, chemistry, biology, computer science, and other fields. To substantiate my answer, I shall proceed along two lines: On the one hand, I shall invoke general principles and concepts of synergetics. On the other hand, and perhaps still more importantly, I shall present concrete models that allow us to represent experimental results in great detail. But again, these models are formulated from a unifying point of view – that of synergetics.

Thus, in the end, the main results of this book are its detailed application-oriented models. I hope that the book will help to pave the way to a deeper understanding of the mechanisms of brain activity and thus prove to be of practical importance (as some of its results already have been in the past).

I came across possible links between concepts of synergetics and human perception at a rather early stage of synergetics. In my book on synergetics (1977) I interpreted the perception of ambiguous figures, such as the one on the cover of this book, as the outcome of bistable states of order parameters in synergetic, i.e. self-organizing systems. This line of thought will be represented in Chap. 16 of the present book. Another line of thought

(*Haken* 1979) that proved very fruitful is the idea that pattern recognition by humans, or animals, or machines is a specific kind of pattern formation again in self-organizing systems. This concept will be followed up again in Chap. 16. These and similar ideas led me to the general proposal of treating the brain as a synergetic system that produces its macroscopic features by self-organization (*Haken* 1983). As a specific tool to study self-organization, I suggested looking at qualitative changes on macroscopic scales. As an explicit example, I mentioned the change of gaits of horses, which show, on the one hand, well-definded behavioral patterns and, on the other, pronounced transitions between these patterns.

Another point of view that I promoted in the above article was that the brain operates close to instabilities, where again and again only few modes of activity show up. According to the concepts of synergetics, these modes are governed by order parameters that act on the individual parts, for instance in neurons, via the slaving principle. (These concepts will be explained in this book.) It was very fortunate for me that later in 1983, *Scott Kelso*, a neurophysiologist, visited me and told me about his finger movement experiments (*Kelso* 1981, 1984). He demonstrated that there are specific transitions between two kinds of finger movements that occur in a well-defined, but entirely involuntary fashion. This experiment presented me with the challenge of modeling it in terms of synergetics. The model (*Haken, Kelso, Bunz* 1985) could not only successfully represent a number of experimental findings, but, jointly with general results of synergetics, predict some further ones, such as the effects of hysteresis, critical fluctuations, and critical slowing down (*Schöner, Haken, Kelso* 1986). This model turned out to be a safe basis for the explanation of a number of further experiments and served as a starting point for more elaborate models on movement coordination. I mention here, as a few important examples, the experiments by *Schmidt, Carello, Turvey* (1990) on the coordination of leg movement even between different persons, the experiments by *Kelso* and co-workers, and those by *Beek, Peper, van Wieringen* (1992), and others on finger tapping. I shall come back to these problems in the course of this book. A central theme in synergetics is the coordination of the actions of individual parts by means of order parameters and the slaving principle. This will lead me to the analysis of pedalo experiments performed by *Körndle* (1992). It turns out that, during the learning process, the complex movement pattern is, eventually, governed by a single order parameter, which obeys a simple standard nonlinear equation.

Movements patterns, or more generally speaking, behavioral patterns, are macroscopic, indirect manifestations of brain activity. A more direct, again macroscopic, manifestation are the electric and magnetic fields produced by the brain. To substantiate my thesis that the brain operates close to instability points, I shall represent the Friedrich–Uhl analysis of EEGs in the case of *petit mal epilepsy* (1995) and the Fuchs–Kelso–Haken (1992) analysis of MEG experiments performed by *Kelso* et al. (1992).

I shall then proceed to a far more complicated phenomenon, namely vision. I shall present a model of vision (*Haken* 1987) that can be extended to the perception of ambiguous figures (*Ditzinger, Haken* 1989, 1990). This extended model allows contact to be made with a number of detailed experiments. Furthermore, this model serves as a basis for treating stereopsis. My model of vision is based on the idea that pattern recognition is nothing other than pattern formation and I shall elucidate this analogy more closely.

A further point of view will be that a study of vision teaches us a lot about mechanisms of cognition. To substantiate this, I shall draw detailed analogies between decision making and pattern recognition, and show how a number of phenomena observed in decision making, such as unique, oscillatory, or random solutions, and hysteresis effects, can be understood by means of this analogy.

The book concludes with a brief study of networks of brains and a critical summary and outlook.

Actually, the book covers more topics than those just mentioned. In order to facilitate its reading by newcomers to synergetics, or to brain research, or to both these disciplines, I have included a number of chapters that provide the reader with the necessary basic knowledge. This background material is mainly concentrated in Part I on Foundations. The following three parts on behavior, on EEG and MEG, and on cognition may be read practically independently of each other so that, for instance, a reader interested mainly in cognition may start with Part I and then proceed immediately to Part IV.

* Marks chapters or sections that are mathematically somewhat more involved.

Part I **Foundations**

1. Introduction

1.1 Biological Systems Are Complex Systems

All biological systems are highly complex. Most of them consist of very many cells which are themselves complicated systems. In addition, these systems show complex behavior. One of the most striking features of animals is the cooperation of many cells that manifests itself, for instance, in the coordination of muscles in locomotion and in other movements. Thus around the turn of the 20th century the famous physiologist *Sherrington* coined the term: synergy of muscles. Such high coordination may also be observed in breathing, heart beat and blood circulation. At a still higher level, in the human brain many cells cooperate in a purposeful manner to produce perception, thinking, speech, writing, and other phenomena, including emotions. In all these cases, new qualities emerge at a macroscopic level, qualities that are absent at the microscopic level of the individual cells. One of the greatest puzzles of biology is certainly this high degree of integration which accompanies the linking between the microscopic and macroscopic level. How strong this integrative power of our brain can be is demonstrated by Fig. 1.1, which shows a painting by Giuseppe Arcimboldo. At first sight we recognize a face, but a closer look reveals just an arrangement of fruits and vegetables. We recognize that face not because, but in spite of its individual parts! This example may be used metaphorically in what follows, where the individual parts to be considered will be mainly the nerve cells and not entities such as fruits.

How is integration brought about? In earlier times we could see pictures illustrating the idea that there is a human within a human's brain to produce steering or organization effects. In the 20th century, the famous neurophysiologist *Eccles* jointly with the philosopher *Popper* (1977) wrote a book on *The Self and its Brain*, where, in a way, they interpreted the Self as a programmer and the Brain as a computer. In my book I shall adopt quite a different attitude. Rather than assuming that integration is brought about by organizing centers, by programmers, or by some kind of computer programs, I shall develop the idea of self-organization. Synergetics can be considered as the most advanced theory of self-organization and I want to study how this theory can be applied to a variety of phenomena found in biological systems, in particular to those associated with brain activities, behavior and cognition. I shall consider a biological system as a giant system based on the laws of

Fig. 1.1. Painting by Giuseppe Arcimboldo (1527 – 1593)

physics. We shall see, however, that the laws in biology cannot be uniquely derived from physical laws. There are other additional laws connected with the emergence of new qualities. In this way, it will turn out that synergetics is not in conflict with physics but, on the other hand, it must not be identified with any physicalism. While synergetics is a discipline, self-organization is a phenomenon.

1.2 Goals of Synergetics

Complex systems are composed of many individual parts, elements, or sub-systems that quite often interact with each other in a complicated fashion. One classical recipe for coping with such systems is that due to *Descartes*. According to him one has to decompose a complex system into more and more elementary parts until one arrives at a level at which these parts can be understood. Quite clearly, molecular biology pursues this line. On the other hand, by means of the interaction of the elements of a system, new qualitative features are brought about at a macroscopic level. Thus undoubtedly there remains an enormous gap in our understanding of the relations between the microscopic and the macroscopic level. It is the goal of synergetics to bridge this gap. At the same time we shall see that in most cases structures are not produced by an organizing hand but are brought about by the systems themselves. That is why we shall speak of self-organization. There is still another difficulty with what we might call Descartes' approach. In order to describe the individual parts, an enormous amount of information is needed, but nobody can handle it. Therefore, we have to develop adequate methods to compress information. A simple example of how this goal can be reached is provided by our temperature sense. As we know, a gas, such as air, is composed of myriads of individual molecules, but we do not notice their indi-vidual motion. Rather, we somehow integrate over their motion and feel only a certain temperature. Similarly, in most cases single words represent whole classes, or categories, or objects, or complicated actions.

Can we develop a general theory that allows us to adequately compress information quite automatically? As we shall see, such information compres-sion takes place in situations where a system changes its macroscopic state qualitatively. In the inanimate world, there are a number of such abrupt changes, called *phase transitions*. Examples are provided by freezing, where liquid water goes over into the state of solid ice, or the onset of magnetism, or the onset of superconductivity. As we shall see, biology abounds with similar qualitative changes, though at a far more sophisticated level.

1.3 The Brain as a Complex System

The human brain consists of about 100 billion (10^{11}) neurons and there may be up to 10^4 connections per neuron. In addition, each neuron by itself is a intricate system. The neurons are linked together in a highly complex manner. In order to illustrate the enormous number of neurons, let us blow them up so that a hundred neurons fit into a thimble, of say one cubic centimeter. We would then need a house of ten meters width, ten meters length and ten meters height to accommodate all these thimbles.

The research on brain, behavior, and cognition has many facets and, to be honest, these facets are inexhaustible. Thus we have to ask what questions are purposeful. These questions, in turn, depend on our level of scientific study, which itself depends on experimental techniques and theoretical concepts as well as on mathematical procedures. But in addition it depends on taste, fashions, our previous training, etc. In view of the great complexity of the brain, we have to look for models, paradigms, or metaphors. But at what level and with what precise meaning shall we use metaphors? We shall address these questions in the following.

1.4 Traditional Versus Synergetic Interpretations of Brain Functions

To give the reader a feeling for how our considerations based on synergetics differ from traditional approaches, we shall anticipate some of the basic results of our book. In the left column of Table 1.1 we list the traditional concepts, while the corresponding concepts based on synergetics are listed on the right.

Table 1.1. Comparison between traditional and synergetic interpretations of brain functions

Traditional	Synergetic
cell	network of cells
individual	ensemble
grandmother cell	collective of cells
steering cell	collective of cells
localized	delocalized
engram	distributed information
programmed computer	self-organized
algorithmic	self-organized
sequential	parallel and sequential
deterministic	deterministic and chance events
stable	close to instability points

Let us consider the contents of Table 1.1 in more detail by comparing the left- and right-hand sides line by line. The traditional experimental and theoretical study of brain functions rests on the single cell, while in synergetics we focus our attention on the action of a whole network of cells. Thus instead of treating the individual, we consider an ensemble. This difference between points of view can most clearly be seen when we discuss the question of the grandmother cell in traditional theories. According to this idea, we recognize our grandmother by means of an individual cell in our brain that identifies her. In the synergetic approach, recognition of patterns is achieved by the action of an assembly of cells. Similarly, steering of motion is attributed to a steering cell in the traditional approach, while in synergetics it is the outcome of the action of an assembly of cells. Quite clearly, while in the traditional approach the actions are strictly localized, they now become delocalized and may be distributed over quite extended areas of the brain. In accordance with this view, instead of engrams we have to look for distributed information. (These points of view are shared, in particular, by connectionism; cf. Sect. 18.4.)

A basic difference between our standpoint and that of other schools of thought occurs when we consider the widely held view that the brain acts as a programmed computer based on algorithms. In our view, the brain acts by means of self-organization, where we may, or may not, invoke algorithms. In the traditional approach incoming information is processed sequentially. In the new approach, incoming information is mainly processed in parallel. The concept of a programmed computer implies that the whole system works in a deterministic fashion. As we shall see in our book, the actions of biological systems are determined both by deterministic and by chance events. A further basic difference shows up with respect to stability. In the traditional approach it is assumed that the brain is in a stable state. This is the basis of many experiments on brain functions. We shall provide evidence that the brain is acting close to instability points.

I hope that Table 1.1 shows that it might be worthwhile following up these new lines of thoughts and, especially, to see what experimental evidence we have to suggest such changes in our theoretical understanding of brain function.

2. Exploring the Brain

In this chapter I give a brief survey of the most prominent experimental methods in brain research. The various physical methods will be particularly emphasized.

2.1 The Black Box Approach

The black box approach is well-known in electrical and radio engineering. In a number of cases, when a new device is to be constructed by putting individual parts together, it is sufficient for the engineer to know some characteristic features of the individual parts, namely the relationship between input and output. In this approach the internal structure of the individual part is of no concern. Similarly, one may consider a human or an animal as a black box that responds to specific stimuli in a specific manner. In other words, in this approach one studies the behavior. The idea of this methodology, called behaviorism, is mainly due to *Skinner*, who, for instance, constructed special cages to study the response of animals to stimuli such as food or punishment. In this approach internal states of the brain are entirely ignored and, moreover, asking questions about such states was considered unscientific. This might have been plausible with respect to animals, which cannot speak, but with respect to humans who can communicate, one can take advantage of an important tool, namely introspection. Actually, nowadays things are changing, and there are even studies in progress to gain insight into the mental states of animals. At any rate, behaviorism per se is no more at the focus of modern research, though the study of different kinds of behavior is still an important tool, as we shall see later in this book. Let us mention some examples of such behavioral studies. Movements of humans and animals are studied by the discipline of movement science. Such studies are also of interest to sports science. Human behavior at the mental level is studied by psychology and psychiatry. In a way, the study of language by linguistics can also be considered a black box approach, because one is not concerned with how the brain produces language but merely with the abstract structure of this mental product. The psychological study of visual and auditory perception may be considered as some kind of bridge between the external and the internal world, depending on the way we interpret the corresponding experiments, i.e., it depends upon the extent to which one invokes introspection.

2.2 Opening the Black Box

When we open the skull, we see white-grey matter, whose form is strongly reminiscent of a walnut (cf. Figs. 2.1 and 2.2). Not many details are visible even under the microscope. Things change, however, if certain stains are used, as was first done by *Golgi*. Then under the microscope a network containing nodes becomes visible. These nodes are the neurons and may have various shapes. For instance, the neuron in Fig. 2.3a has the shape of a pyramid and is called *pyramidal cell*. Other cells are called *Purkinje cells* after their discoverer (Fig. 2.3b). All in all there are now about twenty or more different types of neurons known.

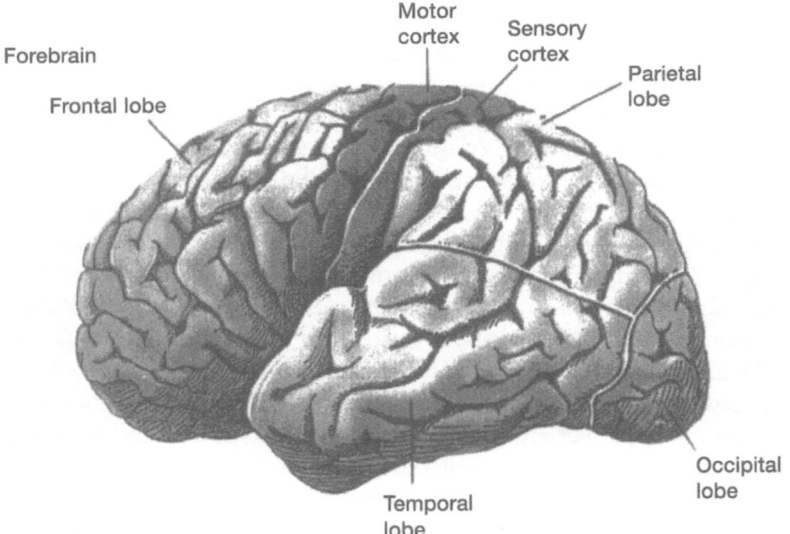

Fig. 2.1. The brain seen from the left side

2.3 Structure and Function at the Macroscopic Level

It has turned out in the course of the study of the brain that specific functions can be attributed to certain areas in the brain. (We should mention, however, from the very beginning that there may be shifts of functions between areas so that this functional map of the brain may undergo changes, for instance, when the brain has been damaged and some recovery takes place.) Brain

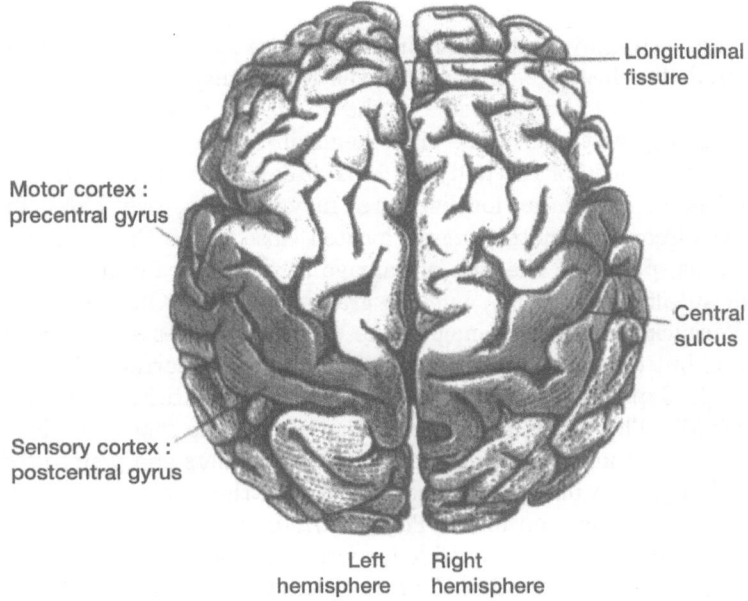

Longitudinal
fissure

Motor cortex :
precentral gyrus

Central
sulcus

Sensory cortex :
postcentral gyrus

Left   Right
hemisphere   hemisphere

Fig. 2.2. The brain seen from above

(a)   (b)

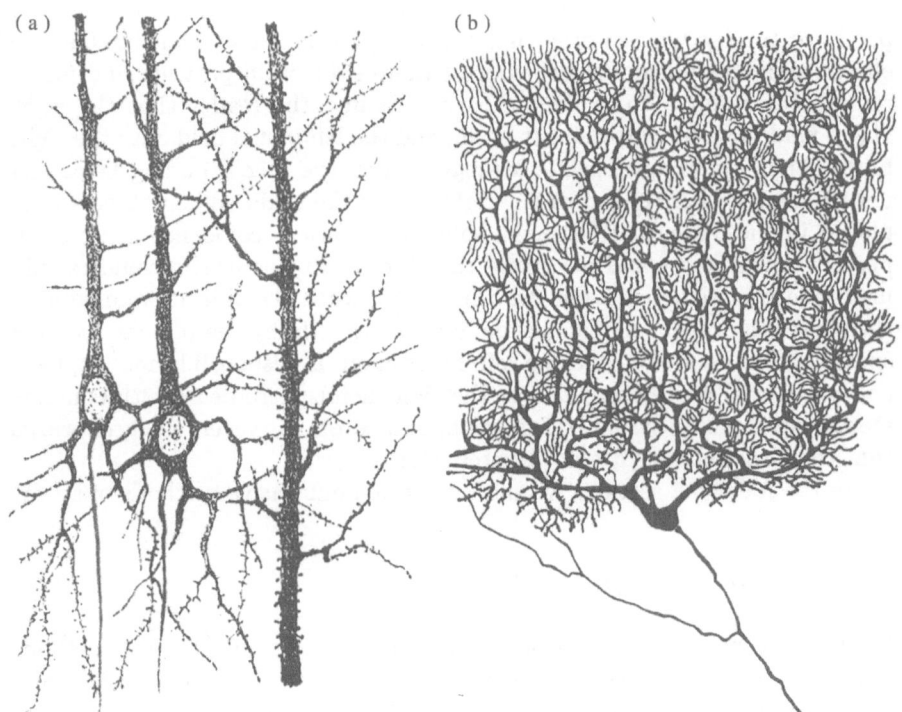

Fig. 2.3. (a) Pyramidal cells, (b) a Purkinje cell (after *Bullock* et al., 1977)

injuries and strokes were the first indicators of where such functional areas are located. For instance, a stroke in the left hemisphere of the brain can lead to paralysis of the right side of the body, e.g. of a leg, an arm, and a hand. Similarly a stroke in the right hemisphere may cause a paralysis in parts of the left side of the body. In 1861 *Brocca* found that a stroke in the left hemisphere in a rather localized area may impair speech. Later, in 1874, *Wernicke* discovered another center located close to the Brocca center that also governs speech. In the case of damage to the Brocca center, the person can still speak meaningfully, but the grammar is lost. On the other hand, if the Wernicke center is damaged, the person produces seemingly correct sentences, but ones that are meaningless. Another important discovery was made by the Japanese medical doctor *Inoue* during the Russian–Japanese war of 1904–1905. The Russians had developed a new rifle that gave the bullets a bigger momentum. Such bullets hit Japanese soldiers and penetrated their heads. Though in a number of cases the eyes of these soldiers were not hit, they nevertheless became blind. This led *Inoue* to the conclusion that the rear part of the brain is responsible for visual perception.

Let us now briefly mention the role of the two hemispheres of the brain (cf. Fig. 2.2). These two hemispheres are connected by bundles of nerve fibres. This bridge is called the *Corpus callosum*. It had been found that epileptic seizures could be positively influenced when the *Corpus callosum* was severed partly or totally. On the other hand, such operations had considerable side effects. *Sperry* studied the behavior of these split-brain patients and found remarkable results. To explain them we remind the reader that the right visual field projects onto the left brain, and the left visual field onto the right brain. Thus by an appropriate arrangement in the corresponding fields one may activate either the left or the right half of the brain. When objects were placed in the left visual field, the split-brain patient could not consciously perceive them and could not name the objects. But he or she could readily manipulate these objects. On the other hand, when the objects were lying in the right visual field, they could be named correctly by the patient. Loosely speaking, we may say that the left hemisphere is responsible for language and sequential processes, whereas the right hemisphere deals with complex scenes, with music, visual imagination, and so on. However, research shows that these distinctions are not entirely strict.

In the following sections we shall describe noninvasive methods of study.

2.4 Noninvasive Methods

In this section we shall be concerned with a number of physical methods that allow research workers to explore a variety of properties of the brain. These methods are X-ray tomography, electro-encephalograms, magneto-encephalograms, magnetic resonance imagery and PET scans. We shall explain these methods in the following:

2.4.1 X-Ray Tomography

Since *Röntgen's* discovery that X-rays penetrate biological tissue, X-rays have become an important tool in medical diagnosis. A considerable step forward was achieved with computer tomography (CT). The principles of computer tomography were independently developed by *Kulmack* and *Hounsfield*. In this approach a biological object, for instance the brain, is X-rayed at a series of different angles. The resulting patterns are then processed in a computer, which allows one to construct three-dimensional images. A tumor in the brain might thus be located, whereby not only its position but also its shape may become visible.

2.4.2 Electro-encephalograms (EEG)

The experimental study of electric fields in the brain has a long history. As early as in 1875 *Caton* stated:[1] "Bringt man Elektroden an zwei Punkten der Oberfläche des Gehirns ... an, so fliessen schwache Ströme wechselnder Polarität durch den Verstärker." But it was not until 1929, that *Berger* introduced the study of electro-encephalograms of humans.

For the measurement of EEGs one or several electrodes are placed on the skull and the electrical potential between these electrodes and a reference electrode, for instance also located on the skull, can be measured as a function of time. Depending on the mental activity, quite different types of curves are found. For further analysis, at least in general, these curves are then frequency-filtered so that only certain frequency bands are investigated. When a person is in a resting state with his or her eyes closed, the so-called α-waves around 10 Hz, occur. In sleep, various phases may be distinguished. One of them is the REM-phase which is connected with dreams and rapid

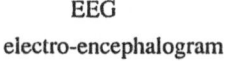

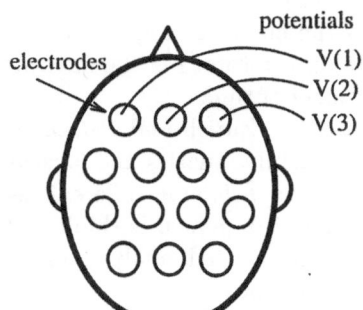

Fig. 2.4. Positioning of the electrodes on the skull

---

[1] If one places electrodes at two points on the surface of the brain ..., then weak currents of oscillating polarity flow through the amplifier.

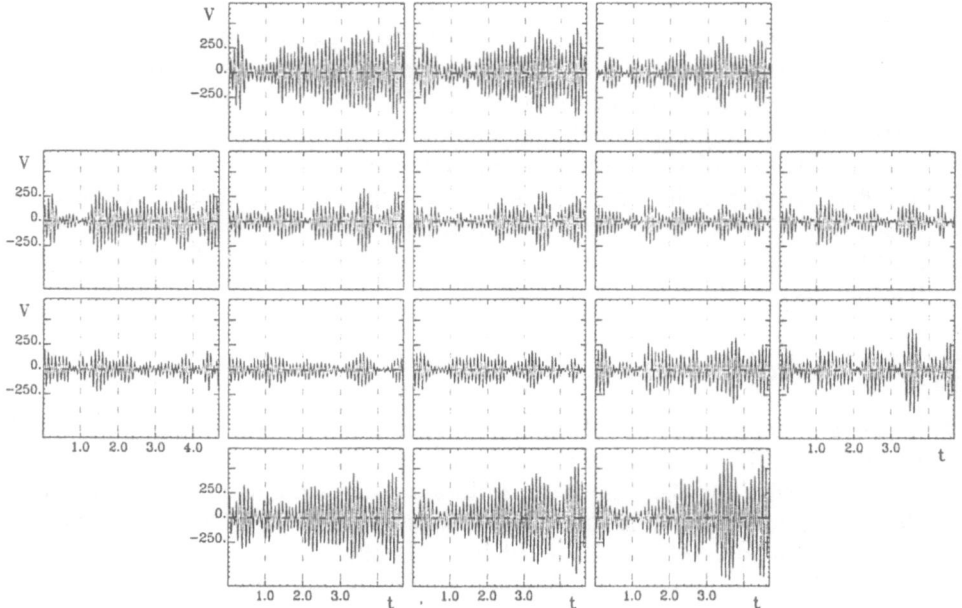

Fig. 2.5. The voltage difference derived from each electrode and a reference electrode is plotted against time for each electrode. The position of the boxes corresponds to that of the electrodes on the skull (after *Lehmann*, private communication)

eye movements (hence the abbreviation REM). When several electrodes are used (Fig. 2.4), one may look at spatio-temporal patterns. Figure 2.5 shows the time-series corresponding to the individual electrodes, where the arrangement of the boxes corresponds to that of the electrodes on the skull. By taking the activities at the same time in each box, we may construct a map of local activity at the positions of the individual electrodes. When we interpolate between these positions, we obtain one of the circles of Fig. 2.6. Repeating this procedure for a series of times, we obtain the total Fig. 2.6 showing the evolution of the spatial pattern over the course of time.

A number of researchers, such as *Lehmann* in Zürich and *Petsche* in Vienna, have studied the connection between mental activities and locations of high electric activity in the brain. According to *Lehmann* different centers can be shown to be active depending on whether a person is thinking of an abstract or a concrete object. Later in this book we shall analyze some EEG patterns and relate them in particular to chaotic processes.

2.4.3 Magneto-encephalograms (MEG)

Even very weak magnetic fields can be measured by physical devices called

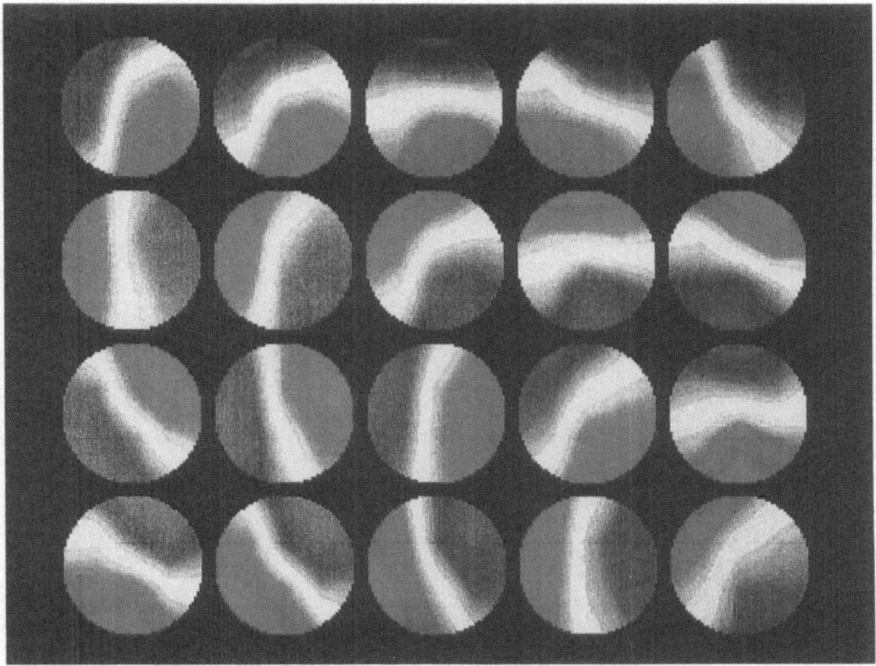

Fig. 2.6. Reconstruction of the spatio-temporal signal from the recordings shown in Fig. 2.5. The time runs from left to right in the first row, starts again in the second row at the left and runs to the right, and so on (after *Fuchs* et al., 1987)

vice; this is based on the Josephson effect. Let us describe a recent example of such a study. (For details see Chap. 15) Here an array of 37 squids was put on the scalp in a region that covers parts of the motor and sensory cortex (Fig. 2.7). In these experiments done by *Kelso* and co-workers, the test person was exposed to a periodic acoustic signal and was asked to tap a finger between successive signals. Most remarkably it turned out that when the frequency of the signal was increased, the person could not maintain the syncopating mode, but suddenly switched to a synchronized mode. The results are schematically shown in Fig. 2.8. Is this transition from one behavioral mode to another one detectable in the MEG? Figure 2.9 shows the time-series of the signals stemming from the individual squid locations before the transition happened, while Fig. 2.10 shows the corresponding signals after the transition. At first sight, it seems difficult to interpret these data, but in Chap. 15 we shall show how we can analyse them in detail and that they beautifully reflect the above mentioned transition.

Quite generally it may be said that the information obtained by MEG experiments is different from that of the EEG despite the fact that both signals have their origin in the electrical activity of the neurons. Among the important differences are: The EEG picks up the electrical potential between

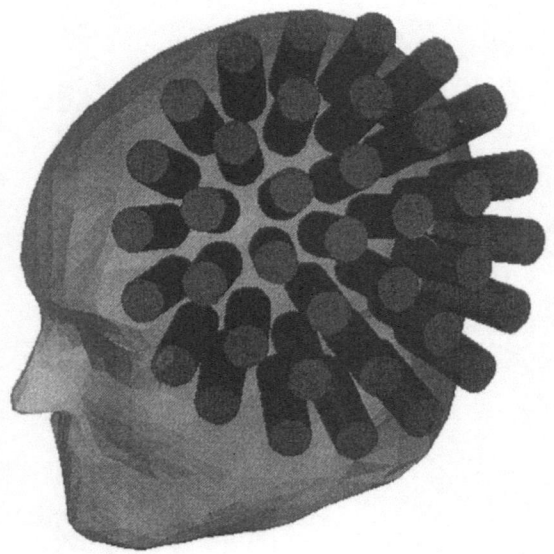

Fig. 2.7. Arrangement of the squid detectors in Kelso's experiment (after *Kelso* et al., 1992)

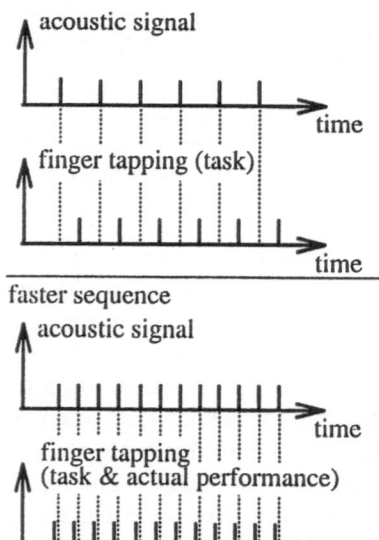

Fig. 2.8. *Upper part*: The finger tapping occurs during the intervals between the acoustic signals. *Lower part*: If the frequency of the acoustic signals is increased, the finger tapping occurs synchronously. The shift between the dotted and solid line is just for illustration. In reality the bars of the finger tapping and the acoustic signal coincide

a point at the skull and a reference point. This potential does not have a fixed value at a specific point in space and time, but is unique only as a difference between two points in space. This reference electrode problem has been addressed extensively in the EEG literature, but up to now no common standard exists. The quantity measured by the squid detectors is the tangential component of the magnetic field, a physical quantity that is defined uniquely at each point in space and time. Thus, experimental results

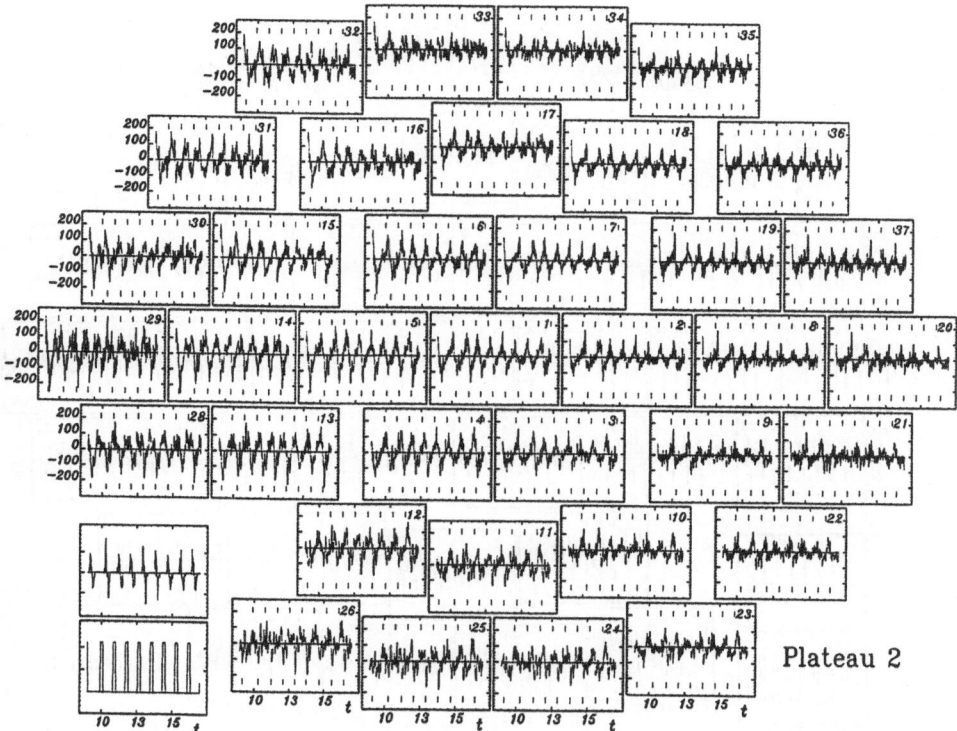

Fig. 2.9. Recording of the magnetic field by the individual squid detectors. Again time-series in analogy to Fig. 2.5 are shown. These signals are recorded before the transition from syncopation to synchronization occurs (after *Kelso* et al., 1992)

are free of artifacts arising from differently chosen reference electrodes as is the case in the EEG.

A second important difference between the EEG and the signal registered by squids is that the former are subject to volume conduction distortions introduced by skull and tissue intervening between electrode and source while the latter are not, i.e., the skull and scalp are transparent to magnetic fields generated by electric currents in the brain. (Little seems to be known about the origin of these currents at the cellular or intercellular level.)

2.4.4 Positron Emission Tomography (PET)

In those parts of the brain that are active, the glucose metabolism is enhanced. The increase of glucose concentration can be studied by positron emission tomography. In these experiments glucose is marked by a radioactive isotope that emits positrons. An emitted positron cannot travel very far, because it quickly hits an electron with which it recombines, i.e., positron and electron annihilate each other. As a result of this annihilation process

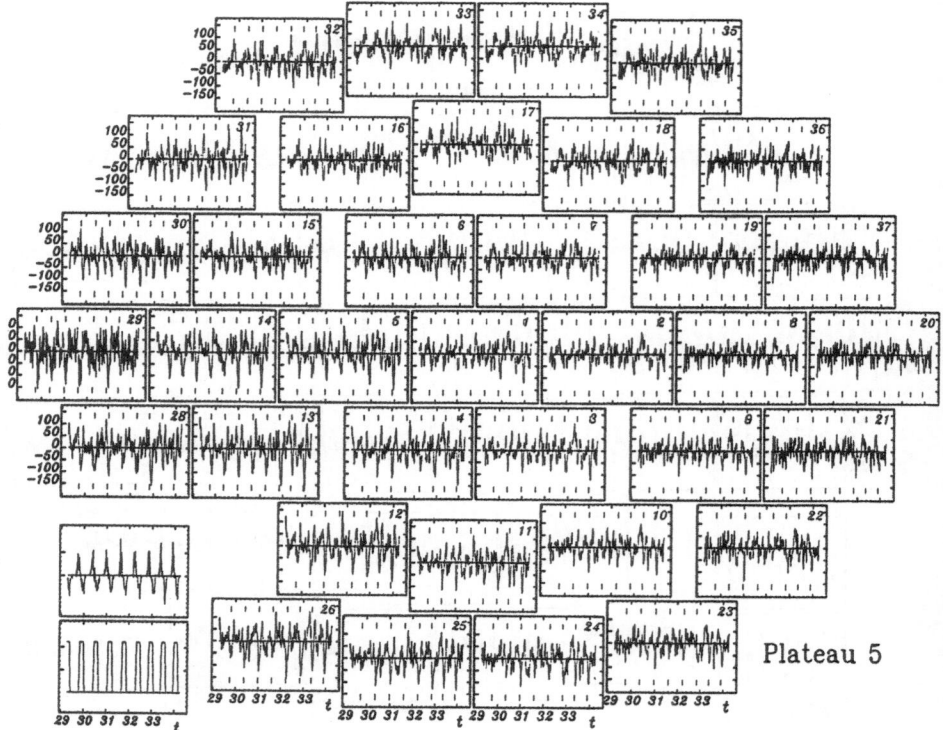

Fig. 2.10. The same as Fig. 2.9, but after the transition to synchronous finger tapping (after *Kelso* et al., 1992)

2 γ-quanta are emitted in opposite directions. When γ-ray detectors are positioned around the skull, by means of these detectors and a subsequent computation, the emission centers can be located to within a few millimeters. So far blood flow has turned out to be the most reliable indicator of local mental activity . Such a relationship had already been foreseen by *Roy* and *Sherrington* in 1890, who said that in an automatic mechanism ... "the blood supply to the brain is connected with local activities of the brain". Active parts of the brain require a larger supply of blood. On the other hand, the neurons, at least on short time scales, act without oxygen, i.e. anaerobically. Therefore, the veins carry a higher concentration of oxygen than usual. In practice, the experiments are done as follows: Water, i.e. H_2O, is injected into the arm vein. In H_2O oxygen is used as a radioactive isotope O^{15}, i.e. a positron emitter. After about a minute, this H_2O has arrived in the brain. It has a half-life of about two minutes, i.e., it decays very fast and does not leave any radioactive material in the brain. When one compares the pictures obtained from positron emission tomography for a brain that is inactive and for a brain tackling a specific task, not much difference can be seen. However, pronounced differences become visible when the two images

are subtracted from each other. In addition, these experiments are repeated and the measurements with respect to different test persons are averaged or the measurement is done several times on the same test person.

The PET method was used by *Petersen* et al. in 1988 (cf. *Raichle*, 1994) to study the organization of language. A noun was given, then the person had to find a word and to pronounce it; for instance, the word *hammer* was given and the test person found the word *hits*. In detail the experiment was done as follows: The persons had to watch a TV screen on which a cross bar appeared and the noun was given either on the TV screen or through ear phones. Then the test person had to say which noun he or she had heard and finally the word to be said. As expected, the optical perception was encoded in the rear part of the brain (primary visual cortex), while the acoustic perception took place in the temporal lobe. The spoken nouns were produced in the motor cortex, but not in Brocca or Wernicke areas, i.e., in a way, no thinking was involved in the production of the noun. On the other hand, when two tasks had to be combined, namely conscious recognition of the meaning of the word and the choice of answer, the left frontal lobe and the temporal lobe were involved. An interesting change of this pattern was observed after fifteen minutes of learning: the areas for the nouns became the same as those which generated the words.

2.4.5 Magnetic Resonance Imaging (MRI)

This method is based on the phenomenon of nuclear magnetic resonance which had been studied by *Bloch* and *Purcell* in physics. As we know, an atom is composed of nuclei with electrons orbiting around them. Many nuclei possess a magnetic moment that is connected with their spin. In this way, the nuclei may be visualized as little elementary magnets having a north and a south pole. The simplest nucleus is that of the hydrogen atom and is called a proton. It possesses a spin $\frac{1}{2}$, which in a constant applied magnetic field can be aligned either in the direction of the field or opposite to it. In one direction it acquires a higher energy, in the opposite direction a lower energy (Fig. 2.11, upper part). When one applies an additional magnetic field that oscillates in time, the elementary magnet may start to change its orientation (Fig. 2.11, lower part). The magnetic field must fulfil a crucial requirement so that this happens: Its frequency ν must be equal to the energy difference $E_2 - E_1$ divided by Planck's constant, h. In this case one speaks of a resonance, which explains the term *nuclear resonance* attached to this method. The angle to which the orientation of the elementary magnet is changed is proportional to the strength of the applied field and its duration. Thus by an appropriate timing one can let the elementary magnet (spin) reach any final orientation. If the spin is flipped from the vertical to the horizontal direction, i.e. by an angle of $\pi/2$, one speaks of a $\pi/2$-pulse of the applied oscillating magnetic field (Fig. 2.12, upper part). If the spin direction is to be changed from its up- to its down-direction, one has to apply a π-pulse (Fig. 2.12, lower part). The effect

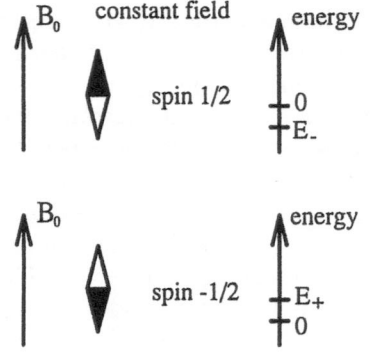

B$_0$ constant field   energy

spin 1/2   0
                  E$_-$

B$_0$ energy

spin -1/2   E$_+$
                  0

alternating field B

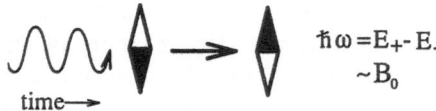

$\hbar \omega = E_+ - E_-$
$\sim B_0$

time→

Fig. 2.11. *Upper and middle part:* In a constant magnetic field B_0 a spin $-\frac{1}{2}$ can adopt one of two positions, namely up or down, with corresponding lower or enhanced transition energy. *Lower part:* An alternating magnetic field, supplied in addition to a constant field B_0, may flip the spin provided the resonance condition $h\nu \equiv \hbar\omega = E_+ - E_-$ is fulfilled. This transition energy difference is proportional to the applied constant magnetic field B_0

1. Saturation-recovery
         (sequence of $\pi/2$-pulses)

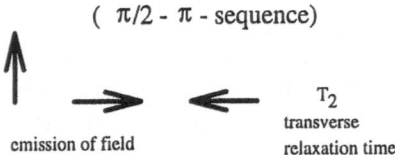

2. Spin-echo-sequence
         ($\pi/2$ - π - sequence)

T$_2$
transverse
relaxation time

emission of field

3. Inversion-recovery
         (π - $\pi/2$ - sequence)

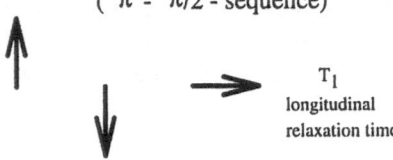

T$_1$
longitudinal
relaxation time

Fig. 2.12. Schematic diagram of different kinds of magnetic resonance processes

of the $\pi/2$-pulse is of particular interest. As is shown in quantum mechanics, under the action of the constant magnet field, the spins rotate in the plane perpendicular to the magnetic field. They rotate coherently, and being like little magnets, they may emit an alternating coherent magnetic field that can be recorded. Thus the protons (or other nuclei) *respond* to the pulse. But the elementary magnets are disturbed by their interaction with their surroundings. As a consequence, in the course of time their *coherent* motion gets lost and their emitted magnetic field drops to zero after a time that is called *transverse relaxation time* or T_2. The emission of the magnetic field also stops when the spins return from their motion within the horizontal plane to their magnetically more favorable vertical orientation. This reorientation takes a time that is called *the longitudinal relaxation time*, T_1. Depending on the kind and sequence of pulses applied to the spins, one may measure T_2 or T_1. From the size of the emitted magnetic signal one may draw conclusions about the concentrations of spins, whereas by measuring the decay times T_1 or T_2 of this field, one may draw conclusions about the chemical surrounding of the spins. Without trying to go into too many details, we just list the three most important procedures (Fig. 2.12):

1) Saturation − recovery:

   a sequence of $\pi/2$-pulses

2) Spin-echo sequence

   a sequence of $\pi/2 - \pi -$ pulses

   measures transverse relaxation time, T_2

3) Inversion-recovery

   sequence of $\pi - \pi/2$ - pulses

   measures longitudinal relaxation time, T_1.

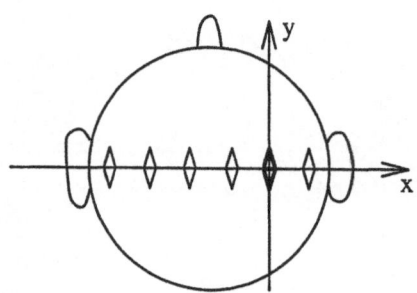

Fig. 2.13. We look at the top of a head from above. The strength of the constant magnetic field B_0 that points in the y-direction and increases its strength along the x-axis. The transition energy, or correspondingly frequency, of the individual spins changes along the x-axis

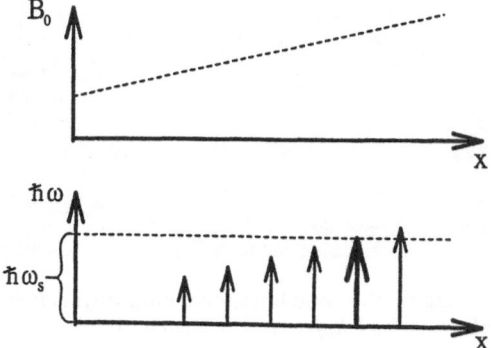

Fig. 2.14. *Upper part*: Variation of the magnetic field B_0 along the x-axis. *Lower part*: The transition energy $\hbar\omega$ of the individual spins as a function of their location x. ω_s is the frequency of the applied alternating magnetic field

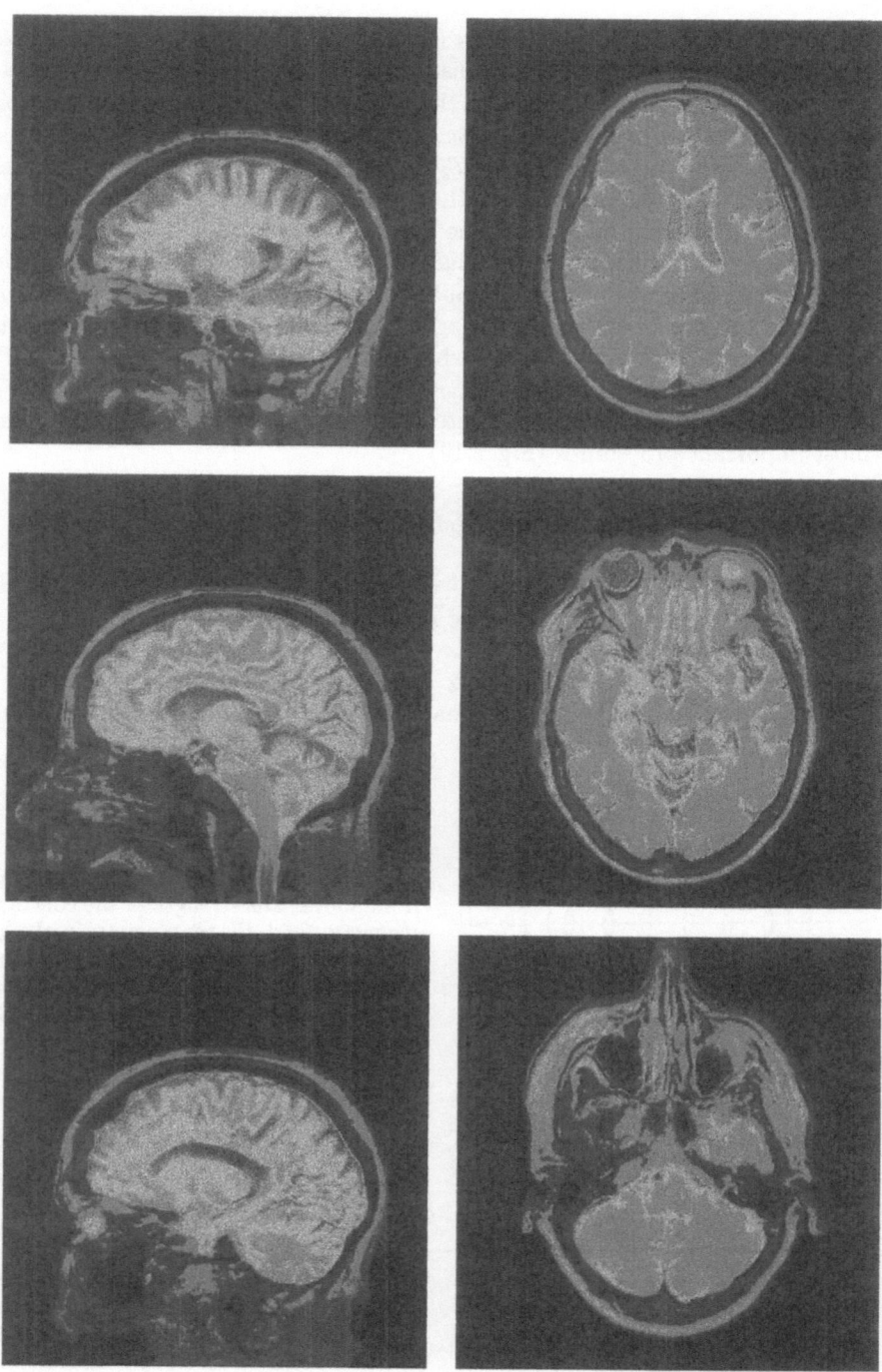

Fig. 2.15. Sequence of MRI pictures along vertical and horizontal cuts. *Left column*: sagittal; *right column*: transversal (after *Daffertshofer* and *Schwartz*, 1994)

The concept of nuclear magnetic resonance was applied in medicine by *Lauterbur*, who used resonances with respect to protons. How can one localize protons in a body by this method? The basic idea is as follows: One applies to the body a time-independent but spatially variable magnetic field (Figs. 2.13, 2.14). As a consequence, the resonance condition becomes space-dependent. One may now probe the spatial dependence of the energy distribution, $E = \hbar\omega$, of the individual spins by sending a time-dependent magnetic field into the probe. The field will effect only those protons (spins) that meet the resonance condition. Since one knows the position of these protons because of the inhomogeneous magnetic field, one can now measure their local concentrations. Examples are shown in Fig. 2.15. By nuclear magnetic resonance not only hydrogen concentrations can be measured but also concentrations of nuclei of oxygen. While brain activity increases the consumption of glucose, it does not increase that of oxygen. But because brain activity requires additional blood, one finds an increased oxygen concentration in the small veins that empty the neurons. Oxygen is transported by hemoglobin. In 1935 *Pauling* could show that the magnetic properties of hemoglobin are changed according to the amount of transported oxygen. In 1990 *Bougava* and others showed that by means of nuclear magnetic resonance (NMR), or in its new terminology by MRI, these small magnetic deviations can be discovered. The spatial resolution is one to two millimeters. The measurements can be done in real time, but the resolution ranges from 100 milliseconds to some seconds. This may still be too long to explore certain important mental activities.

2.5 Structure and Function at the Microscopic Level

At the microscopic level we deal with the structure and function of neurons. A neuron is mainly composed of a cell body and the axon through which electric signals are emitted and then, eventually, transferred to other neurons. (Actually, a neuron is a highly complex system in which a number of components can be identified. Among these components are microtubuli which according to a speculation by *Hameroff* (1987) play a fundamental role in memory and even in consciousness. We shall return to such speculations briefly at the end of this book, but shall be concerned in this chapter with the hard facts only.) In addition, a neuron possesses some branches called *dendrites*, through which it receives signals from other neurons. One also speaks of a dendritic tree. The axon itself eventually splits into a dendritic tree. Signals from one cell to another one are transmitted at the meeting points between dendrites. These meeting points have a special structure and are called synapses (Fig. 2.16). Each synapse contains a number of vesicles, which store specific molecules called *neurotransmitters*. When an electric signal emitted through the axon arrives at the synapse, the vesicles open and

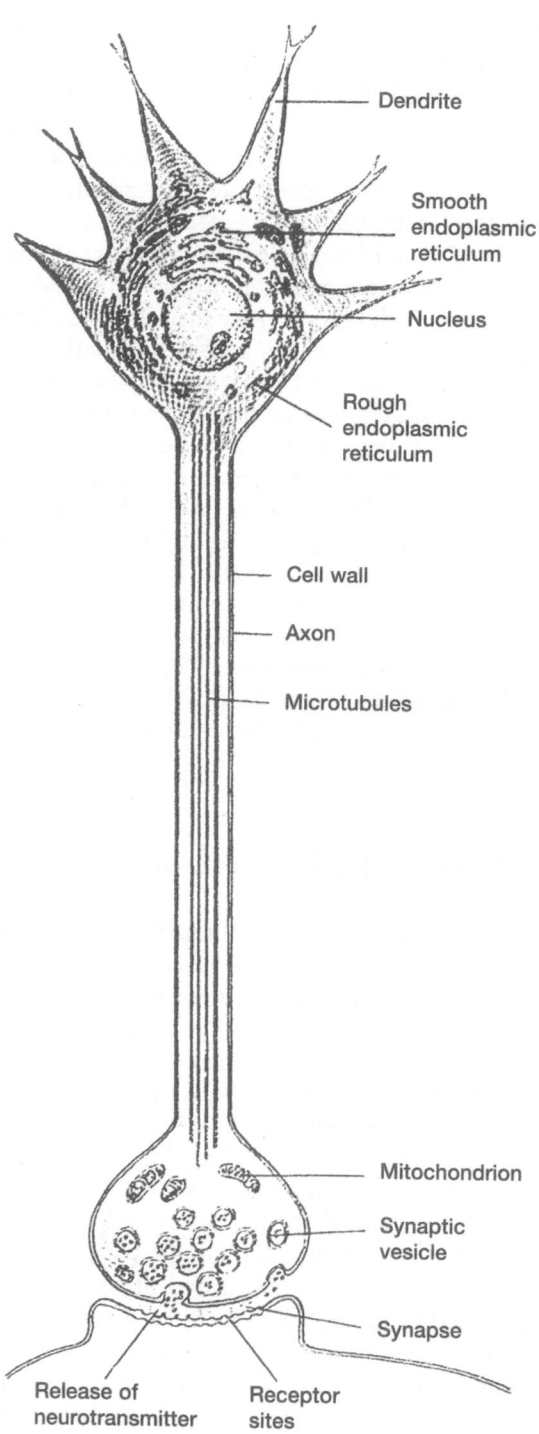

Dendrite

Smooth
endoplasmic
reticulum

Nucleus

Rough
endoplasmic
reticulum

Cell wall

Axon

Microtubules

Mitochondrion

Synaptic
vesicle

Synapse

Release of   Receptor
neurotransmitter   sites

Fig. 2.16. A typical nerve
cell

release neurotransmitters which diffuse through the synaptic cleft and eventually lead to a new signal at the other side. Nowadays quite a number of different neurotransmitters are known, including acetyl choline, serotonin, dopamin, GABA, and others. Serotonin is known to dampen activities, and when its level is too low, this may lead to depressions. Dopamin, on the other hand, stimulates brain activity, but when too high or highly fluctuating concentrations are reached, schizophrenic behavior may result. In the membrane at the synaptic cleft, there are special centers, so-called *receptors*, that are specific for the absorption of neurotransmitters. For instance, there are serotonin receptors. When drugs are administered, they may block certain receptors. For instance, caffein blocks serotonin receptors, while a drug used for treatment of schizophrenia, namely Haloperidol, blocks dopamin-2 receptors. At this point, interesting relationships will emerge with respect to synergetics, because one has to understand why a simple change of drug concentration and thus a simple change in the blocking of specific receptors may change the macroscopic behavior of a person. For instance, when drinking coffee, we become activated and excited, whereas Haloperidol not only dampens activities but may also lead to a qualitative change of behavior.

While some neurons emit signals which activate other neurons, there are also kinds of neurons that emit signals which *inhibit* the action of other neurons. Thus we have to distinguish between activating and inhibiting neurons. When a neuron receives signals from other neurons, activators or inhibitors, the neuron sums up these signals and forms a post-synaptic potential. This transformation from incoming to outgoing signals is done in such a way that a signal can only be emitted when the sum of the incoming signals exceeds a specific threshold.

What kinds of signals can be emitted by a neuron? Independently of its type, a neuron can emit only one kind of signal, namely a short pulse, of about one millisecond duration. This electric pulse then travels along the axon. The degree to which a neuron is excited is encoded by the rate at which these signals are emitted. The higher the activation or the higher the post-synaptic potential, the greater the rate at which these pulses are emitted. In technical terms one may speak of pulse-code modulation. It is most remarkable that the cells belonging to quite different areas in the brain, for instance, areas processing audition, or visual inputs, use the same code.

An important relationship between sensory inputs and the firing rate of neurons was found by *Hubel* and *Wiesel* (1962), who studied the firing rate of specific neurons in the visual cortex of anesthesized cats. These cats had their eyes open and could perceive pictures. A typical pattern was a bar in a specific direction that was moved in some direction. The firing of a specific neuron depended strongly on the orientation of that bar. If the bar, for instance, was in horizontal direction, the neuron did not fire at all; if the bar had an orientation of, say, 45 degrees to horizontal axes, the cell fired moderately, and, finally, when the bar pointed in the vertical direction, the cell fired

very strongly. *Freeman*, when studying the olfactory bulb of rats, found that
the firing of different neurons can become correlated. Such experiments are
most impressive when done on the visual field, as was first demonstrated by
Singer, *Gray* and co-workers and by *Eckhorn* and co-workers. These authors
found that when two bars moving in the same direction and having the same
orientation were perceived by visual fields belonging to two different neurons,
these neurons fired synchronously for a short time.

2.6 Learning and Memory

The structures and processes underlying learning and memory at the neural
level are not yet well-known. According to widely accepted hypotheses due
to *Hebb*, learning is achieved by a strengthening of synapses between neurons
that are again and again *simultaneously* active, and a corresponding weaken-
ing, if one or both are inactive during the same time. The connection between
the acquisition of behavioral patterns and changes at the neuronal level have
been studied in particular in lower animals, such as sea-slugs Aplysia and
Hermissenda, for instance, by *Kandel*. Two kinds of changes of behavioral
patterns may occur, namely habituation and sensitization. These changes of
macroscopic behavioral patterns were correlated to changes in the formation
of new receptor channels, but only over several hours.

3. Modeling the Brain. A First Attempt: The Brain as a Dynamical System

3.1 What Are Dynamical Systems?

Dynamical systems have been studied in mathematics and physics for centuries. Examples are the pendulum and the solar system composed of the sun and the planets orbiting around it. Liquids may also be conceived as dynamical systems. They are composed of myriads of molecules that obey the laws of mechanics (or quantum mechanics). At a more abstract level, we may also treat chemical reactions as processes occurring in the dynamical system of the interacting molecules.

Let us briefly discuss what all these examples have in common. To describe the motion of the pendulum mathematically, we need its amplitude x and its velocity v. Because its amplitude and velocity change in the course of time, we write more explicitly $x(t), v(t)$ in order to indicate their time dependence. In the case of the solar system, we introduce the positions or coordinates $x_1, ..., x_n$ and the corresponding velocities $v_1, ..., v_n$, where the indices distinguish between the different planets. In a liquid, instead of treating the molecules individually, we introduce their densities and the velocities of small volume elements that still contain many molecules. To treat chemical reactions, we may introduce the number of molecules or their densities. In all these cases, the individual quantities will vary in time, at least in general. In order to treat these problems from a unifying point of view, we introduce the variables q and distinguish them by a label. For instance, in the case of the pendulum, we may put $x = q_1, v = q_2$. Quite generally, we introduce a set of such variables and define a *state vector* by means of

$$q = (q_1, q_2, ..., q_n). \tag{3.1}$$

In order to describe a system and, in addition, to make predictions about its evolution in the course of time, we use equations of motion. An example is provided by Newton's equations of mechanics.

Let us consider a single particle of mass m. Its momentum p is related to its velocity v by the relation

$$p = mv. \tag{3.2}$$

Under the action of a force F, the momentum changes according to the equation

$$\frac{dp}{dt} = F. \tag{3.3}$$

An explicit example of F is provided by a spring to which the particle is attached. When we identify the position of the particle x with the elongation of the spring, the force F has the form

$$F = -kx,$$

where k is Hooke's constant. Thus F may depend on the position and, occasionally, also on the momentum. To make contact with the notation (3.1), we put $x = q_1$ and $p = q_2$. Recalling that, of course, $v = dx/dt \equiv dq_1/dt$, we may cast (3.2) and (3.3) into the form

$$\frac{dq_1}{dt} = \frac{q_2}{m} \tag{3.4}$$

$$\frac{dq_2}{dt} = F(q_1, q_2), \tag{3.5}$$

where we have indicated that F may depend on q_1 and q_2.

These considerations may be generalized considerably. When a system is described by a set of adequately chosen variables $q_1, ..., q_n$, we may cast the equations of motion into the form

$$\frac{dq_j}{dt} = N_j(q_1, ..., q_n) \quad , \quad j = 1, ..., n. \tag{3.6}$$

We leave it to the reader as a little exercise to make the identifications between (3.4), (3.5) and (3.6). By use of the state vector (3.1) and by defining the vector $\mathbf{N} = (N_1, ..., N_n)$, we may write the equations (3.6) together as a single equation of the form

$$\frac{d\mathbf{q}}{dt} = \mathbf{N}(\mathbf{q}). \tag{3.7}$$

The left-hand side describes the temporal change of the state vector, the right-hand side depends on the present state $\mathbf{q}$ of the system and tells us how the temporal change is determined by the present state. The form of the equations (3.6) and (3.7) is not restricted to mechanics. In fact, a huge variety of processes can be described in this frame work, which is the starting point of dynamical systems theory. There are at least two limitations, however: eqs. (3.6) and (3.7) do not cover quantum mechanical processes and the inclusion of stochastic processes requires additional considerations. Thus (3.6) and (3.7) refer to classical, deterministic processes. The central problem of dynamical systems theory is the analytical, numerical, or qualitative solution of (3.6) and (3.7). For later applications we mention that one can also study dynamical systems by means of discrete maps. Here we follow the path, say, of a particle only at discrete time steps $t_n, n = 1, 2, ...$, and study how the state of the system at time t_{n+1} depends on the state at the previous discrete time t_n.

3.2 The Brain as a Dynamical System

As we have seen in Chap. 1, numerous chemical and electrical *processes* are going on in the brain. At least from a formal point of view, we may thus conceive of the brain as a giant dynamical system. But there are some fundamental difficulties. First of all we must address the following questions: How can we precisely define the variables q_j? Do these variables refer to the constituents of the individual neurons? What kinds of molecules taking part in the chemical process must we consider? Since there are about 10^{22} molecules per cubic centimenter, we must formulate at least such many equations for brain activity. But we don't even know the parameters, e.g. the kinetic constants! At this level of consideration, we are surely confronted with an insoluble task. So let us try the neuronal level. Here again we have to formulate the equations of motion. Let us assume that we know all the connections and parameters. (In fact we are still far away from knowing these, because the neurons are quite complicated and differ from each other.) Since we are dealing with about one hundred billion neurons, we have to solve 10^{11} equations. That is a hopeless task even for the most advanced computers. Quite clearly, an approach based on the treatment of the brain as a dynamical system is not feasible. Nevertheless, a number of attempts have been made along to these lines, which can only be done using enormous simplifications. One such simplification is the neuro-computer in which the properties of neurons and of their links are greatly simplified and only very limited numbers of model neurons have been treated, say, a dozen or a hundred instead of a hundred billion (cf. Chap. 18). Quite obviously, there is a need to treat brain activity and behavioral patterns by means of different concepts. This will be the main concern of the present book, where we shall treat the brain as a *synergetic system*. Basic to this idea is the concept that, by the cooperation of individual parts, new qualities emerge via self-organization. In order to be able to explain this new approach, we have to make use of the basic principles of synergetics, which will be explained in the next chapter.

4. Basic Concepts of Synergetics I: Order Parameters and the Slaving Principle

4.1 Factors Determining Temporal Evolution

In this chapter we adopt a rather broad view. We shall consider the complex systems that may occur in a variety of disciplines, such as physics, chemistry, biology, and medicine, but also economy and ecology. In all the cases considered, the systems are composed of many components, subsystems, elements, or parts. Some of the first steps of our analysis will be strongly reminiscent of dynamical systems theory, but later on decisive differences will appear. We distinguish the components by an index $j = 1, \ldots, N$, where N may be a very large number (Fig. 4.1) and denote the activity of the component j by q_j. Such activity may, for instance, be the firing rate of a neuron, but there are many other interpretations of q_j depending on the kind of system. In order to characterize the activity of all the parts of a system, we must list the individual components. It is convenient to combine them into the state vector

$$q = (q_1, q_2, \ldots, q_n). \tag{4.1}$$

In a number of cases, the components may be considered as being continuously distributed, for instance, we may treat a fluid as a continuum. In such systems we shall replace the index j by the spatial coordinate $x, y, z = x$ and

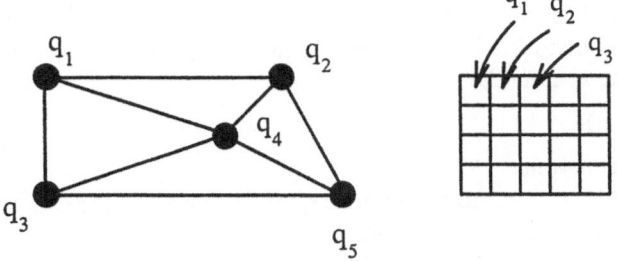

Fig. 4.1. Examples of systems described by the variables $q_1, q_2, \ldots$. *Left*: A network with nodes whose activities are described by these variables. *Right*: An arrangement of cells, for instance in a tissue, where the state of each cell with index j is characterized by a variable q_j. Such a variable may refer, for instance, to the concentration of a particular chemical in such a cell

q_j by $q(x)$. An example of $q(x)$ is provided by the density of molecules in a fluid where the density varies as a function of the spatial coordinate x. In many cases, the state of a system will change in the course of time, i.e., the state vector q (4.1) becomes a function of time t

$$q = q(t). \tag{4.2}$$

In order to study this temporal change in synergetics, we adopt the following attitude: We assume that the temporal change of q may be determined by a number of factors, namely

1) by the present state q of the system
2) by the connections between the components q_j
3) by control parameters α and
4) by chance events.

Let us discuss these different factors in more detail.

1. Present state. An example is provided by the overdamped motion of a point mass fixed to a spring. There is a certain equilibrium position of the mass. When we elongate the spring and let the point mass move, under usual conditions it will undergo an oscillatory process. When we suppose, however, that the whole process is going on, say, in some kind of molasses, the friction may be so strong that no oscillations can occur. In such a case, we obtain overdamped motion. If we denote the deviation from the equilibrium position by q, the equation of motion can be cast into the form

$$\dot{q} = -kq \tag{4.3}$$

where k is a constant. Clearly, the temporal change of q (left-hand side of (4.3)) depends on its present state, namely q (right-hand side of (4.3)).

2. Links between the q_j. An example is provided here by a spring that connects two point masses (1) and (2) with positions q_1 and q_2, respectively (Fig. 4.2). If a is the equilibrium distance between the point masses, then the force acting on point mass 2, is given by

$$F = -k(q_2 - q_1 - a), \tag{4.4}$$

where k is the spring constant.

This example shows that two quite different quantities enter in the equations of motion, namely on the one hand the variables q_j and, on the other hand, specific constants, such as k and a. This also holds in far more complicated cases, for instance, the links can be the synaptic strengths between

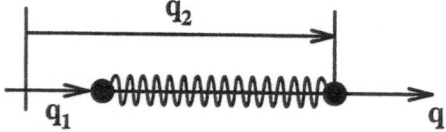

Fig. 4.2. Two point masses at positions q_1, q_2 coupled by a spring whose equilibrium length is a

two neurons. In continuously distributed media, such as fluids, or chemical distributions, where we deal with space-dependent concentrations $q(x)$ of
molecules, the links may have the form of a gradient or may depend on still
higher spatial derivatives of $q(x)$.

3. Control parameters. Let us consider an example to which we will return
later, namely a fluid in a vessel that is heated from below and cooled at
its upper surface (Fig. 4.3). Because of the cooling and heating, a temperature difference between the lower and upper surfaces, $\alpha = T_1 - T_2$, will be
established. As experiments show, the temperature difference has quite a remarkable influence on the behavior of the fluid. If the temperature difference
is smaller than a critical value α_c, the fluid remains at rest. If, however, the
temperature difference α exceeds that critical value, quite suddenly a macroscopic motion of the fluid becomes visible, for instance, in the form of rolls
as is shown in Fig. 4.4. We may say that the macroscopic behavior of the
system is controlled by the *control parameter* α. Similar dramatic changes
may be observed in chemical reactions. When, in a chemical reactor to which
chemicals are continuously added and removed by an overflow, the concentration of some chemical is increased beyond a critical value, an oscillation
may suddenly set in. The chemicals may change their color, say, from red to
blue to red, and so on. In this case, the concentration of the added chemical
serves as a control parameter. Even in the most complex system, namely the
brain, we may identify control parameters. These may be the concentrations
of neurotransmitters, such as serotonin or dopamin, or the concentrations of
administered drugs, such as Haloperidol, caffein, etc. In a number of cases
the concentrations of hormones may also be considered as control parameters. Why these chemicals may be considered as control parameters in the
brain will be discussed later in this book.

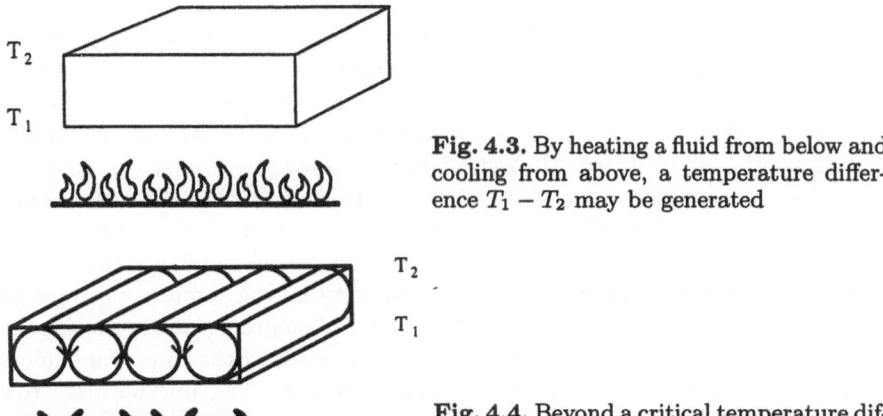

T_2

T_1

Fig. 4.3. By heating a fluid from below and
cooling from above, a temperature difference $T_1 - T_2$ may be generated

T_2

T_1

Fig. 4.4. Beyond a critical temperature difference, a fluid may develop a roll pattern

Some caution must be used, however, when we apply the concept of control parameters to biological systems. In physical and chemical systems, we fix the value of the control parameter(s) by imposing experimental conditions from the outside, for instance, by the amount of heating in the case of a fluid. In biological systems the control parameters are quite often produced by the system itself and are thus in a way variables. But these variables change slowly compared to the actions they trigger. We shall elucidate this interrelation between control parameters and other variables in biological systems later in a number of explicit examples.

4. Chance events. Finally, we have to discuss chance events or random events. Many fields of science are dominated by the idea that there are no chance events, but rather that all processes are entirely deterministic. There are, however, a number of important processes which involve chance events. They are abundant in quantum physics, which deals with the behavior of atoms and molecules. An example is provided by the radioactive decay of the nucleus of an atom. When a nucleus is radioactive, or has been made radioactive by nuclear collisions, we cannot predict when it will undergo radioactive decay by emitting an elementary particle, say an electron, or a γ-ray. The same is true for the spontaneous emission of light by atoms. When an atom is excited, it may emit a light wave, or, in quantum mechanical terms, a photon, but we are unable to predict the emission time. Precisely speaking, all elementary processes in chemical reactions are of quantum mechanical and, therefore, of a random nature. According to our understanding of physics, these chance events are fundamental and cannot be predicted by the development of a more detailed theory.

There is yet another kind of randomness, namely that of thermal fluctuations, e.g., density fluctuations in gases, fluids, or solids, or electric current fluctuations in semiconductors and metals. Such fluctuations occur even if we neglect the quantum nature of molecules, i.e., if we treat them as classical particles. In this case we are seemingly dealing with chance events (of fluctuations) because of our lack of knowledge of the precise positions of the molecules. Thus, except in quantum theory, the question of whether we should speak of chance events or not may depend on our level of description. For instance, we may treat the motion of gas atoms according to classical mechanics, i.e., according to a fully deterministic theory. Nevertheless, we may treat the fluctuations of density as if they were chance events that can be described by statistical theories. There is good reason to believe that chance events also occur in our brain. Such events may be the spontaneous opening of vesicles in neurons, or the random firing of neurons, or the occurrence of tremor. But with our present state of knowledge of the important microscopic processes in the brain, it is not clear whether their fluctuations are of a fundamental, quantum mechanical nature, or merely depend on our level of description.

4.2 Strategy of Solution

4.2.1 Instability, Order Parameters, Slaving Principle

Quite obviously, the systems that we sketched in the preceding section, are very general and many that we are going to treat are highly complex. Therefore, finding a general recipe for how to treat these systems theoretically seems to be hopeless, and we are confronted with the same problems we discussed in Sect. 3.2. To find a way out of this dilemma, we shall be guided by experiments and by the basic idea of synergetics: look for qualitative changes at macroscopic scales! Indeed, there is a large class of phenomena in which complex systems change their macroscopic states qualitatively. We mentioned the example of a fluid in Sect. 4.1. When a control parameter α (in that case the temperature difference) is changed beyond a critical value, the system suddenly forms a new macroscopic state that is quite different from the formerly homogeneous state in which the fluid was at rest. There are numerous other examples in physics and chemistry (Figs. 4.5 and 4.6) and quite similar phenomena are observed in pattern formation in biology (Fig. 4.7). Let us look more closely at the example of the fluid in which the roll pattern forms.

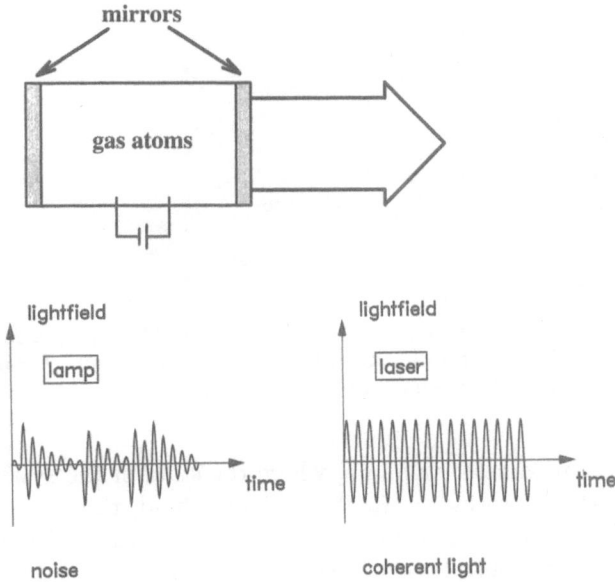

Fig. 4.5. *Upper part*: Typical set up of a gas laser. A glass tube is filled with gas atoms and two mirrors are mounted at its end faces. The gas atoms are excited by an electric discharge. Through one of the semi-reflecting mirrors, the laser light is emitted. *Lower part, left*: In a conventional lamp, the light field generated by the gas atoms consists of individual wave trains that represent noise. *Right*: In the laser, a highly ordered, i.e. coherent, light wave is generated. (For more details, see Sect. 4.2.2.)

Fig. 4.6. In certain chemical reactions circular waves emerge spontaneously, as shown schematically in this figure

Fig. 4.7. This tropical fish shows a pronounced stripe pattern. The emergence of such a pattern can be explained by a mechanism first proposed by *Turing* (1952). (For a treatment in the framework of synergetics see *Haken* (1993))

We start with a control parameter value α_0, where the state of the system is known. We denote this state by q_0. In the case of a fluid, the state q_0 is the resting state. Then we change the control parameter until suddenly a new state occurs. After the control parameter has been changed, the fluid starts its motion, or, in other words, the rotation speed of the rolls increases in the course of time. This increase indicates an *instability*: The old resting state has become unstable. The newly developing motion grows out of a small fluctuation, such as the density fluctuations of the molecules of the fluid or its local velocity fluctuations. The first step of an analysis consists in a study of the behavior slightly above the instability point. As it turns out, the system may undergo quite different collective motions, a few examples

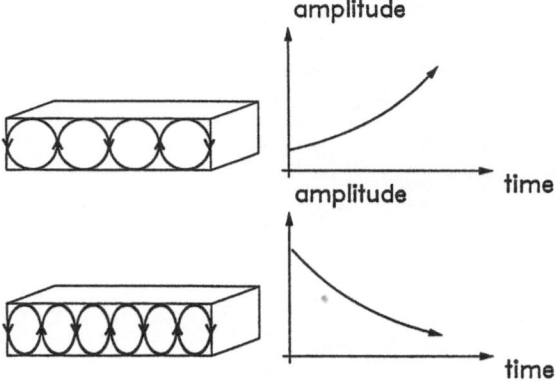

Fig. 4.8. *Left*: Two different roll configurations in a fluid. *Right*: The behavior of the amplitudes of these configurations in the course of time. While in one case the amplitude increases, in the other it decays. (For a definition of the amplitudes, see Fig. 4.9.)

of which are shown in Fig. 4.8 (left-hand side). The amplitude (Fig. 4.9) of some of these configurations tends to grow, whereas others decay even if they have been initially triggered by some fluctuation (Fig. 4.8, right-hand side). The amplitude ξ of the growing configuration is of particular interest for the further analysis, because it determines the evolving macroscopic patterns, for instance, a roll pattern or other patterns (Fig. 4.10). When pattern formation starts, the amplitudes are initially small. This allows us to study the growth

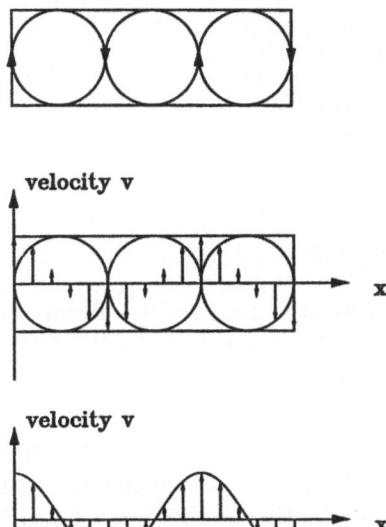

$$v = \xi \sin(2\pi x/\lambda)$$

Fig. 4.9. *Upper part*: A roll configuration as shown in Fig. 4.8. *Middle part*: When we draw a horizontal middle plane through the liquid, we may plot the local velocity versus the space point x, as shown by the arrows. *Lower part*: In a more precise description, the velocity v changes continuously along the x-axis according to the formula $v = \xi \sin(2\pi x/\lambda)$, where λ is the wavelength

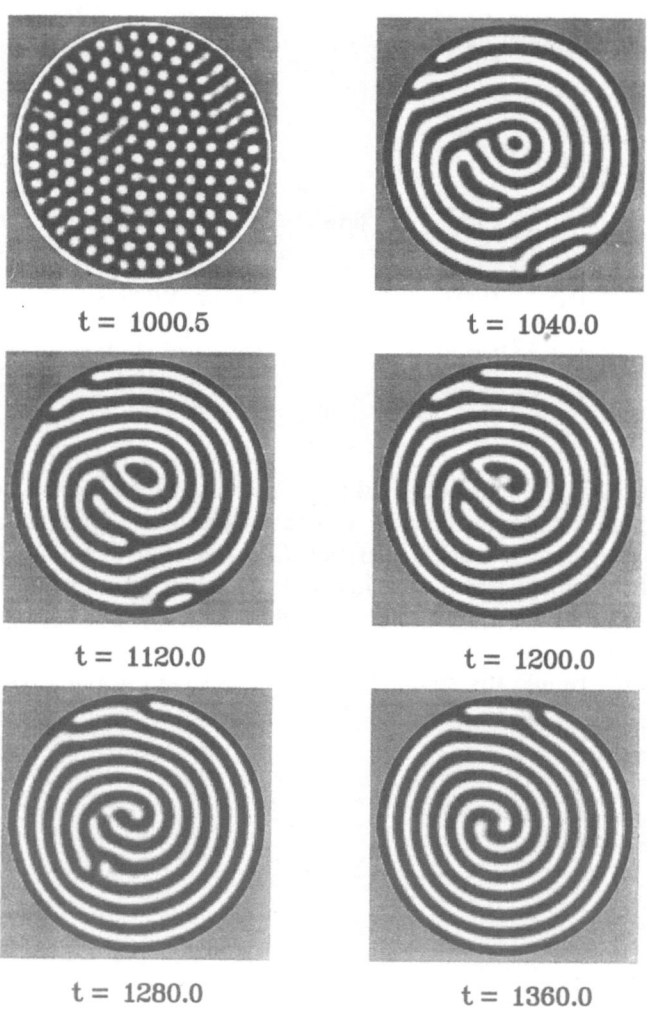

t = 1000.5   t = 1040.0

t = 1120.0   t = 1200.0

t = 1280.0   t = 1360.0

Fig. 4.10. An example of the transition from a hexagonal pattern in a liquid to a spiral pattern. This transition occurs when the walls are also heated (after *Beste-horn* et al. (1993))

and decay of these individual configurations independently of each other. When the amplitudes grow further, the configurations start to influence each other. For instance, they may compete with each other so that only one wins the competition and suppresses all the other configurations. In other cases, they may coexist or they may even stabilize each other. The amplitudes of the growing configurations are called *order parameters*. They describe the macroscopic order, or, more generally speaking, the macroscopic structure of the system. The state q of the system can be described by a superposition of all configurations, i.e., the growing and the decaying ones.

If a system has many components, there are correspondingly many individual configurations. That means the information needed to describe the behavior of the system is not reduced by a decomposition into its configurations. But now a central theorem of synergetics comes in. It tells us that not only the behavior of the growing configurations, but also that of the decaying configurations is uniquely determined by the order parameters. As a consequence, the total space-time behavior of the state q is governed (or enslaved) by the order parameters. This is the *slaving principle* of synergetics (Fig. 4.11). Since, in general, the number of order parameters is much smaller than the number of components of a system, in this way we achieve an enormous reduction of the degrees of freedom, or in other words, an enormous information compression takes place. In a way, the order parameters act as puppeteers that make the puppets dance. There is, however, an important difference between this naive picture of puppeteers and what is happening in reality. As it turns out, by their collective action the individual parts, or puppets, themselves act on the order parameters, i.e., on the puppeteers. While on the one hand the puppeteers (order parameters) determine the motion of the individual parts, the individual parts in turn determine the action of the order parameters (Fig. 4.12). This phenomenon is called *circular causality*. The principle of circular causality allows us to interpret the slaving principle in yet another fashion. Because the individual parts of the system determine or even generate the order parameters which in turn enslave the individual parts, the latter determine their behavior cooperatively. It is tempting to describe this phenomenon in anthropomorphic terms as *consensus finding* by the individual parts. Thus enslavement and *consensus finding* are two sides of the same coin.

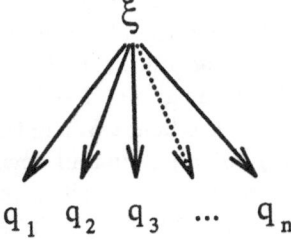

Fig. 4.11. Visualization of the slaving principle. One or several order parameters enslave the behavior of the subsystems described by the variables $q_1, q_2, \ldots$

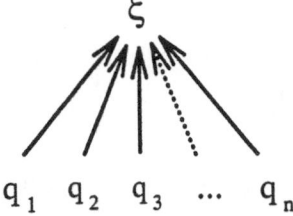

Fig. 4.12. The subsystems described by variables $q_1, q_2, \ldots$ act on the order parameter and even generate it

4.2.2 The Laser Paradigm or Boats on a Lake

Let us illustrate these relationships by another example from physics, namely the light source laser (Fig. 4.5). This device produces a unique kind of light, namely coherent light. Let us consider the example of a gas laser. A glass tube contains a gas of laser-active atoms. By means of an electric discharge, the individual atoms of the gas are energetically excited. After excitation, they can act like a miniature radio antenna, but emitting light waves instead of the radio waves of conventional antennas. If the rate of excitation is increased, more and more light waves can be emitted. They are reflected by the mirrors, which serve to trap the light wave in the laser device for some time. Thus the concentration of light waves grows. Beyond a critical density of light waves, a new process sets in, one that was first introduced by *Einstein*. Namely, when a light wave hits an excited atom, it may force the atom to give away its energy to exactly this wave so that the light wave is enhanced (Fig. 4.13). This process is called *stimulated emission*. The enhanced light wave may hit a further excited atom and quite obviously in this way a light avalanche may start. The laser has reached its unstable state. In exact analogy to the fluid, where different configurations may appear (Fig. 4.8), different kinds of light wave may be emitted and may tend either to grow or to decay. A competition between the growing waves sets in and is, eventually, won by one of them. The amplitude of this wave is the order parameter. To discuss the slaving principle in a more explicit fashion, we remind the reader that in physics the light field is represented by its electric field strength. According to the slaving principle, the order parameter(s) enslave the individual parts; in the present case, the individual parts are the gas atoms. Each atom may be considered as being composed of a nucleus and one specific electron orbiting around it. As is well-known, electric fields act on charged particles such as electrons. When the light wave oscillates, it acts in an oscillating fashion on the electron and forces it to move in phase. This process can be visualized as follows: Consider a lake on which boats are floating (Fig. 4.14). An electric field may be put in analogy to a water wave and the individual electrons to the boats. When the water wave propagates across the lake, the boats will be pulled up and down according to the wave. Similarly, electrons move according to the oscillations of the electric field strength. This is the slaving principle in action. Under the influence of the electric field, the electron starts an oscillation. On the other hand, it is known from electrodynamics that oscillating charges, in our

atom   atom   atom   avalanche

Fig. 4.13. By the process of stimulated emission of light by excited atoms, a light avalanche is built up

a)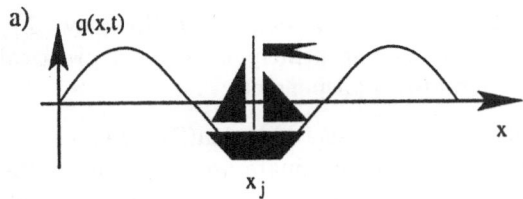

b)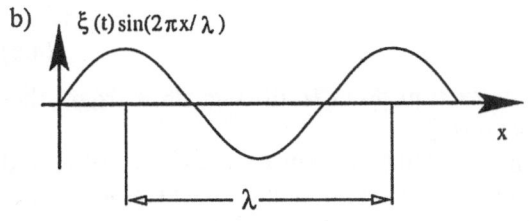

Fig. 4.14a–c. Visualization of the slaving principle by the movement of a boat on a lake. (a) Position of the boat on the wave on a lake. (b) Description of the wave amplitude (4.5) as a function of space x and time t. The up-and-down movement of the wave is described by the time-dependent function $\xi(t)$. (c) By means of the wave in (b), the height of the boat at position x_b is uniquely determined

c)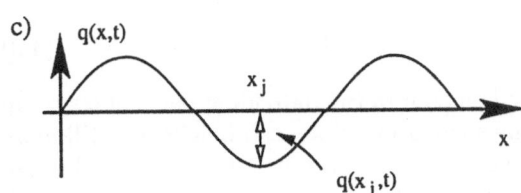

case oscillating electrons, generate an electric field. Again we find the circular causality principle.

4.2.3 The Slaving Principle

Our picture of boats floating on a lake allows us to illustrate the mathematical content of the slaving principle in some detail. We may describe a standing wave on the lake in the form

$$w(x,t) = \xi(t)\sin(2\pi x/\lambda), \tag{4.5}$$

where λ is the wavelength (Fig. 4.14). The amplitude $\xi(t)$ plays the role of the order parameter. Since the wave goes up and down, $\xi(t)$ depends on time in an oscillatory fashion:

$$\xi(t) = a\sin(\omega t), \tag{4.6}$$

where a is a constant amplitude and ω the (circular) frequency. Let us consider a boat at position x_j (Fig. 4.14). We denote its displacement along the vertical axis at time t by $q(x_j,t)$. Quite clearly, this displacement is given by

$$q(x_j,t) = \xi(t)\sin(2\pi x_j/\lambda). \tag{4.7}$$

The variable of the enslaved subsystem, namely the boat with number j, depends on the size of the order parameter, $\xi(t)$, at the same time. In addition, the factor $\sin(\pi x_j/\lambda)$ occurs and establishes a specific link between the order parameter ξ and the variable of the subsystem, x_j. In this way (4.7) captures an essential ingredient of the slaving principle. In many practical cases generalizations of (4.7) are, however, indispensible.

1) The spatial part, which in (4.7) is represented by $\sin(2\pi x_j/\lambda)$, can be a more complicated function of the spatial coordinate and depends on the problem treated. Denoting this function by $v_u(x_j)$, we may replace (4.5) and (4.7) by the more general form

$$q(x_j, t) = \xi(t)v_u(x_j). \tag{4.8}$$

(We have equipped v with the index u in order to distinguish it from other functions that we shall introduce now.)

2) In the problems of synergetics, which are always nonlinear, (4.7) and (4.8) represent only a first (though leading) approximation. Therefore, higher order terms in ξ must be added so that (4.8) is replaced by

$$q(x_j, t) = \xi(t)v_u(x_j) + \xi^2(t)v_2(x_j) + \xi^3(t)v_3(x_j) + \dots . \tag{4.9}$$

The relationship between the behavior of the individual parts of a system and the order parameters, i.e. the slaving principle, can be given a still more general and abstract form. It does not matter whether the parts are discrete or continuously distributed. Let us again denote the variable (or a set of variables) that describes the part j by q_j and consider several order parameters $\xi_1, \xi_2, \dots, \xi_M$. Then the slaving principle states

$$q_j = f_j(\xi_1, \dots, \xi_M). \tag{4.10}$$

Or in words: The state q_j of part j is uniquely determined by the order parameters. In particular this holds true if the order parameters are time dependent. We shall call a configuration of a system that is described by a state vector of the form $q = (q_1, \dots, q_N)$ with q_j given by (4.10) a *collective mode*, or, for short, a *mode*. Since, in general, the number of order parameters M is much smaller than the number of the parts of a system, (4.10) is a mathematical expression of the *information compression* mentioned above.

To finish this little excursion into the mathematical description of the slaving principle, let us consider a *propagating* instead of a standing wave. The amplitude of such a wave at space point x and time t may be written as

$$w(x, t) = \xi(t)\sin(2\pi x/\lambda - \phi(t)) \tag{4.11}$$

where $\phi(t)$ is a *phase*. Quite clearly, the displacement of a boat is not only determined by the amplitude ξ, but also by the phase ϕ. *Phases can be order parameters!*

At first sight, the illustration of the slaving principle in the cases of the fluid and of the laser seems to be rather different, because in the first case we have been speaking about configurations or collective motions of the fluid whereas in the case of the laser we have been considering the motion of individual electrons in the laser atoms. However, mathematical analysis, which we shall not dwell on here, shows that both descriptions are equivalent and depend on the kind of problem.

Both cases have a fundamental common feature, namely in both cases we can clearly distinguish between the role of order parameters and of the enslaved variables or components. When we perturb the order parameters and then drop the perturbation, the order parameters relax only slowly. When we perturb the individual parts, they relax quickly, i.e., order parameters and parts (components) are distinguished by their time scales. If we wish to apply the principles of synergetics to the brain, we must thus ask whether the time-scale separation holds. Let us consider the neurons as individual enslaved components, and macroscopic actions, such as perception or movement control, as the macroscopic quantities described by order parameters. The individual neurons have time constants of one millisecond, whereas perception, etc. takes place at time scales of about hundred milliseconds. The time-scale separation is certainly well fulfilled in the brain.

4.2.4 The Central Role of Order Parameters

Let us summarize the results of this section. We considered a class of systems that have the following properties: When one, or maybe several, control parameters are changed, the system enters an instability. In other words, it leaves its former state and starts to form a qualitatively new macroscopic state. Close to the instability point, different kinds of collective configurations occur; some of them grow, whereas others decay after their generation by fluctuations. By a study of the growing and decaying states, we may distinguish between the unstable and stable configurations and are thus led to the configurations which are governed by the order parameters. The order parameters determine the behavior of the individual parts via the slaving principle. Thus the behavior of complex systems can be described and understood in terms of order parameters. At the same time, we need no longer consider the action or behavior of the individual parts, but may instead describe the total system by means of the order parameters. The slaving principle underlying this relationship thus leads to an enormous information compression, which we shall use as a tool to study brain, behavior, and cognition. In quite a lot of cases, the number of order parameters is very small and we may discuss the behavior of the system in terms of these order parameters. For instance, in the laser the number of atoms, may be, say, 10^{18}, whereas the number of order parameters is just one, namely the electric field strength of the winning mode. In addition we may state that via this order parameter the motion of the individual electrons becomes highly correlated. Similarly, in the brain we are dealing with myriads of neurons, but with much fewer behavioral patterns (which are still numerous, however!). As we shall see more clearly in Chap. 5, order parameters are abstract quantities. In many cases, they acquire their specific meaning via the slaving principle.

One warning should be added at the end of this section. Because the principles of synergetics have been illustrated by means of examples from physics, namely liquids and lasers, one may prematurely draw the conclusion that synergetics represents a physicalism. This, however, is not true at all, because in synergetics we start from abstract mathematical relationships, which then are applied to numerous systems including those of physics. But because physical systems are still comparatively simple as compared to biological systems, they provide the nicest way to illustrate the meaning of the mathematical principles of synergetics.

In this section we have discussed the contents of the slaving principle more or less qualitatively, but in a way that is sufficient for an understanding of the subsequent parts of the book. Readers who are interested in the mathematical proof of the slaving principle (which is actually a theorem) are referred to my book *Advanced Synergetics* (see References). The proof requires that the system considered is close to an instability point. General arguments as well as numerous practical applications of the slaving principle have shown, however, that in quite a number of cases it also holds in systems farther away from their instability points. For further information on these, see the works listed in "References and Further Reading".

4.3 Self-Organization and the Second Law of Thermodynamics

Our previous discussion clearly shows what we understand by self-organization of a system. Namely, quite generally, when certain external or internal control parameters are changed, there are certain situations where the system itself does not change slightly, but undergoes a dramatic change of its macroscopic state as is witnessed by the spontaneous formation of patterns in lasers and liquids. It is superfluous to say that biology and medicine abound with such qualitative changes that range here from the formation of spatial patterns or structures (morphogenesis) to dramatic changes in behavior, for instance, in psychosis. It is important to note that the control parameters do not anticipate the evolving macroscopic patterns. For instance, a liquid is heated quite uniformly; nevertheless the interaction between its molecules produces quite distinct patterns.

The spontaneous formation of patterns by self-organization seemed to be in conflict with the second law of thermodynamics. According to this law, in so-called closed systems macroscopic order should disappear and be replaced by a homogeneous state that may, however, show a chaotic motion at its microscopic level. For instance, the motions of gas atoms are quite chaotic, but at the macroscopic level a gas appears to be practically uniform. This tendency towards a maximally chaotic state at the microscopic level and a structureless state at the macroscopic level is formulated by the statement that the entropy of a system increases to its maximum value. However, this statement only holds for closed systems which do not exchange energy or matter with their surroundings. The biologist *von Bertalanffy* recognized that biological systems are *open* systems, whose structures and functions are maintained by an influx of energy and matter either in the form of sunlight and material as used by plants, or in the form of nutrients and oxygen as used by animals. *Von Bertalanffy* coined the term *Fliessgleichgewicht* (flux equilibrium) to characterize this state of living matter. All systems treated in synergetics may be considered as open systems and thus fulfill a necessary condition for self-organization.

5. Dynamics of Order Parameters

5.1 One Order Parameter

In the last chapter we saw that even the behavior of complex systems may, close to their instability points, be governed by very few variables, namely the order parameters. In a number of important cases, we deal with very few order parameters and we shall start here with the discussion of the behavior of a single order parameter. To describe the behavior of such an order parameter below and above the instability point of a system, a mechanical model has proved to be very useful, because it allows a simple interpretation. We identify the size of the order parameter ξ with the coordinate of a ball that slides down a hill towards the bottom of a valley. In technical terms, we are dealing with the overdamped motion of a particle under the gravitational force, but under the constraint that the particle moves on a parabola (Fig. 5.1, upper part). According to the overdamped motion, the speed with which the particle moves is proportional to the size of the slope. The steeper the slope, the higher the velocity. Quite clearly, at the bottom of the valley the slope is equal to zero and the particle has come to rest. Typical order parameters are subject to fluctuations. These fluctuations may be visualized as the kicks of soccer players hitting the ball. Quite often in actual games, the players hit the ball entirely at random, a beautiful picture of the impact of fluctuations! Because of these kicks, the ball will not remain at rest, but will go to the left- or to the right-hand side, then slide down, will be kicked up again, and so on. The situation just described applies to the region below threshold or, in other words, below the critical value of the control parameter.

If the control parameter is changed so that it reaches its critical size, where the instability sets in, the landscape is deformed (Fig. 5.1, middle part). The valley becomes very flat. When the ball is now displaced from the origin, because the driving force is practically zero, it will take a very long time for the ball to relax to the equilibrium point at $\xi = 0$. This phenomenon is called *critical slowing down*. When we remember that the ball is still subject to fluctuations and the restoring force is very small, we realize that the ball may move far away from its equilibrium point, until it hits the rather steeply rising slopes. Thus the ball undergoes a random motion, or, in other words, it displays large fluctuations. In technical terms, we are speaking of *critical fluctuations*.

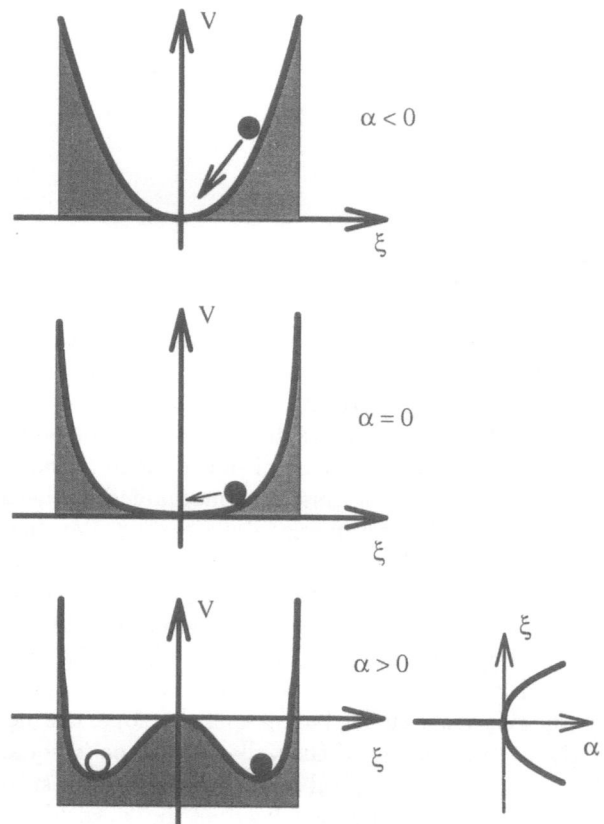

Fig. 5.1. This figure allows us to visualize the behavior of the order parameter ξ for different values of the control parameter α, where in the present case the critical value is $\alpha = 0$. *Upper part*: Below the critical value of the control parameter, the order parameter behaves as a ball that slides down a hill towards the bottom of the valley. We shall call this picture a *potential landscape*. Middle part: At the critical point, the bottom of the valley becomes very flat. *Lower part (left)*: Above the critical value of the control parameter, the formerly stable position $\xi = 0$ has become unstable and is replaced by two stable positions. *Lower part (right)*: Here we plot the equilibrium position of ξ as function of α. For $\alpha < 0$ there is only one stable position for ξ corresponding to the bottom of the valley in the upper part of this figure. Above threshold, i.e. for $\alpha > 0$, there are two stable branches corresponding to the stable positions at the bottoms of the valley in the lower part of this figure and one unstable branch corresponding to $\xi = 0$

Finally, when the control parameter is set to a value above its critical point, the landscape is qualitatively deformed, and instead of the one former valley, two valleys appear (Fig. 5.1, lower part). Each bottom of the two valleys represents a stable position. Since there are two stable positions, we speak of *bistability*. In the present case, the landscape is symmetric with respect to the vertical V-axis, but the ball can occupy only one valley. It has thus to choose between two possibilities that are symmetric. We are therefore speaking of *symmetry breaking*. Since the former position in Fig. 5.1 at $\xi = 0$ has become unstable and is now replaced by a new stable position, we also speak of a *symmetry-breaking instability*. The phenomena of critical slowing down, critical fluctuations and symmetry breaking had been observed previously during so-called phase transitions of systems in thermal equilibrium. Such phase transitions take place, for instance, when water freezes to ice, or when a magnetic material becomes magnetic, or when conductors become superconducting. We have found that these phenomena also occur in systems far from thermal equilibrium which are kept in their active state by a continuous flux of energy into the system and by dissipation of energy. An example is the laser, which we discussed in Sect. 4.2.2. Because we are dealing with systems far from equilibrium, we may speak here of *nonequilibrium phase transitions*. In the realm of synergetics, we may consider critical fluctuations and critical slowing down as strong indications for processes of self-organization, because the fluctuations exerted on the order parameter stem from the random, i.e. incoherent, movement of the individual parts of a system. When the system becomes destabilized at the critical point, these fluctuations grow until the system reaches a new stable state in one of the two valleys at the corresponding value of the order parameter. When the control parameter is further changed, the valleys soon become very steep so that the fluctuations become extremely small. This phenomenon has been studied in great detail in the case of the laser. We shall see below that the same phenomenon also occurs in biological movement coordination.

To conclude of our discussion of the behavior of a single order parameter, we show another plot of the behavior of the order parameter (Fig. 5.1, lower part, right-hand side). In this case, we plot the equilibrium position ξ versus the control parameter α. For $\alpha < \alpha_c$, the order parameter value is $\xi = 0$. For $\alpha > \alpha_c$, we obtain three branches. The branch at $\xi = 0$ corresponds to the position that has now become unstable in Fig. 5.1, lower part, left-hand side, whereas the upper and lower branches refer to the positive and negative stable points in Fig. 5.1, lower part, left-hand side. Since the figure in which ξ is plotted versa α has the shape of a fork, one speaks of the phenomenon of *bifurcation*, which is dealt with in mathematics by the branch of bifurcation theory. A decisive difference between our considerations in synergetics and those of bifurcation theory must be noted. In the latter approach the impact of fluctuations is entirely neglected so that one of the important indicators of self-organization is neglected here. In addition, in synergetics we study the relaxation behavior of the order parameter, for instance, its critical slowing down; another phenomenon that is not treated in conventional bifurcation theory. For those interested in mathematics, we write down the order parameter equation belonging to Fig. 5.1 in the form

$$d\xi/dt = \alpha\xi - \beta\xi^3 + F(t), \tag{5.1}$$

where the fluctuating force $F(t)$ is usually assumed to be Gaussian distributed and having white noise. In (5.1), α plays the role of the control parameter, where $\alpha_c = 0$. Equation (5.1) can also be written by means of a *potential V* in the form

$$\frac{d\xi}{dt} = -\frac{\partial V}{\partial \xi} + F(t), \qquad (5.2)$$

where

$$V = -\frac{\alpha}{2}\xi^2 + \frac{\beta}{4}\xi^4. \qquad (5.3)$$

$V = V(\xi)$ is precisely the curve that is plotted in Fig. 5.1, where $\alpha < 0$ in the upper part, $\alpha = 0$ in the middle part, and $\alpha > 0$ in the lower part of this figure. Another important possibility is shown by the sequence of landscapes in Fig. 5.2, which stem from an equation of the form

$$d\xi/dt = \alpha\xi - \beta\xi^2 - \gamma\xi^3 + F(t) \qquad (5.4)$$

with control parameters α and β. When we start in the upper left row of Fig. 5.2 and change the control parameter, the ball first remains at its equilibrium position $\xi = 0$. Only at the value α_3 can it jump to its new equilibrium position. However, when we now change the control parameter in the reverse order, the ball stays at the equilibrium position at $\xi = \xi_1$ and only

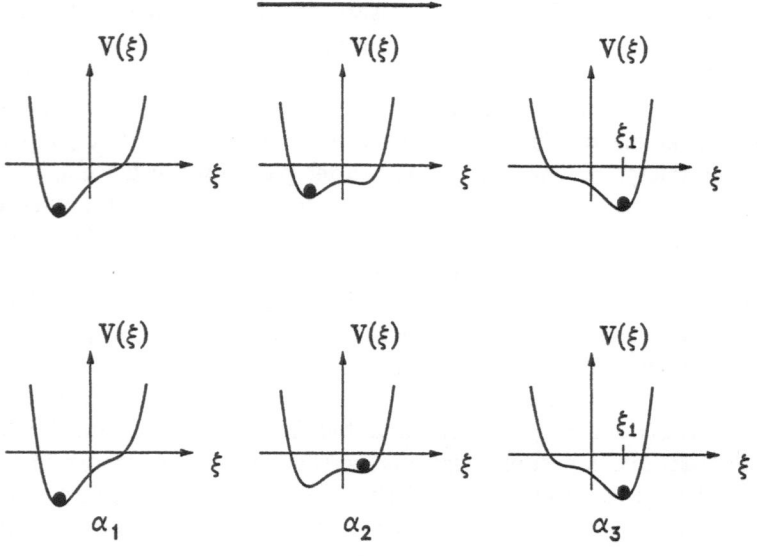

Fig. 5.2. Visualization of the effect of hysteresis. If a control parameter is changed, the potential landscape is deformed as shown in the upper part of this figure. In the lower part, the change of the control parameter occurs in the reverse direction. Note that at $\alpha = \alpha_2$ the order parameter occupies two different positions depending on the history

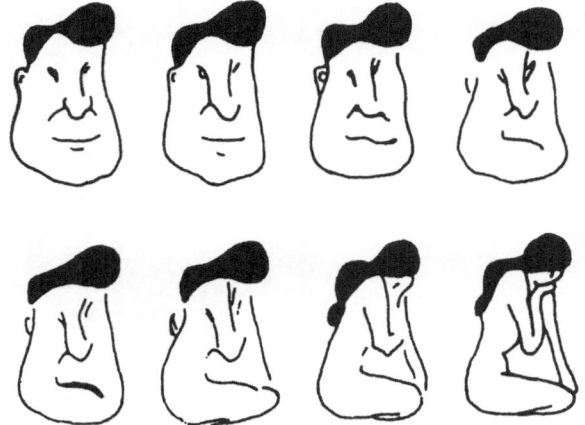

Fig. 5.3. Hysteresis in visual perception. Look first at the first row from left to right and then at the second row from left to right. Then do the same in the reverse order

at the control parameter value $\alpha = \alpha_1$, does the ball return to its original position. Thus in the middle part of the upper and lower row of Fig. 5.2, the ball will occupy *different* positions at the *same* control parameter value. Quite evidently, the actual position depends on the history. In this case one speaks of *hysteresis*. When the bottom of a valley becomes very flat, critical fluctuations and critical slowing down occur once again.

We shall see in the next chapter that the phenomena of critical fluctuations, critical slowing down, and hysteresis may also be observed in biological movement coordination, thus giving us a strong hint that the phenomenon of self-organization occurs here. The phenomenon of hysteresis can easily be observed by the reader in the case of visual perception (consider Fig. 5.3). When we look at this figure, starting from the left and following the first row, we recognize the face of a man that is also recognized when we look at the second row from the left. Only in the middle of this row does a switch to a woman suddenly occur. When we consider the picture in the reverse direction, the switch from the woman to the face of a man occurs at a different, i.e. later, position.

5.2 Two Order Parameters

Let us start with a special case in which the temporal change of the order parameters ξ_1, ξ_2 can be visualized as motion in a landscape. An example of such a motion is shown in Fig. 5.4. When we again look at overdamped motion, the ball will slide down to the bottom of the closest valley. When we plot the time-dependence of ξ_1 and ξ_2 we are led to Fig. 5.5. At each instant of time, ξ_1 and ξ_2 have specific values. Quite often, we wish to represent these values on a single diagram. This is done in Fig. 5.6, where each point on the trajectory shows us, by the projections on the ξ_1 and ξ_2 axes, which values ξ_1 and ξ_2 have at a particular time. As we can see quite clearly from

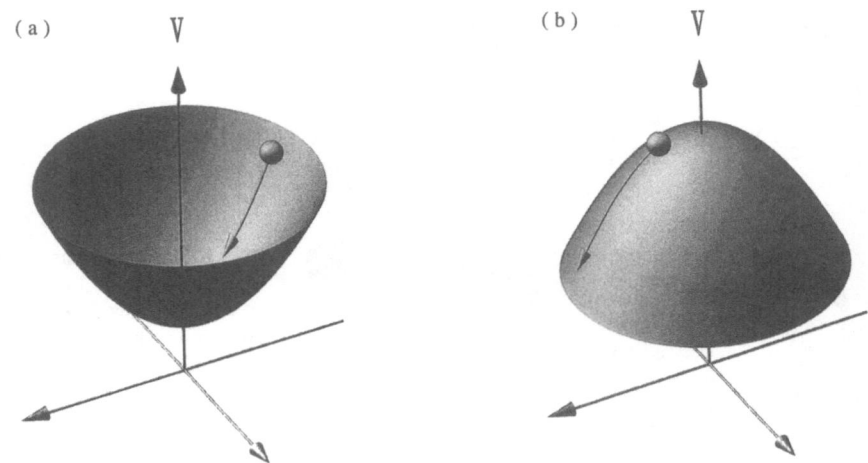

Fig. 5.4. Two examples of potential landscapes: (a) giving rise to a stable fixed point; (b) giving rise to an unstable fixed point

the landscape of Fig. 5.4, the ball will follow different paths depending on its initial starting point. Correspondingly, when we let the ball start from different initial values in the ξ_1, ξ_2 plane, it will follow different trajectories. The whole picture of these trajectories is called a *flow*. An example of a flow for a case where the ball comes to rest at $\xi_1 \neq 0, \xi_2 \neq 0$ is shown in Fig. 5.7. The point at which the ball comes to rest is called a *fixed point* and, more precisely speaking, a *stable fixed point*. When we place the ball at the top of one of the hills, it will stay there in unstable equilibrium. The slightest push, however, will cause it to roll down in any of the directions, depending on the direction of the initial push. When we plot such a flow, then we obtain Fig. 5.8. The point in the middle is called an *unstable fixed point*. We mention that the concept of fixed points and flows is not connected with the picture of a landscape; it is far more general.

Let us consider another typical special case, namely that of *limit cycles*. We again take an example from mechanics, namely a pendulum, as shown in Fig. 5.9, upper part. In mathematics and in theoretical physics we often attach a specific diagram to the motion of the pendulum, namely that of Fig. 5.9, lower part. In it we plot the elongation or position x along the horizontal line, the abcissa. Along the vertical line, the ordinate, we plot the velocity v. Quite clearly, at the right turning point, the position has its maximum value and the velocity is zero. Then the pendulum starts its movement to the left, its velocity increases, but because the velocity points to the left-hand side, it is counted as negative. In this way, the pendulum follows the lower part of the ellipse shown in Fig. 5.9. Then at the lowest point $x = 0$, the pendulum has its maximum speed. When it moves further to the left-hand side, the speed decreases and, finally, the left-hand turning point is reached

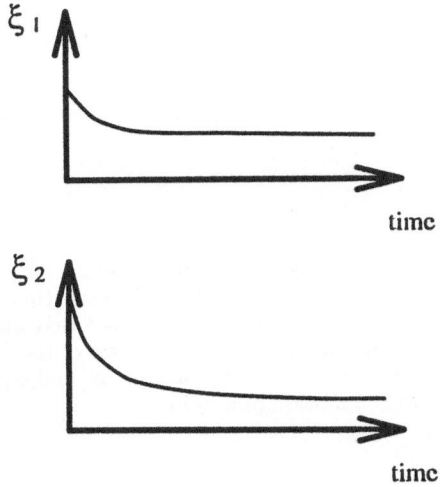

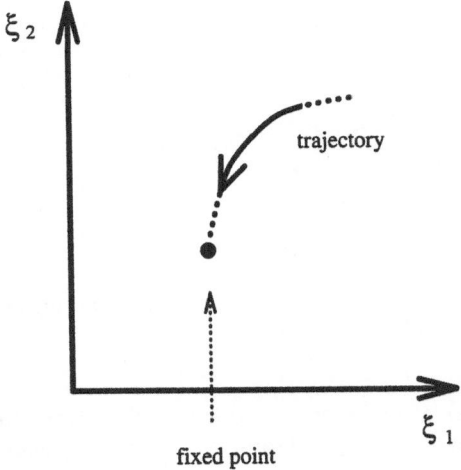

Fig. 5.5. Relaxation of the order parameters ξ_1 and ξ_2 towards a stable state in the course of time

Fig. 5.6. Here the two diagrams of Fig. 5.5 are combined into a single one, in which the trajectory is followed by ξ_1 and ξ_2 in the course of time

with a maximum negative elongation at a vanishing speed. Then the speed increases to the right and we now follow the upper part of the ellipse. In this way, the motion of the pendulum is described as the motion of a point on an ellipse in the so-called phase plane. When we let the pendulum start from a large elongation and zero speed, it will again follow an ellipse but with a bigger diameter. The whole picture, however, is oversimplified, because we have ignored friction. Because of friction, the mechanical energy of the pendulum will be dissipated as heat and the motion of the pendulum will slow down. When we plot this motion including friction in the phase plane, we obtain Fig. 5.10. The pendulum eventually comes at rest at $x = 0, v = 0$. Such a point is called a *stable focus*.

ξ_2

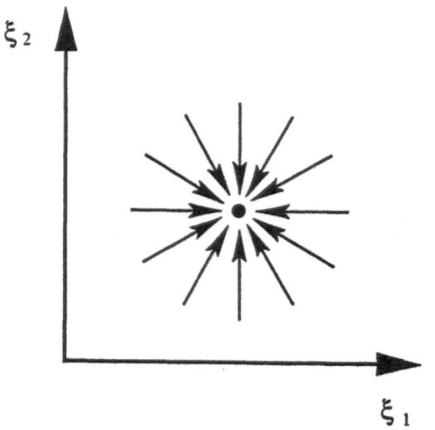

ξ_1

Fig. 5.7. When the initial state of ξ_1 and ξ_2 is chosen differently each time, a whole set of trajectories arise that all approach a fixed point. The set of these trajectories is called a flow in analogy to the motion of a fluid

ξ_2

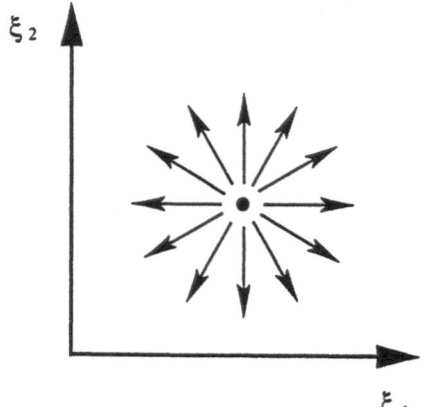

ξ_1

Fig. 5.8. Trajectories around an unstable fixed point

In former times, a pendulum was a necessary component of a clock. In order to overcome friction, one had to give the pendulum regular little pushes to overcome friction and thus to keep it moving. This was done in such a way that whenever the motion of the pendulum was disturbed, it returned to one and the same closed curve, which is called a *stable limit cycle*. A limit cycle may be described either in phase plane or by plotting ξ_1 and ξ_2 individually versus time, as shown in Fig. 5.11. Fluctuations (or pushes) acting on the motion on a limit cycle may have two effects; they may act in a direction vertical to the limit cycle, or parallel to it. Since motion on the limit cycle proceeds in the course of time similarly to the propagating wave of (4.10), we may speak of a phase. In other words, the fluctuations in tangential direction cause *phase fluctuations*.

So far we have been speaking on the motion of order parameters in the case where we keep the control parameter fixed. When we change a control parameter, similar processes as shown for one order parameter may occur,

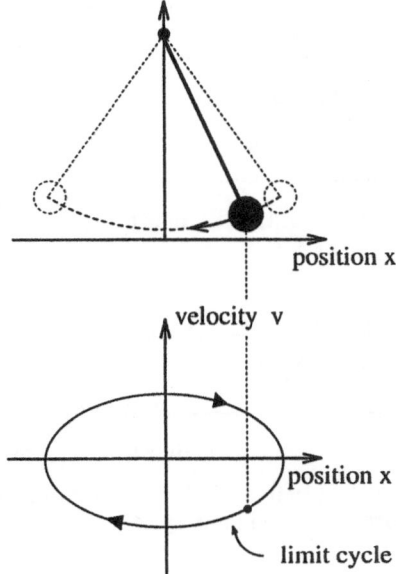

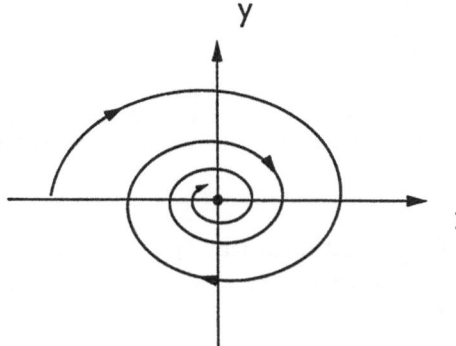

Fig. 5.9. *Upper part:* A pendulum. *Lower part:* A representation of the motion of the pendulum in a plane with the axes position and velocity (compare text)

Fig. 5.10. In the case of a damped pendulum, the ellipse is replaced by a spiral that approaches the origin

namely one stable fixed point may become unstable and will be replaced by two new stable fixed points. Again the phenomenon of bifurcation occurs, or when fluctuations are taken into account we are dealing with a nonequilibrium phase transition. When two order parameters are present, more general transitions may occur also, for instance, a stable fixed point becomes unstable and will then be surrounded by a stable limit cycle. This would happen, for instance, when we kick the pendulum of a clock that was formerly at rest such that it will begin the usual motion of a clock pendulum.

As is well known, physiology abounds with oscillators. The brain has many oscillatory actions, which we shall discuss later in more detail. In vision, too, we can observe oscillations, for instance, when looking at ambivalent

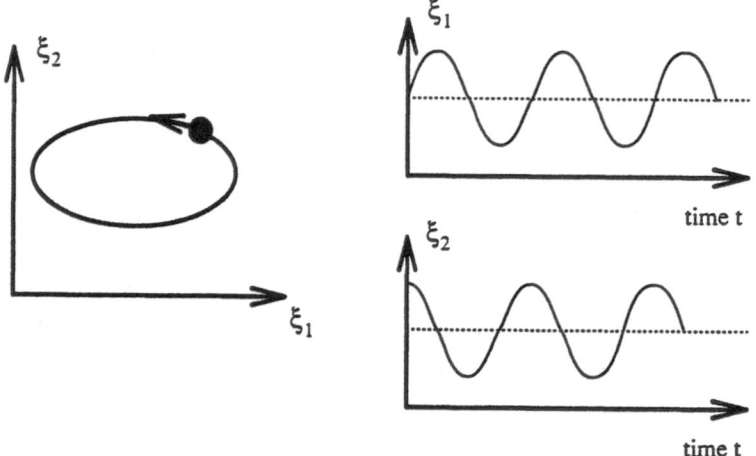

Fig. 5.11. Representation of a limit cycle. *Left*: Limit cycle in the ξ_1, ξ_2-plane. *Right*: Periodic motion of ξ_1 and ξ_2 is associated with the limit cycle

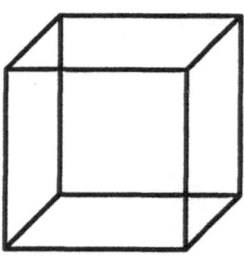

Fig. 5.12. Oscillations in visual perception: the Necker cube. A side that is parallel to the plane of this page may be perceived as front side *or* as rear side of a cube. The interpretation *front* and *rear* oscillates in the course of time

figures, as shown in Fig. 5.12. Finally, for the mathematically interested reader we write down some typical order parameter equations. An example of the dynamics around a stable fixed point at $\xi_1 = 0, \xi_2 = 0$ is described by the equations

$$d\xi_1/dt = -\gamma_1 \xi_1 \tag{5.5}$$

and

$$d\xi_2/dt = -\gamma_2 \xi_2. \tag{5.6}$$

An example of a limit cycle is provided by the equations

$$d\xi_1/dt = \omega \xi_1 + \omega_2 \xi_2 + \gamma[a - (\xi_1^2 + \xi_2^2)]\xi_1 \tag{5.7}$$

and

$$d\xi_2/dt = -\omega_2 \xi_1 + \omega_2 \xi_2 + \gamma[a - (\xi_1^2 + \xi_2^2)]\xi_2. \tag{5.8}$$

A closer investigation of (5.7) and (5.8) shows that the first two terms on the right-hand sides give rise to oscillatory motion, whereas the last terms give rise to a stabilization on a limit cycle with the radius $\xi_1^2 + \xi_2^2 = a$.

5.3 Three and More Order Parameters

The most convenient way to study the behavior of three order parameters is in three-dimensional phase space, i.e. a space that has the coordinates ξ_1, ξ_2, ξ_3. Again we may study the behavior of the flow, now, of course, in three dimensions, and investigate specific behaviors of the flow. First of all, the flow may approach a fixed point, or it may start from a fixed point. Secondly, we may have stable or unstable limit cycles.

Finally, for a long time it was believed that the only further interesting object of a typical behavior of the flow is a so-called torus: Imagine you are sitting on a merry-go-round and are waving a flag in a circle that is perpendicular to the motion of the merry-go-round. When the merry-go-round stands still, the flag is moved, so-to-speak, on a limit cycle. But when the merry-go-round and you yourself are also moving on a limit cycle, the flag moves along a spiral. After one turn of the merry-go-round, the flag may be at precisely the same position as before. It moves again on a limit cycle. In the general case, however, its position will be different and the spiral never joins onto itself. In this case, the spiral fills a torus – an object with the shape of a doughnut (Fig. 5.13). It was a great surprise for the mathematics and physics communities, but not only for them, when it turned out that there is still another strange object: a chaotic attractor. We shall come back to discuss this in Chap. 13.

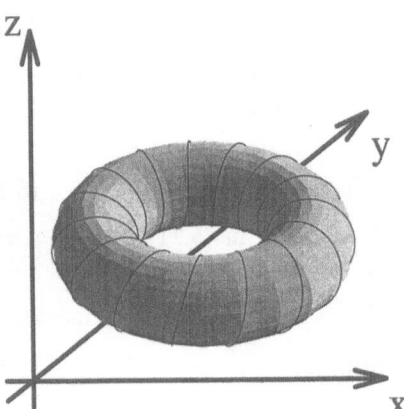

Fig. 5.13. Diagram of a torus

5.4 Order Parameters and Normal Forms *

As we have seen above, in particular close to its instability points, a system's dynamics can be described, at least in many cases, by one or several order parameters. In many important cases, close to the transition point of the system, the order parameters are still small quantities. This allows one to keep only a few leading powers in their equations of motion. For instance, in the case of one order parameter such a typical equation is given by (5.4). If systems possess, for instance, only one order parameter, their equations are the same in spite of the fact that the microscopic elements may be quite different. For instance, the dynamics of a single-mode laser and of a roll pattern in a fluid are governed by precisely the same order parameter equations. One may even go a step further by transforming order parameters so that equations that at first sight look different describe the same dynamics. An important tool in achieving this identification of seemingly different kinds of dynamics is provided by normal form theory. It is definitely not our goal to provide the reader with a complete introduction to that theory, but we wish to give him or her a feeling forwhat is involved.

Let us consider the simple case where the dynamics of an order parameter, which we shall denote by x, is governed by a potential function

$$V(x) = ax^2 + bx + c. \tag{5.9}$$

The relevant dynamics, i.e., the overdamped motion of a particle in this potential field V, is then given by

$$\dot{x} = -\frac{\partial V}{\partial x} = -2ax - b. \tag{5.10}$$

We transform the coordinate x to a new coordinate $\tilde{x}$ by means of the linear transformation

$$x = \alpha\tilde{x} + \beta. \tag{5.11}$$

This transformation means that we shift the origin of the coordinate system along the abcissa and scale x in a new way. Inserting (5.11) into (5.9), we find, of course,

$$V = a\left(\alpha^2\tilde{x}^2 + 2\alpha\beta\tilde{x} + \beta^2\right) + b\left(\alpha\tilde{x} + \beta\right) + c. \tag{5.12}$$

It now our goal to choose the coefficients α and β in (5.11) in such a way that (5.12) acquires a simple form. To this end we require that

$$a\alpha^2\tilde{x}^2 = \tilde{x}^2 \tag{5.13}$$

or, in other words, that α is chosen according to

$$a\alpha^2 = 1 \quad \text{or} \quad \alpha = 1/\sqrt{a}. \tag{5.14}$$

We further require that the term linear in $\tilde{x}$ vanishes

$$(2\alpha\beta a + b\alpha)\,\tilde{x} = 0, \tag{5.15}$$

which is fulfilled if we choose β according to

$$2\beta a + b = 0 \quad \text{or} \quad \beta = -b/(2a). \tag{5.16}$$

Finally, we introduce the abbreviation V_0 via

$$a\beta^2 + b\beta + c = V_0. \tag{5.17}$$

We then may define a new potential function $\tilde{V}$

$$\tilde{V}(\tilde{x}) = V(x) - V_0 = \tilde{x}^2. \tag{5.18}$$

Quite obviously, (5.18) is simpler than (5.9), because all coefficients a, b, c have disappeared or have been replaced by unity. $\tilde{V}(\tilde{x})$ is the normal form of $V(x)$. The equation of motion now becomes very simple, namely

$$\dot{\tilde{x}} = -2\tilde{x}. \tag{5.19}$$

Another example is provided by a potential function of the form

$$V(x) = ax^4 + bx^3 + cx^2 + dx + e, \tag{5.20}$$

which again by a linear transformation can be reduced to

$$\tilde{V}(\tilde{x}) = \tilde{x}^4 + \tilde{b}\tilde{x}^3 + \tilde{c}\tilde{x}^2. \tag{5.21}$$

If we know from other considerations that the potential $\tilde{V}$ is symmetric, i.e.,

$$\tilde{V}(-\tilde{x}) = \tilde{V}(\tilde{x}), \tag{5.22}$$

we may drop the cubic term in (5.21) and obtain as normal form

$$V(\tilde{x}) = \tilde{x}^4 + \tilde{c}\tilde{x}^2. \tag{5.23}$$

Normal forms can also be obtained for potentials that depend on several variables and Thom's catastrophe theory may be viewed as a theory that establishes such normal forms. Normal forms can also be established for equations which are not derivable from a potential. But we shall not dwell on this rather difficult problem here, preferring simply to give the reader a feeling of what is meant by normal forms and to show that normal forms may subsume several different kinds of potential under one single potential. If symmetries are invoked in addition, the number of coefficients can be further reduced.

After this introduction to the basic concepts of synergetics, we are now in a position to treat some important biological problems from a new point of view.

Part II **Behavior**

6. Movement Coordination – Movement Patterns

The subtitle of this book contains the word *behavior*. The study of human behavior is a huge field of research. Under the concept of behavior we may subsume simple movements up to highly complex behavior, such as dealing with other humans or with specific situations. The intention of my approach is to be as operational as possible and to study those experimental phenomena that can be treated in a quantitative manner. This is why I shall choose in this and the following chapters movement coordination. In the spirit of synergetics, I shall be concerned with qualitative changes (that can be measured quantitatively!). It should be noted, however, that qualitative changes can be observed in quite a number of different kinds of human behavior. To take a most striking example, let us think of cases studied in psychiatry. Here in the case of schizophrenia we find well-defined transitions between normal behavior and psychotic episodes. The same is true for transitions between depressions and manic episodes. The most striking feature of these phenomena is the existence of well-defined behavioral patterns; for instance, we can readily distinguish between a normal and a psychotic state of a person, though there are a variety of manifestations of, say, schizophrenia. At any rate we may state that behavioral patterns are quite clearly defined and coherent within themselves. It appears as if the behavior of a person is governed by – what I would like to call – a single order parameter. This picture may be oversimplified, but the main issue of this chapter is to demonstrate that such well-defined transitions occur in movement coordination, and my claim is that from here on we may extrapolate to far more complicated kinds of behavior and transitions between them. In a way the problem will be not so much to model these more complicated transitions, but to develop adequate means to quantify human behavior in complex circumstances. We shall elucidate this kind of problem from a different point of view in Chap. 17, where we shall deal with decision making, which may again be considered as a certain kind of human behavior. Studies of the circumstances under which transitions between behavioral patterns occur are, of course, of practical importance in many respects. Let us again take psychosis as an example. Can one predict when such an event will happen or are there any indicators that a psychosis is starting? When we take seriously the analogy between movement patterns and more general behavioral patterns, I believe one can claim that such indicators exist.

6.1 The Coordination Problem

When humans or animals such as vertebrates move their limbs, dozens of muscles must cooperate in a highly organized fashion. Physiologists have realized for a long time that an explanation of this high degree of coordination is a deep puzzle. The famous British physiologist *Sherrington* (1906) coined the word of *synergy of muscles*. In Russia, *Bernstein* (1967) devoted much thought to this problem and considered it under the assumption that in movement only few degrees of freedom occur. In the US *Gibson* (1979) studied the relationship between the surroundings and subjects and spoke of *affordance*. In Germany *von Holst* (1935, 1939, 1943) did detailed experiments on coordinated motion, for instance, of fins of fish and of the centipede. Here, for instance, he found that when he cut off most of the feet of this animal so that only six feet were left, this centipede moved like an insect; with four feet left, it moved using the gaits of common quadrupeds such as horses. In retrospect, this looks like strong support for self-organization. If we suppose that there is a computer program in the brain of the centipede, it can hardly be imagined that this program would be preprepared for all kinds of losses of feet and for steering the motion of the remaining feet.

Thus we wish to discuss the question of whether the general concepts of synergetics can provide us with a window into the study of movement coordination. As we have seen in Chap. 4, the central strategy of synergetics is based on the idea of studying complex systems at those points where their macroscopic behavior changes qualitatively. First of all we know that humans and quadrupeds have specific movement patterns in locomotion, for instance, horses display different kinds of gait, such as walking, trotting and galloping. So the question arises of whether we can investigate the transitions between gaits in detail and draw conclusions in terms of synergetics. According to its methodology, we have to identify one or several order parameters, one or several control parameters, and look for indications of self-organization, such as critical slowing down and critical fluctuations.

6.2 Phase Transitions in Finger Movement: Experiments and a Simple Model

It was very fortunate for the development of my ideas that *Kelso* (1981, 1984) performed some detailed experiments on the coordination of finger movements. He told subjects to move their fingers in parallel (Fig. 6.1, left), and then the test persons were either asked verbally or forced by a metronome

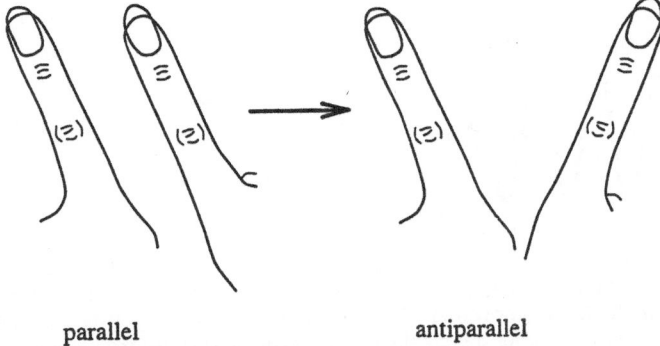

parallel   antiparallel

Fig. 6.1. Parallel and antiparallel finger movement

to move their fingers more quickly. At a certain critical frequency, ω_c, a sudden entirely involuntary change of the finger movement pattern occurred, namely from parallel to antiparallel, or, in other words, to symmetric movement (Fig. 6.1, right). Because the change of finger movement occurred at a specific frequency, that frequency ω suggests itself as a control parameter.

Let us discuss the results of this study in more detail. To this end, we introduce the displacement of the finger tips x_1, x_2 (Fig. 6.2). Figure 6.3 shows a schematic plot of the results of *Kelso*. Time is plotted to the right-hand side. The ordinate shows the elongations x_1, x_2 by means of solid or dashed lines. While the experiment was going on, the frequency of the metronome was slowly increased. Figure 6.3 clearly shows that the curves that originally exhibit a phase-shift, eventually coincide, indicating that the transition to Fig. 6.1 (right) had taken place. In the following I wish to develop the Haken-Kelso-Bunz (1985) model. In order to introduce an adequate order parameter, let us dwell a little on the mathematical description of oscillatory motion. Harmonic oscillations are described by cosine or sine functions as shown, for instance, in Fig. 6.4. The cosine curve can be shifted along the ωt-axis by introducing the phase ϕ, according to the relation

left   right

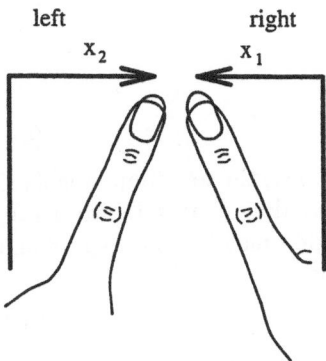

Fig. 6.2. Coordinates describing the displacement of the finger tips

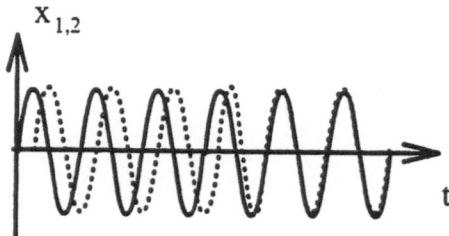

Fig. 6.3. Schematic representation of the results of the experiments on the displacement of the finger tips x_1 (*solid line*) and x_2 (*dotted line*) as a function of time, when the frequency of the finger movement is continuously increased. In the actual experiment of *Kelso* the amplitudes x_1, x_2 decreased with increasing frequency ω

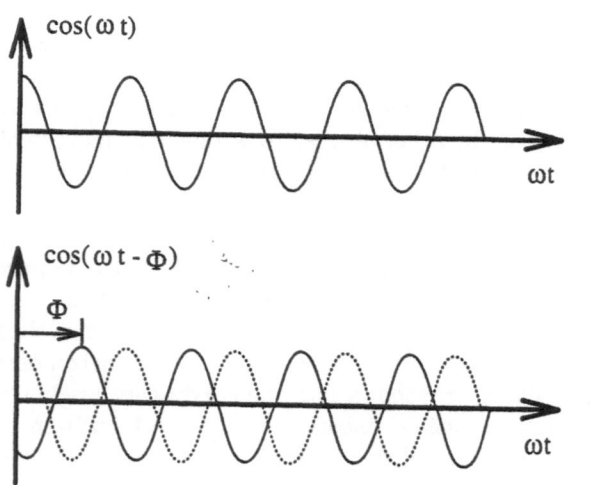

Fig. 6.4. Visualization of the phase shift ϕ. *Upper part*: Cosine function without phase shift. *Lower part*: Cosine function with phase shift. The unshifted function is represented by the dotted curve

$$\phi = \omega t_0, \tag{6.1}$$

where t_0 is a shift in time that corresponds to a shift of the maximum of the cosine curve by an amount ϕ according to Fig. 6.4. The displacement of the finger tips may then be written in the form

$$x_1 = r_1 \cos(\omega t - \phi_1) \tag{6.2}$$

and

$$x_2 = r_2 \cos(\omega t - \phi_2) \tag{6.3}$$

where r_1, r_2 are the amplitudes of the oscillation, i.e., the maximum displacement of the finger tips, and ϕ_1 and ϕ_2 the corresponding phases. Quite clearly, the transition from Fig. 6.1 (left) to Fig. 6.1 (right) may be understood as a change of the relative phase, which is defined by

$$\phi = \phi_1 - \phi_2, \tag{6.4}$$

from $\phi = \pi$ to $\phi = 0$. Note that because of the periodicity of the cosine function, $\phi = \pi$ is equivalent to $\phi = -\pi$. Quite clearly, the transition from the parallel to the antiparallel finger movement is a qualitative change on a macroscopic scale. In addition, this change is described by the relative phase (6.4). Thus ϕ becomes our candidate for the order parameter. Before proceeding, we mention that for the following model the right-hand sides of (6.2) and (6.3) need not be purely cosine functions. They may be replaced by any periodic function, i.e., any function with the property

$$f(\omega t + 2\pi) = f(\omega t) \tag{6.5}$$

that has qualitatively the same shape as a cosine function. Thus instead of (6.2) and (6.3) we may also consider

$$x_1 = r_1 f(\omega t - \phi_1) \tag{6.6}$$

and

$$x_2 = r_2 f(\omega t - \phi_2). \tag{6.7}$$

Let us now discuss the derivation of an equation for the order parameter ϕ. Generally, for a single order parameter, we may expect an equation of the form

$$\dot{\phi} = -\frac{\partial V}{\partial \phi}, \tag{6.8}$$

where V is a potential function of the type discussed in Sect. 5.1. Two important general features of V can easily be derived. First of all, because of the periodicity of the cosine functions, the whole problem is periodic when ϕ is changed by 2π or multiples thereof. Thus we require that V be periodic

$$V(\phi + 2\pi) = V(\phi) \tag{6.9}$$

Furthermore, the left- and right-hand fingers play a symmetric role, i.e. the potential V must be symmetric. As is shown in mathematics, the most general form of such a potential is given by

$$V(\phi) = a_1 \cos \phi + a_2 \cos 2\phi + \dots + a_n \cos n\phi + \dots, \tag{6.10}$$

where n is an integer that may in principle approach infinity. For what follows, we will be satisfied by studying the most relevant features given by the simplest form of V. First, in (6.10) we keep only the first term

$$V(\phi) = -a \cos \phi \tag{6.11}$$

and assume

$$a > 0. \tag{6.12}$$

We obtain a potential as shown in Fig. 6.5. It has only one minimum at $\phi = 0$, i.e., there would be only one stable movement possible. In fact, however, at least at low frequencies ω, there are two stable movements possible, namely the parallel *and* the symmetric finger movements. In order to capture this *bistability*, we add a second term to (6.11), and obtain

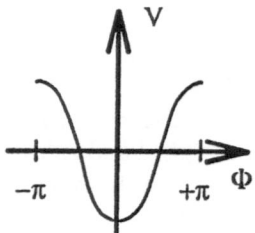

Fig. 6.5. Representation of the potential (6.11)

$$V(\phi) = -a\cos\phi - b\cos 2\phi. \tag{6.13}$$

Inserting (6.13) into (6.8), we find the equation of motion of the order parameter ϕ in the form

$$\frac{d\phi}{dt} = -a\sin\phi - 2b\sin 2\phi. \tag{6.14}$$

The corresponding shape of the potential is shown in Fig. 6.6, which must be read row by row, starting each time from the left. The individual parts of that figure show the shape of the potential for a decreasing value of the parameter b/a. The position of the ball at $\phi = \pi$ (or $= -\pi$) corresponds to a *parallel* finger movement of the subject. When the parameter b/a decreases, this minimum becomes flatter and eventually disappears entirely. The ball will, of course, now roll down to the position at $\phi = 0$, which corresponds to the symmetric finger movement. Thus this simple model mirrors the ob-

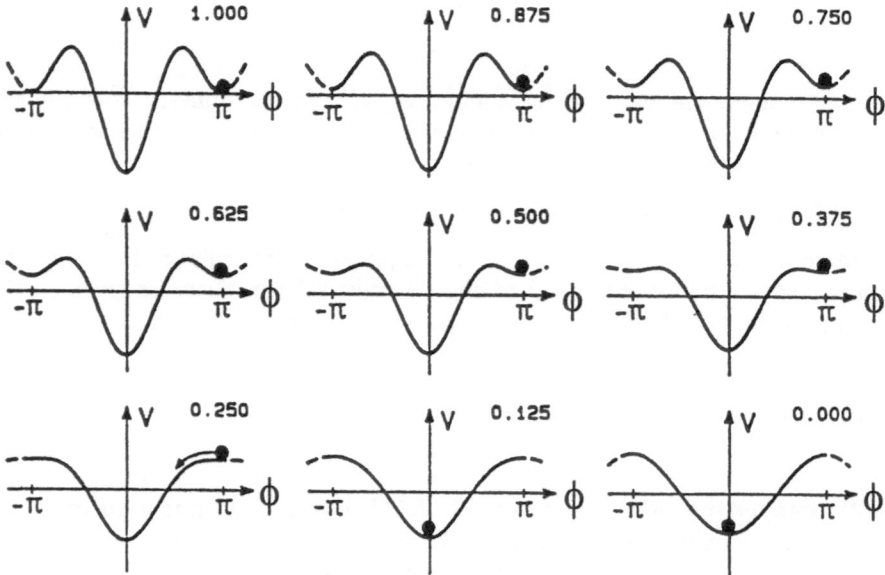

Fig. 6.6. Representation of the potential (6.13) for decreasing values of b/a (see text)

served transition quite well. But in addition it allows an important prediction, namely that of *hysteresis*. When a subject moves his or her fingers in the symmetric mode at a frequency $\omega > \omega_c$, this movement corresponds to $\phi = 0$. When the frequency is now decreased, we run through the sequence of the individual parts of Fig. 6.6 in the reverse direction as before with one decisive difference, namely the ball will certainly not spontaneously jump up again; it will rest at the position $\phi = 0$. This phenomenon is actually observed. When the subjects are asked to move their fingers more slowly, they stay in the initial symmetric state even at a slow frequency.

Let us return to the original sequence and study where the upper minimum disappears. To this end we perform a little algebra which is somewhat boring but, on the other hand, requires only a few lines. The position of the minima may be determined from the requirement that the slope of the potential V vanishes. Using the explicit form (6.13) of V, we find

$$a \sin \phi + 2b \sin 2\phi = 0. \tag{6.15}$$

The second term may be transformed by using a formula from trigonometry and we may cast (6.15) into the form

$$\sin \phi (a + 4b \cos \phi) = 0. \tag{6.16}$$

The left-hand side of (6.16) vanishes if either

$$\sin \phi = 0 \tag{6.17}$$

or

$$a + 4b \cos \phi = 0. \tag{6.18}$$

In the first case, we obtain the solutions

$$\phi = 0, \quad \phi = \pm \pi \tag{6.19}$$

which simply correspond to the minima of the potential. The solution of (6.18) can be written in the form

$$\cos \phi = -a/(4b). \tag{6.20}$$

The values of ϕ resulting from this equation tell us where the maxima of V lie. The transition happens when this maximum vanishes, which means that (6.20) has no solution for any real ϕ. This occurs when the cosine function is required to become bigger than 1 or smaller than -1, or, in other words, when

$$\left| \frac{a}{4b} \right| > 1. \tag{6.21}$$

Solving (6.21) for b, we obtain a critical value b_c provided we consider b_c as the variable that may be changed by the experimental conditions and keep a fixed

$$| b_c | < | a | /4. \tag{6.22}$$

Since, according to the experimental results, the finger movement amplitudes r_1 and r_2 decrease with increasing ω, one is led to cast the potential V in the form

$$V = -a\left(\cos\phi + \frac{1}{4}\frac{r(\omega)^2}{r_c^2}\cos(2\phi)\right) \tag{6.23}$$

where r_c is the critical amplitude at which the transition occurs. It must be stressed that (6.23) is only a guess at this moment and must be substantiated later, in particular by appropriate experiments.

Before we go on to include fluctuations, we discuss an alternative that was proposed in the literature (cf. References to the next section).

6.3 An Alternative Model?

The experimentally observed exchange of stability points from $\phi = \pi$ to $\phi = 0$ provided us with a starting point for the formulation of the model outlined above. Subsequently in the literature another model was proposed that is based on the first term of the expansion (6.10), namely on $V = a\cos\phi$. For $a > 0$, the potential has the form of Fig. 6.7 (left). When we assume that a depends on the control parameter ω, we may consider a change from positive to negative a depending on that control parameter. Because we assume that it changes continuously, a must pass through zero, which leads to the situation depicted in Fig. 6.7, middle part. Eventually, for negative a, we obtain Fig. 6.7 (right). The transition of the potentials of Fig. 6.7 (left) to the potential of Fig. 6.7 (right) shows, indeed, the required exchange of stability. Thus at first sight, this model not only seems equivalent, but also appears to be simpler than that of the preceding section. So experiments have to decide which model is the appropriate one. First of all, this alternative model does not exhibit any hysteresis. When we pass through the transition region of Fig. 6.7, the fluctuations will immediately drive the system out of its old state to the new state. In addition, the fluctuations become extremely large (critical fluctuations). Quite evidently on the flat curve of Fig. 6.7, the fluctuations will drive the ball, which represents the value of the order parameter, from

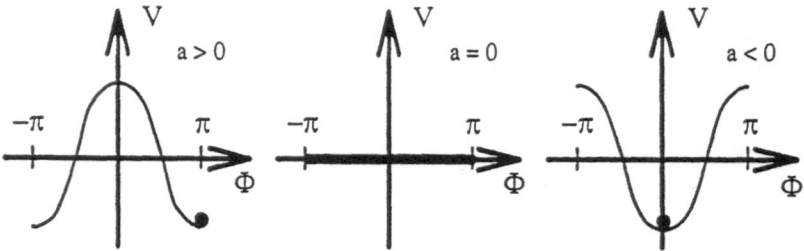

Fig. 6.7. Representation of the potential $V = a\cos\phi$ for different values of the control parameter a

$\phi = -\pi$ to $\phi = +\pi$. As we shall see, such large fluctuations are not observed. All these facts force us to dismiss this model.

6.4 Fluctuations in Finger Movement: Theory *

As we have seen in Sect. 5.1, the occurrence of critical fluctuations is an important signature of the process of self-organization. Thus the question arises of whether fluctuations may be observed in finger movement. As a starting point, we look into the mathematical modeling of fluctuations in more detail. To this end, we adopt a procedure that is well known in synergetics, namely we start by writing down an equation for the order parameter which includes the impact of fluctuations. In our case, we are dealing with the order parameter *relative phase* ϕ. To its equation of motion (6.14), we add a fluctuating force $F(t)$ so that the equation we wish to study reads

$$\dot{\phi} = -a \sin \phi - 2b \sin 2\phi + F(t). \tag{6.24}$$

We shall assume that $F(t)$ represents stochastic forces that act on the finger movement and stem from a microscopic level, for instance, from spontaneous firing of neurons. All we need to know for our analysis are two simple features of that fluctuating force:

1) Its average value is zero

$$\langle F(t) \rangle = 0. \tag{6.25}$$

Equation (6.25) means that the value of $F(t)$ vanishes when we average it over many runs of an experiment. In our case it would mean that we repeat the same experiment with the finger movements again and again.

2) Furthermore we shall assume that its correlation function is given by

$$\langle F(t)F(t') \rangle = Q\delta(t - t'). \tag{6.26}$$

In it we multiply the two values of F at different times and average over several (ideally very many) runs of the experiment. Q is the strength of the fluctuating force, the strength, so to speak, of the kicks of the individual football players that push the ball back and forth. δ is the Dirac δ-function whichis depicted in Fig. 6.8. It vanishes practically everywhere where $t - t' \neq 0$ and only at $t - t' = 0$ does it have a very sharp peak. The occurrence of this function signals that the correlation time between the fluctuating forces is very short, at least compared to all the time constants in the process of ϕ.

Equation (6.24) together with (6.25, 26) is the starting point of the *Schöner-Haken-Kelso* (1986) model, whose results I shall present in the following. Equation (6.24) is called a Langevin equation. It may be written in a more concise form by using the explicit form of the potential V

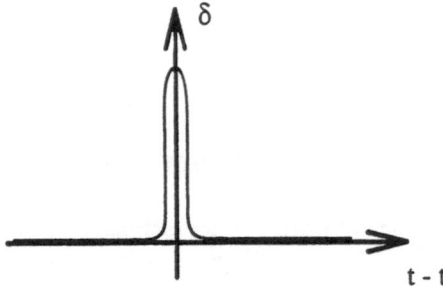

Fig. 6.8. Sketch of the δ-function

$$V = -a\cos\phi - b\cos 2\phi \tag{6.27}$$

and can then be cast into the form

$$\dot{\phi} = -\frac{\partial V}{\partial \phi} + F(t). \tag{6.28}$$

In a number of cases it is more practical to treat, instead of the Langevin equation (6.28), the Fokker–Planck equation. This equation refers to the probability distribution $P(\phi, t)$, where $P(\phi, t)d\phi$ is the probability finding the relative phase in the interval $\phi, ..., \phi + d\phi$ at time t. This probability distribution may change in the course of time. An example is provided in Fig. 6.9. The Fokker–Planck equation reads

$$\dot{P}(\phi, t) = \frac{\partial}{\partial \phi}\left(\frac{\partial V}{\partial \phi}P\right) + \frac{Q}{2}\frac{\partial^2 P}{\partial \phi^2}, \tag{6.29}$$

where the left-hand side is the temporal change of the probability distribution. The first term on the right-hand side is called the drift and the second term the diffusion term. If P no longer changes in time, it has reached its stationary value which is defined by

$$\dot{P}_{\text{stat}} = 0. \tag{6.30}$$

In this case (6.29) can be solved explicitly. Its solution reads

$$P_{\text{stat}}(\phi) = N\exp[-2V(\phi)/Q], \tag{6.31}$$

where V is the potential function that occurs in (6.29) and need not have the special form (6.27). N is the normalization factor

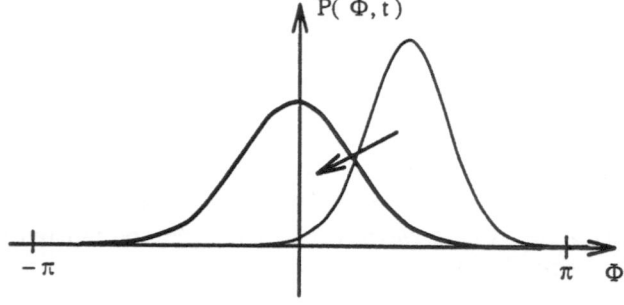

$P(\Phi, t)$

$-\pi$

$\pi \quad \Phi$

Fig. 6.9. Evolution of the probability distribution P in the course of time. The initial distribution is shown on the right

$$N^{-1} = \int_{-\pi}^{+\pi} P_{\text{stat}}(\phi)d\phi. \tag{6.32}$$

In order to make predictions about the behavior of the finger movement, we have to study the solutions of (6.24) or (6.29) in detail. To this end, we start with some simple cases:

a) *Symmetric finger movement.* Here we are studying the behavior of the two index fingers with their relative phase close to $\phi = 0$. We shall assume that the phase fluctuations are not too big. This allows us to replace the sine-functions on the right-hand side of (6.24) by their arguments according to

$$\sin x \approx x, \quad \text{where} \quad x = \phi \quad \text{or} \quad = 2\phi. \tag{6.33}$$

In this way, (6.24) is replaced by

$$\dot{\phi} = -(a + 4b)\phi + F(t), \tag{6.34}$$

where the corresponding potential V is a simple quadratic function in ϕ, namely (cf. Fig. 6.10, dotted curve)

$$V = \frac{1}{2}(a + 4b)\phi^2. \tag{6.35}$$

In this case, according to (6.31) the stationary solution of the Fokker–Planck equation (6.29) reads

$$P_{\text{stat}} = N \exp(-d^2\phi^2), \tag{6.36}$$

where we have used the constant d defined by

$$d = \sqrt{\frac{a + 4b}{Q}}, \tag{6.37}$$

and N is a normalization factor given by

$$N = \frac{d}{\pi \text{erf}(\pi d)}. \tag{6.38}$$

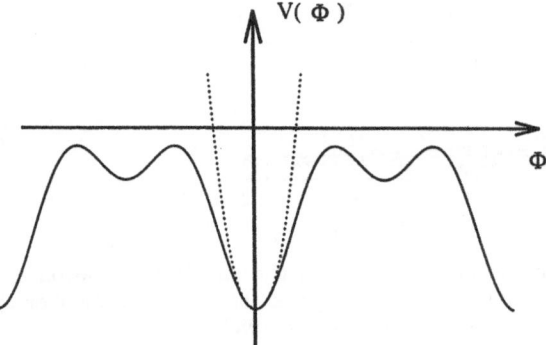

Fig. 6.10. Close to $\phi = 0$ the potential V (6.13) (*solid line*) may be approximated by a parabola (*dotted line*)

The detailed form of N is of no great concern here, and we merely note that erf is the error function. P is obviously a Gaussian distribution, i.e., a bell-shaped function. The importance of (6.36) lies in the fact that we can calculate the mean value of $|\phi|$ and the standard deviation σ explicitly. The mean value is defined by

$$\langle|\phi|\rangle \equiv \int\limits_{-\infty}^{+\infty} d\phi\,|\phi|\,P_{\text{stat}}(\phi). \tag{6.39}$$

It can easily be evaluated by the use of (6.36) and reads

$$\langle|\phi|\rangle = \frac{1 - \exp(\pi^2 d)}{\sqrt{\pi d}\ \ \text{erf}(\pi d)}. \tag{6.40}$$

The standard deviation, σ, is as usual defined by

$$\sigma^2 \equiv \langle\phi^2\rangle - \langle|\phi|\rangle^2 \tag{6.41}$$

and can be evaluated explicitly

$$\sigma^2 = \frac{1}{2d^2} - \frac{\sqrt{\pi}\exp(-\pi^2 d^2)}{d\ \ \text{erf}(\pi d)} - \langle|\phi|\rangle^2. \tag{6.42}$$

A plot of the average value and the standard deviation as a function of the parameter d is shown in Fig. 6.11. The two lines indicate the experimental values for mean and standard deviation so that experimentally realistic values for d can be read off. An important time constant is the relaxation time. This is the time it takes the system to relax to its former state at $\phi = 0$ after having been disturbed. This relaxation time can be read off from (6.34), whose meaning can be visualized as follows: The last term $F(t)$ represents

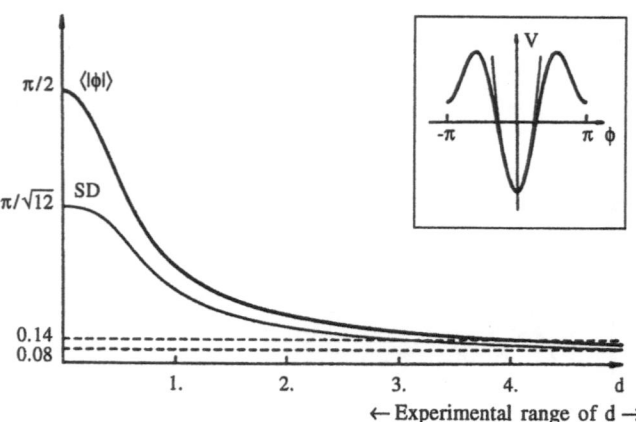

Fig. 6.11. The dependence of the mean phase and the standard deviation as a function of the parameter d (6.37). The insert shows the approximation of the potential by a parabola. After *Schöner, Haken, Kelso* (1986)

individual kicks to the relative phase ϕ. After each kick the relaxation of the phase ϕ is described by the first term on the right-hand side, which has the meaning of a relaxation time as indicated. Quide evidently, this time is given by

$$\tau_{\text{rel}} = \frac{1}{4b + a}. \tag{6.43}$$

We shall look at an experimental test later in Sect. 6.5. Here we continue the theoretical study by investigating the behavior of the antisymmetric, i.e. parallel mode.

b) *Parallel finger movement.* Here the relative phase is close to $\phi = \pm\pi$

$$\phi_{\text{stat}} = \pm\pi. \tag{6.44}$$

As we know from our previous analysis in Sect. 6.2, this state is stable for $|a| < |4b|$. Provided the fluctuations are not too big, we may proceed as in the previous analysis, but now we have to linearize (6.24) around $\phi = \pm\pi$. To this end, we introduce a new variable (cf. Fig. 6.12) by means of

$$\begin{aligned}\delta &= \phi - \pi \quad \text{for} \quad 0 < \phi \leq \pi \\ &= \phi + \pi \quad \text{for} \quad -\pi < \phi \leq 0.\end{aligned} \tag{6.45}$$

The following steps are completely analogous to those of the foregoing part. Equation (6.34) is now replaced by

$$\dot{\delta} = -(4b - a)\phi + F(t). \tag{6.46}$$

The stationary solution of the corresponding Fokker–Planck equation reads

$$P_{\text{stat}}(\delta) = N \exp(-f^2 \delta^2), \tag{6.47}$$

where the parameter f is given by

$$f = \sqrt{\frac{4b - a}{Q}} \tag{6.48}$$

and the normalization constant N by

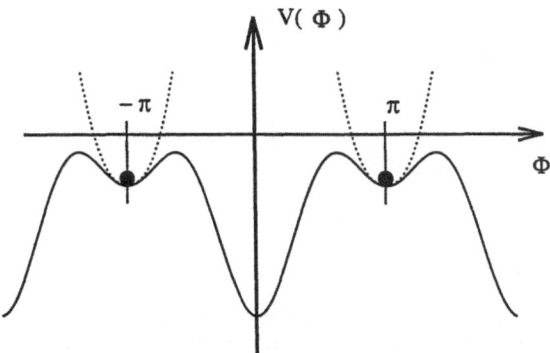

Fig. 6.12. Approximation of the potential V (*solid line*) by two parabolas close to $\phi = -\pi$ and $\phi = \pi$ (*dotted lines*)

$$N = \frac{f}{\pi \operatorname{erf}(\pi f)}.$$
(6.49)

The mean value of the modulus of δ is given by

$$\langle |\, \delta \,| \rangle = \pi - \frac{1 - \exp(-\pi^2 f^2)}{\sqrt{\pi} f \quad \operatorname{erf}(\pi f)}$$
(6.50)

and the standard deviation by

$$\sigma = \frac{1}{2 f^2} - \frac{\sqrt{\pi} \exp(-\pi^2 f^2)}{f \quad \operatorname{erf}(\pi f)} - \langle |\, \delta \,| \rangle^2.$$
(6.51)

In spite of the formal analogy between (6.40,42) and (6.50,51), a deep-rooted difference exists here, because we know that the phase transition occurs at $4b - a = 0$. Quite evidently, at this point f (6.48) vanishes. In this case, the mean modulus of δ goes to π, whereas the standard deviation σ diverges. In other words, we find *critical fluctuations* with $\sigma \propto 1/f^2 \propto Q/(4b - a)$. This divergence is an artefact of the linear approximation (6.46) of (6.24), but even in the case of an exact treatment, σ would become large. From (6.46) we may read off the relaxation time as

$$\tau_{\mathrm{rel}} = \frac{1}{4b - a}.$$
(6.52)

Again we find a divergence for $4b - a \to 0$ at the transition point indicating *critical slowing down.*

The mean absolute phase and its standard deviation for the local model of the antisymmetric mode as a function of f are shown in Fig. 6.13. The approach to the transition point corresponds to $f \to 0$. The f-axis is oriented to the left to illustrate this. The curves have been used to determine from

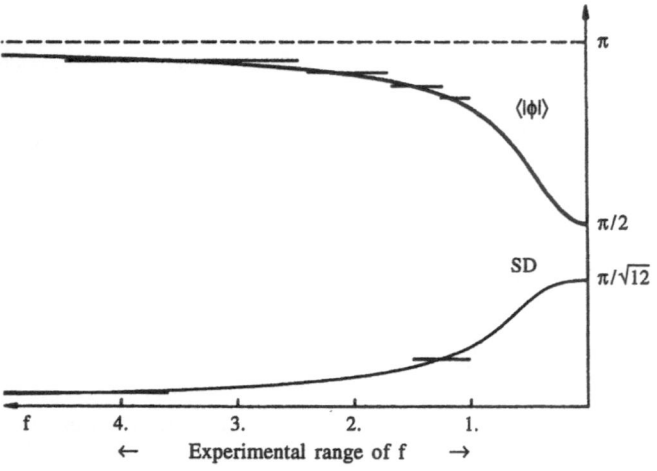

Fig. 6.13. The dependence of the mean phase and the standard deviation on the parameter f (6.48) in the case of $\phi \approx \pm \pi$. After *Schöner, Haken, Kelso* (1986)

experimental values of mean and standard deviation the corresponding values of f. The range of such values is indicated. Since the transition region requires a more careful study, we now consider this region.

In this region, the Fokker–Planck equation has to be solved numerically. The resulting temporal evolution of the probability density for parameter values $a_{cr} = 2.0$ Hz, $b_{cr} = 0.5$ Hz is illustrated by showing it at times $0.0s, 1s, 10s, 2.19s, 3.29s$ and $4.39s$ (Fig. 6.14). All distributions are normalized. In this case, the initial condition was chosen as a stationary distribution of the local model of the antisymmetric mode. One can clearly see from the figure how the probability weight, initially concentrated at π and $-\pi$, flows to the central peak at $\pi = 0$. The resulting distribution is much sharper than the initial one, because $\phi = 0$ is a deeper and steeper minimum of the potential than $\phi = \pm\pi$ was at the last pretransitional parameter plateau. Figure 6.15 shows the mean absolute phase and its standard deviation evolving in time. $\langle | \phi(t) | \rangle$ quite clearly marks the switching from $\pm\pi$ to 0. The standard deviation shows how fluctuations are enhanced during the transient, while they settle to a level even lower than before the transition once the transient has died out.

Let us make the following quantitative observations:

a) We can determine the duration of the transient from the plots of mean and standard deviation. This transient time is approximately $2.5 - 5s$, in accord with the experimental estimate of $1 - 2s$. This result is actually a true prediction.

b) The temporal mean of the standard deviation is defined as the mean up to the time at which the switching has occurred with a 90% probability. With that definition, we find from the data of Fig. 6.15 a standard deviation of 56 degrees. This is the correct order of magnitude when compared to the experimental value of 60 degrees.

Another quantity that can be measured is the mean switching time. To reach an adequate definition, we observe that, during the transient, probabil-

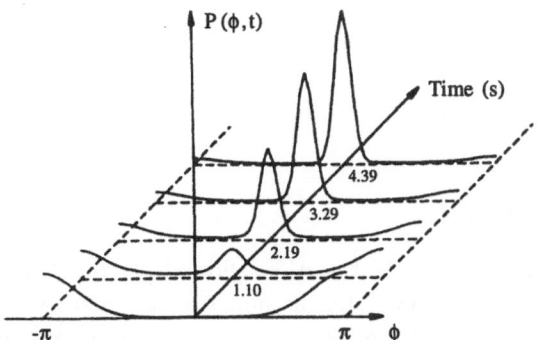

Fig. 6.14. Time evolution of the probability distribution starting from the parallel finger movement and developing into the antiparallel movement. After *Schöner, Haken, Kelso* (1986)

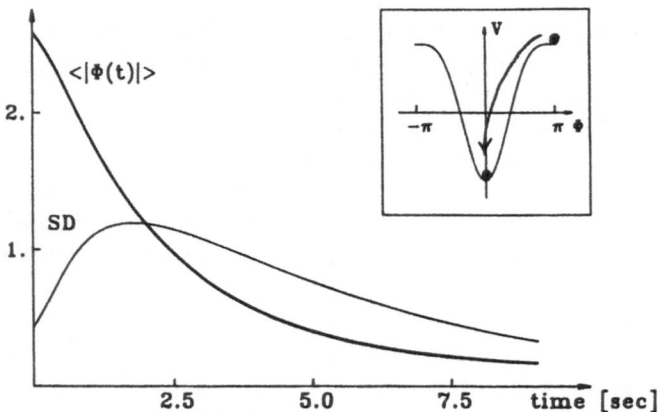

Fig. 6.15. Time evolution of the mean phase and the standard deviation for the transition from $\phi = \pi$ to $\phi = 0$. After *Schöner, Haken, Kelso* (1986)

ity weight, originally concentrated at $\phi = \pm\pi$, flows to $\phi = 0$ and accumulates there. The quantity

$$P_\delta(t) = \int\limits_{-\delta}^{\delta} d\phi P(\phi, t) \qquad (6.53)$$

is, for suitably chosen δ, the probability weight of the new peak and grows during the transient.

$$v(t) = \frac{dP_\delta(t)}{dt} \qquad (6.54)$$

is the velocity of this growth, i.e. $v(t)dt$ is the probability that switching occurs in the time interval $t, ...t + dt$. Thus the switching time can be defined by

$$\tau_{\text{switch}} = \int\limits_{0}^{\infty} tv(t)dt. \qquad (6.55)$$

6.5 Critical Fluctuations in Finger Movements: Experiments

6.5.1 The Experimental Set-Up

The theoretical model presented in the last section allowed us to calculate fluctuations, including critical fluctuations. Let us now turn to their experimental verification by *Kelso, Scholz* and *Schöner* (1986). Two kinds of correlation experiments were done: on the movement of wrists and on the movement of the index fingers. In analogy to the finger movement experiments described in Sect. 6.2, we may also study the transition from a parallel wrist movement to an antiparallel, or, in other words, symmetric one.

For experiments involving oscillatory wrist flexion/extension, the subject's forearms were fixed in a comfortable position while each hand grasped a vertical handle attached to the experimental apparatus that rested on a table top. The axes of the wrist joints were colinear with the axes of the handles. The latter incorporated potentiometers for conversion of wrist rotation angle to dc voltages. The finger movement experiments used a similar experimental set-up except that the forearms were stabilized to restrict movements to the index fingers alone. To reproduce the phenomenon precisely, movements must be restricted to the relevant degrees of freedom, i.e., wrists and fingers, respectively. On a given run subjects oscillated the index fingers bilaterally in the transverse plane of motion, i.e. abduction-adduction. The continuous x, y coordinates of the tip of each finger were measured using infra-red light emitting diodes attached to the fingertips. All data were digitized with a 12-bit A-D converter at 200 samples/s and stored on magnetic tape for later computer analysis.

Here we briefly summarize the results of the *Kelso, Scholz, Schöner* experiments on the finger movement. In these experiments, the frequency of oscillation was systematically increased in 0.25 Hz steps at 4 s intervals according to a metronome pacing stimulus. Data from the finger experiments could, therefore, be time-averaged for each driving frequency. The relative phase and its fluctuations were measured in two ways:

1) by a point estimate. Here one measures the phase of the wrist's oscillatory peak, i.e., its maximum displacement relative to the other.
2) a practically continuous measurement. Here the relative phase was measured every 5 ms.

To find a way to determine the time-dependent phase from the experiments, we use the idea that we may describe the motion of each finger tip similarly to that of a pendulum in the phase plane (cf. Fig. 5.9), where the abscissa is given by the finger displacement x and the ordinate by this finger speed $\dot{x}$. For the finger tip of the right hand with its phase plane coordinates $x_r, \dot{x}_r$ we may define the phase angle ϕ_r, using geometry, by

$$\tan\phi_r = \frac{\dot{x}_r}{x_r}. \tag{6.56}$$

The phase angle introduced in this way coincides with our former definition (6.2) provided we may ignore the time-dependence of the amplitude r. In order to compensate for that time-dependence, we replace x_r and $\dot{x}_r$ by normalized quantities, where $\bar{x}_r$ is the position of the right index finger normalized to the cycle extrema and $\dot{\bar{x}}_r$ is the normalized instantaneous velocity. Making these replacements in formula (6.56) and resolving this relation for ϕ_r, we obtain

$$\phi_{\rm r} = \tan^{-1}\left(\dot{\overline{x}}_{\rm r}/\overline{x}_{\rm r}\right).\tag{6.57}$$

In complete analogy to (6.57), we may determine the phase angle ϕ_ℓ belonging to the finger tip of the left hand. We then take the difference $\phi = \phi_{\rm r} - \phi_\ell$ to obtain the relative phase. In this difference, the non-fluctuating contribution that increases with time, $\omega(t) \approx \omega t$, drops out so that we may use this difference directly as a measure of the phase fluctuations.

6.5.2 Experimental Results

The main results of the experiments concerning the relative phase are shown in Fig. 6.16. They allow us to compare the continuous estimate of relative phase (lower part of Fig. 6.16) and the point estimate of that phase (upper part of Fig. 6.16) for a representative experimental run. The slow component of a phase fluctuation is apparent in both parts, though a finer fluctuational structure emerges from the continuous estimate. Because of the anharmonicities present in the individual finger movement trajectories, the continuous relative phase also contains an oscillatory component. Due to the controlled, stepwise increase of cycling frequency, stationarity could be checked by averaging over a 0.5 s window that was moved through the 4 s of data at each frequency. Stationarity was guaranteed less than 1 s after the parameter change. The modulus of the phase and the standard deviation were, therefore, calculated on a given run for the last three seconds at a given frequency, i.e. 600 data points. The average of the phase modulus and the standard deviation were taken over a set of 10 experimental runs for a representative subject. The results are shown in Figs. 6.17 and 6.18. Figure 6.17 shows the results when the finger movement of the subject was originally parallel at a low frequency and switched to an antiparallel movement at higher frequencies. The solid cycles clearly show this change of the modulus. The open circles are of primary interest to us, because they show how the fluctuations build up, reach a maximum at the transition point and then drop sharply. In this case, the ball representing the relative phase, is pushed out from its upper minimum, which disappears, and jumps to the lower minimum. The

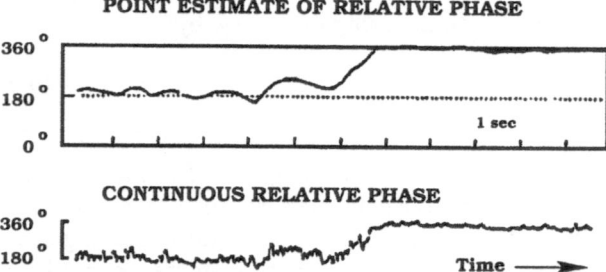

Fig. 6.16. *Upper part:* Point estimate of the relative phase. *Lower part:* Continuous relative phase. (Redrawn after *Kelso, Scholz, Schöner* (1986))

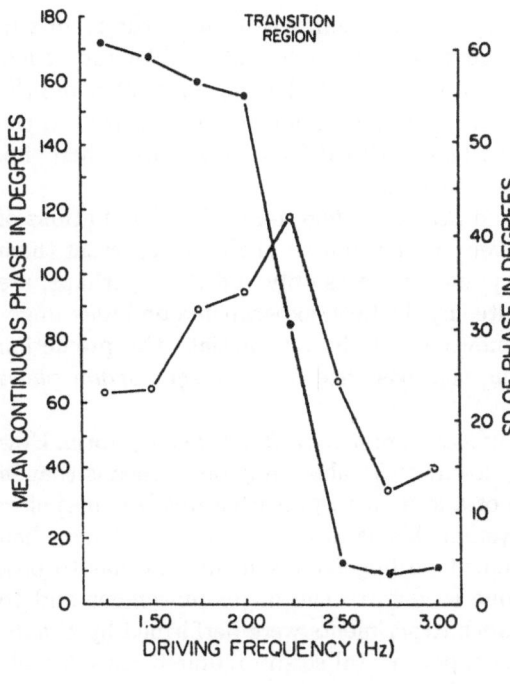

Fig. 6.17. The mean phase (*solid dots*) and the standard deviation (*open circles*) when the fingers are initially moved in parallel and the driving frequency is increased. (Redrawn after *Kelso* et al. (1986))

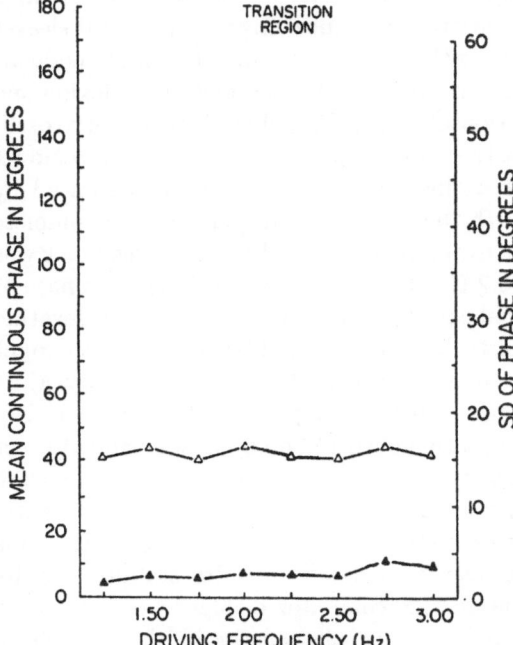

Fig. 6.18. Same plot as Fig. 6.17, but when the finger movement starts in its antiparallel mode and the frequency is increased. (Redrawn after *Kelso* et al. (1986))

significance of this result is underlined by a comparison with the results in which the subject's finger movements were originally antiparallel and at low frequencies. As can be seen clearly from Fig. 6.18, the phase modulus and the fluctuations of the phase remain practically constant in agreement with the theoretical expectation. Quite clearly, no critical fluctuations occur here, i.e. no nonequilibrium phase transition takes place.

The results of *Kelso, Scholz* and *Schöner* (1986) on critical fluctuations of the wrist movement are qualitatively similar, but we shall not represent them here, because they are based on point estimates only and thus, perhaps, are not quite so pronounced. But quite clearly, both experiments on index finger movement and those on wrist movement fully substantiate the prediction of *critical fluctuations* supporting the existence of a *nonequilibrium phase transition.*

Let us now turn to a second indicator, namely *critical slowing down.* How can one measure critical slowing down? Critical slowing down means that in a transition region the response of a system to perturbations becomes slow compared to the reaction of a system that is in a stable region. Thus when one wishes to do such an experiment on finger movements, one has to provide an apparatus that allows one to disturb the finger movement and to track the perturbed movement. Such experiments were performed by *Scholz, Kelso* and *Schöner* (1987), whose experimental set-up required considerable modifications from that used to study the critical fluctuations as described above. Now the subjects had to insert their index fingers into metal sleeves whose axes of rotation coincided with those of the so-called first metacarpophalangeal joints. The metal sleeves restricted finger motion to flexion and extension at this joint in the horizontal plane. The subject's task was to oscillate both fingers rhythmically at the same frequency in one of the two modes of coordination, i.e., parallel or antiparallel. The movement was paced by an auditory metronome pulse with the instruction to perform one complete movement cycle per pulse. In the experimental runs the pacing frequency was increased every 10 s in steps of 0.2 Hz. Trials began either in the in-phase or in the antiphase mode of bimanual coordination. To determine the relaxation time, the finger movement was disturbed by exerting little pushes to the right index finger when the finger flexion movement reached its peak velocity. More technically speaking, a 50 ms torque pulse was applied via the metal sleeves.

In each trial, movement perturbation occurred on up to four nonadjacent frequency plateaus. Perturbations were randomly distributed over a block of trials, such that each of the nine frequency plateaus was perturbed a total of ten times. An estimate of the relaxation time was obtained from the time of torque pulse offset until the relative phase time series stabilized at its pre-perturbation mean value. Interactive computer displays were also used to measure the switching time on frequency plateaus in which a transition occurred. Here the estimate was determined as the time from the beginning of the frequency plateau to the point where the relative phase time series

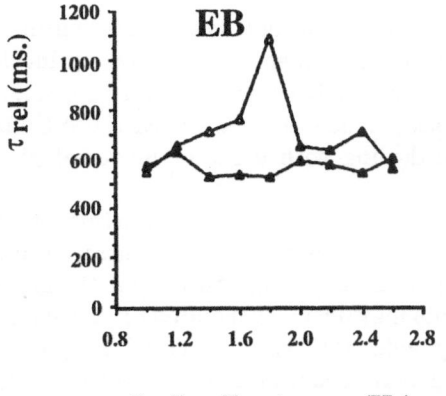

Scaling Frequency (Hz)

Fig. 6.19. Typical behavior of the experimentally found critical relaxation time. Open triangles: The test person starts with parallel finger movements and a transition to the antiparallel movement occurs with increasing frequency. Solid triangles: The test person starts with antiparallel finger movements and remains in that mode. (Redrawn after *Scholz, Kelso, Schöner* (1987))

stabilized at 0° mean value, corresponding to the completion of the transition. The measurements were done for five subjects. A typical result is shown in Fig. 6.19. The open triangles refer to the antiphase mode, i.e., parallel finger movement, the closed triangles to the in-phase mode, i.e., symmetric finger movement. The critical slowing down, i.e., the increase of relaxation time, can clearly be observed for the antiphase mode. Only one out of the five subjects did not show this clear-cut behavior. In agreement with theory, it turns out that the antiphase mode is dynamically less stable than the in-phase mode and that the transition from antiphase to in-phase mode is connected with the loss of stability.

Let us compare these results with the theoretical prediction of the model described in Sect. 6.4. The relaxation times for the motions close to $\phi = 0$ and $\phi = \pi$ were described as

$$\tau_{\text{rel},0} = \frac{1}{4b+a}, \quad \tau_{\text{rel},\pi} = \frac{1}{4b-a}, \tag{6.58}$$

where b and a are the coefficients that occurred in (6.14) and (6.24). This gives a new approach to the determination of the parameters a and b. We shall not dwell, however, on this detailed discussion, because here we simply wanted to elucidate the most salient points.

6.6 Some Important Conclusions

Let us first discuss whether the phase transition observed in finger movements is a unique phenomenon or can be found in other movements, too. A number of coordination experiments were performed, for instance on the relative phase between hand/lower arm and lower arm/upper arm and also on other coordination tasks; see for example *Jeka, Kelso, Kiemel* (1993). Surprisingly, such coordination phenomena occur also in between different persons. In experiments done by *Schmidt, Carello* and *Turvey* (1990), two persons were asked to move their lowerlegs in an antiparallel fashion, where the coordination was mediated by eye-contact. When the frequency was raised, an involuntary transition to the in-phase motion suddenly occurred (Figs. 6.20, 6.21). All these experimental findings are in accordance with the Haken-Kelso-Bunz phase transition model. Thus we are dealing with a rather universal phenomenon in biological coordination.

But why are the results we discussed in the foregoing sections so important for our general understanding of brain and behavior? Let us consider these results in the light of the ideas that the brain is a computer or that the brain is a dynamical system. Since movements may also be generated or steered, at least partly, by the spinal cord, we may understand the spinal cord as a part of our brain which does not change our general conclusions at all. What would we expect from the performance of a computer which runs on a program that in our case we may call *a motor program*, a term used quite often in neurophysiology? We may imagine that there may be some kind of a switch that is operated by means of a control parameter, namely the frequency of the finger movement. In this way, the computer may be switched from one motor program to another. But then two features cannot be understood, namely

1) computer programs are deterministic, so there is no room for fluctuations, in particular in the transition regime.
2) programs are switched immediately or at least with a fixed response time. Thus an increase in relaxation time is not understandable from the point of view of a computer program unless one provides very particular rules. But the experiments on finger movements demonstrate that the relaxation time is a stochastic, i.e. random, variable which would not occur in a computer program.

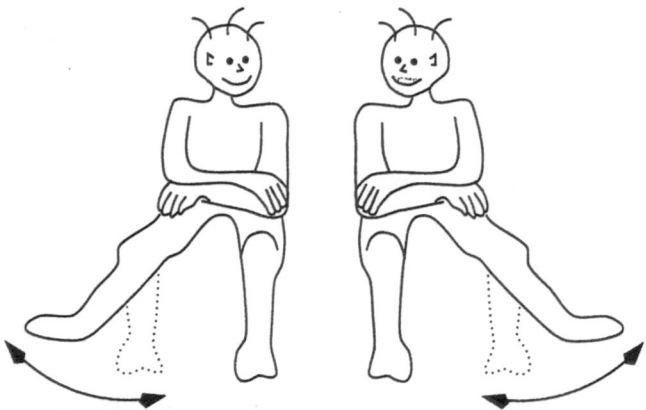

Fig. 6.20. Illustration of the *Schmidt–Carello–Turvey* experiment (see text)

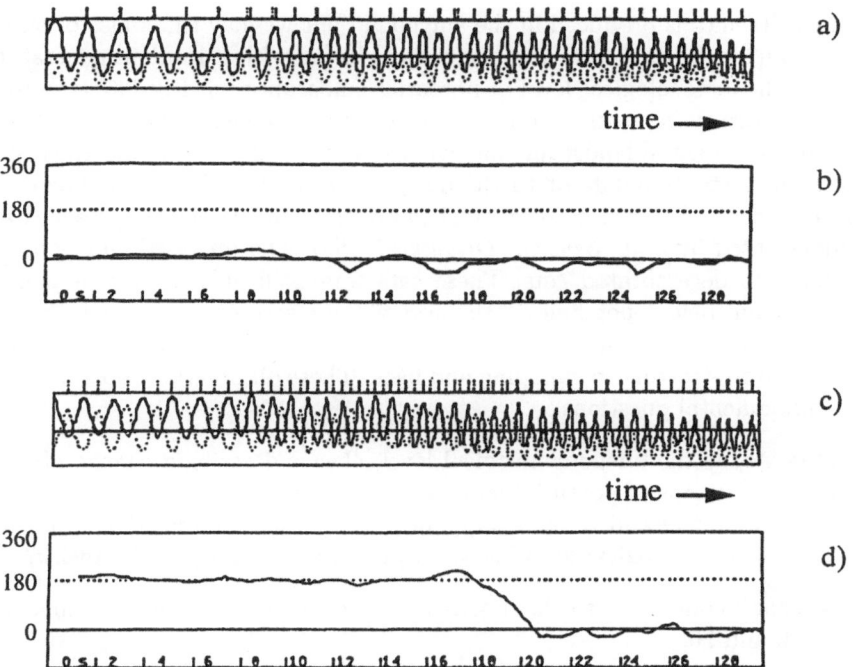

Fig. 6.21. (a) Displacement of feet versus time in in-phase experiments; (b) relative phase; (c) displacement of feet versus time; at the beginning the antiphase state is realized; (d) transition of relative phase. In all these cases the frequency increases in the course of time. (Redrawn after *Schmidt, Carello, Turvey* (1990))

The phenomena of critical fluctuations and critical slowing down can be quite easily explained in terms of *self-organization*, i.e., using the concepts of synergetics, and we remind the reader of our discussions in Chap. 5.

In addition, an important conclusion can be drawn for biological systems. As we have seen, when we wish to switch the behavior of a system from one mode to another, we have to change a control parameter so that, for instance, the upper part of Fig. 5.1 is replaced by the lower part of that figure. But, in addition, an initial random push is needed to drive the system from the former equilibrium state to the new one. It appears that these random pushes are necessary not only in physical systems but also in biological systems. There are strong indications that fluctuations are provided by a biological system in a pronounced fashion, in particular by tremor. This idea is supported by earlier estimates of parameters in the paper by *Schöner, Haken* and *Kelso* (1986). From these discussions, new light is shed on the experimental findings of *von Holst* (1939) on centipedes, where, after most of the feet were removed, the animal was able to develop specific new gaits.

This brings us to the question of how self-organization takes place, or, in other words, what role is played by self-organization in biological motor actions. Tremor is produced in the amygdala, but we may think also of other noise sources. So in a way, the brain provides us with a random generator. At the phenomenological level of potential landscapes (cf. Fig. 6.6), the role of the control parameter becomes evident. When we change the control parameter, a potential landscape may change, even qualitatively, and thus give rise to new stable points or to the disappearance of former ones. The dramatic impact of a change of control parameters in biological systems was demonstrated by *Shik, Severin, Orlovskii* (1966), who provided an electrical stimulus to decerebrated cats. These cats were still able to trot or gallop on a moving belt depending on the size of the electric stimulus, which evidently worked as a control parameter. In a critical region, the cat randomly changed from trotting to galloping and back. These observations pose at least two fundamental questions:

1) May we, at the phenomenological level, always describe a change of stability by means of potential landscapes?
2) Can we understand what is happening at the neuronal level, i.e., how is the phenomenological level based on processes in a neuronal network?

We shall come back to these questions later in the book, in particular in Sects. 16 and 20.

7. More on Finger Movements

7.1 Movement of a Single Index Finger

In Chap. 6 we were concerned with the relative phase of finger movement. In this section we wish to show how to formulate a model that takes into account the oscillatory motion of the individual fingers. As we have seen in Sect. 5.2, stable oscillations can be described by limit cycles. In the literature, a number of equations have been studied that allow the mathematical description of such limit cycles. A highly useful example is that of the Van der Pol oscillator, which was originally developed to describe the behavior of the vacuum tubes used in the early days of radio. As we know, in the realm of synergetics, the material substrate is quite irrelevant for the formulation of these equations, however. The Van der Pol equation refers to a single variable $x(t)$ which, in the present context, may be identified with the displacement of one of the fingers (cf. Fig. 6.2). This equation reads

$$\ddot{x} + \epsilon(x^2 - r_0^2)\dot{x} + ax = 0. \tag{7.1}$$

The meaning of the individual terms in (7.1) is straightforward. $\ddot{x}$ represents an acceleration of the variable x. The first and last term together, i.e.,

$$\ddot{x} + ax = 0, \tag{7.2}$$

describe a harmonic oscillation with solutions of the form $x(t) = r \sin \omega t$ and $x = r \cos \omega t$ (or their superpositions), where the circular frequency ω is related to the constant a by means of

$$\omega^2 = a. \tag{7.3}$$

The middle term in (7.1) is proportional to the velocity $\dot{x}$. An expression of the form $\gamma \dot{x}$ is well known in oscillator theory and represents a velocity dependent damping of the oscillator with the damping constant γ. A comparison between $\gamma \dot{x}$ and the middle term in (7.1) suggests that $\epsilon(x^2 - r_0^2)$ may be interpreted as some kind of damping constant γ that depends on the variable (or displacement) x. If $x^2 > r_0^2$, this constant is positive and the motion will be damped. If, however, $x^2 < r_0^2$, γ becomes *negative* and the variable x will be enhanced. As a consequence, the middle term will serve to stabilize the motion around $x = r_0$, or, in other words, it keeps the motion on a limit cycle with radius r_0.

In order to study the properties of the solution of (7.1) in more detail, we make the hypothesis

$$x = Ae^{i\omega t} + A^* e^{-i\omega t}, \tag{7.4}$$

where the amplitude A and its conjugate complex A^* may still be time dependent. We insert (7.4) into (7.1) and apply two approximations that are well known in oscillator theory. We shall assume that $A(t)$ changes much more slowly than the exponential function $e^{i\omega t}$. This allows us to apply the *slowly varying amplitude approximation* in which we neglect $\dot{A}$ compared to ωA. After having inserted (7.4) into (7.1), we may collect terms that are of the form

$$e^{\pm 3i\omega t} \tag{7.5}$$

or

$$e^{\pm i\omega t}. \tag{7.6}$$

Since the terms (7.5) oscillate more rapidly than the terms (7.6), we may neglect them to a good approximation. This is called the *rotating wave approximation*. Having these two approximations in mind, a little algebra leads us from (7.1) to the equation

$$2\dot{A} + \epsilon(|A|^2 - r_0^2)A = 0, \tag{7.7}$$

where we have dropped the factor $i\omega e^{i\omega t}$ which is common to all terms of the left-hand side of this equation. In the stationary state $\dot{A} = 0$ and we obtain from (7.7) the relationship

$$|A|^2 = r_0^2. \tag{7.8}$$

This tells us that r_0^2 has the meaning of a modulus squared, and we conclude from (7.8) that the amplitude is independent of the frequency ω that occurs in (7.3). This does not quite agree with the experimental findings according to which the amplitude of the finger movement decreases with increasing frequency.

To this end, we study a new model equation, which, as we shall see immediately, will lead us to an amplitude that decreases with increasing frequency. This Rayleigh equation reads

$$\ddot{x} + \epsilon(\dot{x}^2 - \omega_0^2 r_0^2)\dot{x} + ax = 0. \tag{7.9}$$

and differs from (7.1) by the form of its damping term. Here the damping constant $\gamma = \epsilon(\dot{x}^2 - \omega_0^2 r_0^2)$ is positive if the velocity $\dot{x}$ is bigger than a given constant value, $\omega_0^2 r_0^2$, and negative otherwise. Thus in this case (7.9) stabilizes solutions with a given velocity.

Making again the hypothesis (7.4), using the slowly varying amplitude approximation and the rotating wave approximation, and dropping the factor $i\omega e^{i\omega t}$, we may transform (7.9) into

$$2\dot{A} + \epsilon A(3|A|^2 \omega^2 - \omega_0^2 r_0^2) = 0. \tag{7.10}$$

In the steady state, where

$$\dot{A} = 0 \tag{7.11}$$

we immediately find

$$| A | = \frac{\omega_0 r_0}{\sqrt{3}\omega}, \tag{7.12}$$

i.e., the required result that A decreases with increasing ω. In the observed data, A appears to behave as some kind of mixture between (7.8) and (7.12). Therefore, we shall formulate an equation that combines these two effects. To this end we introduce a superposition of the damping terms that occur in (7.1) and (7.9) (cf. *Haken, Kelso, Bunz* (1985)). The resulting equation reads

$$\ddot{x} + \left[\epsilon_1(x^2 - r_0^2) + \epsilon_2(\dot{x}^2 - \omega_0^2 r_0^2)\right] \dot{x} + ax = 0. \tag{7.13}$$

Under the same approximation that has led us in the steady state from (7.1) to (7.8), we obtain

$$| A | = \sqrt{| c | /(\epsilon_1 + 3\epsilon_2\omega^2)}, \tag{7.14}$$

where $c = -(\epsilon_1 + \epsilon_2\omega_0^2)r_0^2$.

The corresponding experiments were performed by *Kay, Kelso, Saltzman* and *Schöner* (1987), who also showed how the parameters occurring in (7.13) can be determined from the experimental data. Kinematic data, i.e., the position of the finger and its velocity, were sampled at 200 samples/second and a detailed analysis of movement amplitude, frequency, and peak velocity were performed. The frequency which enters through $a = \omega^2$ was scaled from 1 Hz to 6 Hz in steps of 1 Hz using a pacing metronome. We note that the stiffness a was the only control parameter. When the frequency ω increased, the amplitude dropped according to the law (7.14). A plot of the trajectory in the position–velocity plane (phase plane) showed typical limit cycle behavior, as is exhibited in Fig. 7.1, where the left column shows the experimental results with increasing frequency, and the right column the results of our model. We can clearly see the decrease of amplitude (horizontal axis) and the simultaneous increase of the peak velocity (vertical axis). In these experiments the accuracy of movement was neither fixed nor manipulated. Only frequency is scaled systematically and amplitude allowed to vary in a natural way, i.e., the test persons were allowed to move their fingers with whatever amplitude they wanted without any conscious interference.

As an aside we mention that there has been surprisingly little research in movements performed under these particular experimental conditions previously (cf. *Freund* (1983)). *Feldman* (1980) reported data from a subject who attempted to keep a maximum amplitude (elbow angular displacement) as frequency was gradually increased to a limiting value. An observed inverse relation was accompanied by an increasing tonic coactivation of antagonistic muscles. In addition, the slope of the so-called *invariant characteristic* (cf. *Asatryan* and *Feldman* (1965); *Davis* and *Kelso* (1982)) – a plot of joint

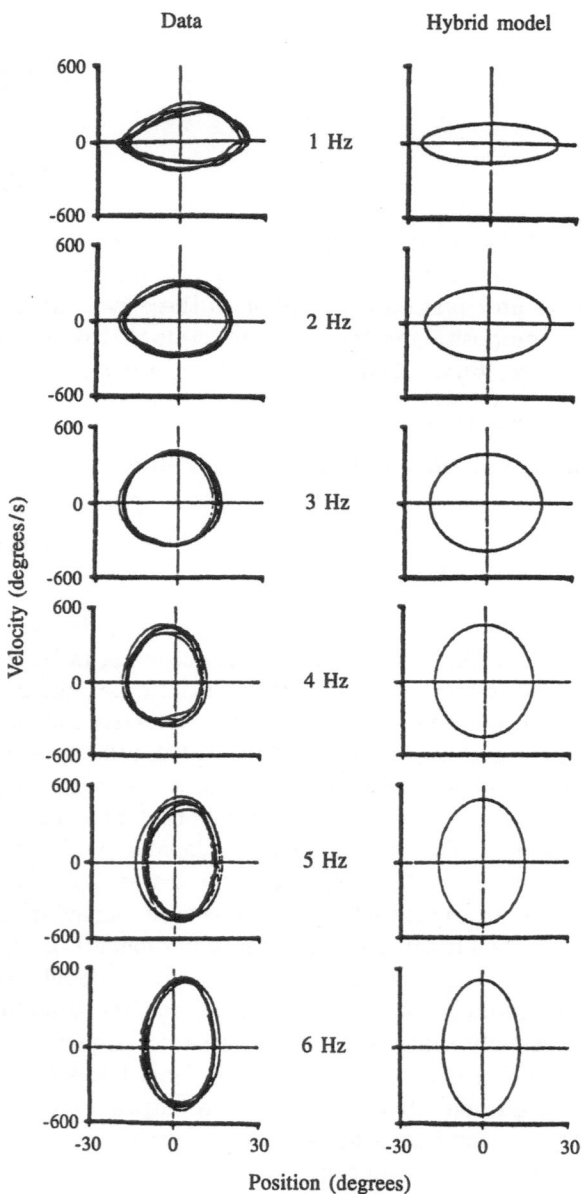

Fig. 7.1. Phase plane trajectories from 1 Hz to 6 Hz. *Left panel:* Representative examples from the collected data set of one subject. *Right panel:* Trajectories of the model (7.13). After *Kay, Kelso, Saltzmann, Schöner* (1987)

torque versus joint angle – increased with rhythmical rate, suggesting that natural frequency (or its dynamic equivalent, stiffness)was a controllable parameter. Other studies have scaled frequency but fixed movement amplitude. Their conclusions were similar to *Feldman's*: Frequency changes over a range were accounted for by an increrase in system stiffness (see *Viviani, Soechting*, and *Terzuolo* (1976)).

The decrease of amplitude with increasing frequency is well-covered by our hybrid oscillator model, as can be seen from Fig. 7.2, in which a comparison of the experimental data with the hybrid oscillator, the Van der Pol oscillator and the Rayleigh oscillator is made. The limit cycle model also provides us with the simple relationship between the peak velocity v_p and the amplitude $v_p = \omega A$. A was taken to be of the same scale as the experimental values measured in degrees. The model parameters were determined by

a) identifying the pacing frequency with ω
b) choosing $c = -0.05\omega_{pref}$ ($= 0.64$ Hz) and finding ϵ_1 and ϵ_2 by a least squares fit of the amplitude frequency relation:

$$c = -0.05\omega_{pref.} \qquad (= 0.641\,\text{Hz})$$
$$\epsilon_1 = 12.457\,\text{Hz}$$
$$\epsilon_2 = 0.007095\,\text{Hz}^3. \tag{7.15}$$

Standard deviations of ϵ_2 and ϵ_1 were $0.001025\,\text{Hz}^3$ and 1.0129 Hz, respectively.

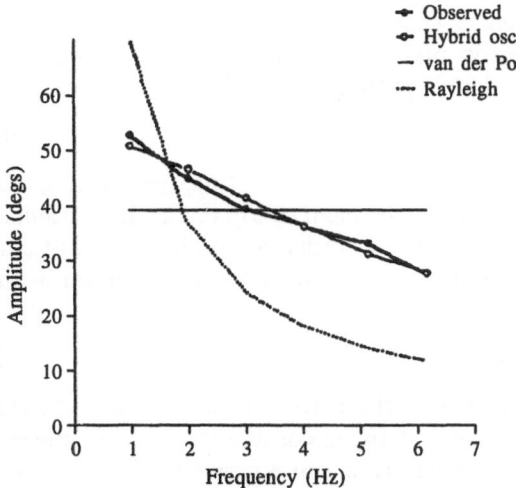

Fig. 7.2. Frequency (in Hz) versus amplitude (in degrees) for the single-handed data and the curves of the best fit for the Van der Pol, the Rayleigh, and the hybrid oscillator (7.13). After *Kay* et al.(1987)

The results of this experimental and theoretical study are in a way astounding. They demonstrate that our body, a highly complex system with its numerous degrees of freedom, behaves as if it were governed by a simple oscillator equation for a single degree of freedom, namely the finger tip displacement. An explanation of this result lies, on the one hand, in the order parameter concept in synergetics, but, on the other hand, must be found in studies on the behavior of the brain and spinal cord. We shall come back to this question later in Chap. 15.

7.2 Coupled Movement of Index Fingers

After our detailed study of the movement of a single index finger, we shall now try to model their coupled movement with the oscillatory motion being taken explicitly into account.

We recall that the displacements of the finger tips are denoted by x_1, x_2. We first write down the general form of the equations sought

$$\ddot{x}_1 + f_1(x_1, \dot{x}_1) = I_{12}(x_1, x_2), \tag{7.16}$$

$$\ddot{x}_2 + f_2(x_2, \dot{x}_2) = I_{21}(x_1, x_2). \tag{7.17}$$

Let us first consider the left-hand sides of these equations. A comparison with that of (7.13) shows the explicit form of f_1 and f_2. The formulation of the left-hand sides of (7.16) and (7.17) required some knowledge of oscillator theory. The concepts of synergetics come in when we discuss the terms on the right-hand sides of (7.16), (7.17). They describe a coupling between the two fingers and must be chosen in such a way that, eventually, the order parameter equation for the relative phase (6.14) emerges. This is, indeed, a very strong guide-line. The simplest hypothesis will be

$$I_{12} = \alpha(x_1 - x_2), \tag{7.18}$$

which is suggested by oscillator theory. As a detailed analysis shows, this hypothesis does not lead to the correct right-hand side of (6.14). In particular, the second term there would be missing. In order to obtain this additional term we need a cubic term. So we may try a nonlinear approximation in the form

$$I_{12} = \alpha(x_1 - x_2) + \beta(x_1 - x_2)^3. \tag{7.19}$$

Again this hypothesis does not lead to the correct phase relation. At this point, it might be useful to give more detailed consideration to the physiological system that we are dealing with. How can a correlation between the finger tips be brought about? A correlation of finger tips needs a signaling from the right finger to the left one, or vice versa, via the nervous system, where we have to leave it open at the present moment whether the signaling goes via the spinal cord only, or involves, for instance, the cerebellum. At any rate, we have to assume that a time delay will occur. Therefore, we make the hypothesis

$$I_{12}(t) = \int\limits_{-\infty}^{t} K(t,\tau)[\alpha_1(x_1 - x_2)_\tau + \alpha_2(x_1 - x_2)_\tau^3]d\tau, \qquad (7.20)$$

where we have used the abbreviation

$$(x_1 - x_2)_\tau \equiv (x_1(\tau) - x_2(\tau)). \qquad (7.21)$$

According to (7.20) the term I_{12} which describes the impact of the finger displacements x_1, x_2 on the movement of finger 1 at time t depends on x_1, x_2 at previous times τ. The amount of the delay is contained in the delay function $K(t,\tau)$. In order to determine $K(t,\tau)$ we impose the following requirements on it:

1) K shall act as a filter that allows the passage of the information carried by the factor in the square brackets in (7.20) [...] only during the time interval $t - t_1 < \tau < t - t_2$. We assume that this interval is short.
2) In order to avoid satiation of nerve cells, quite often the nervous system only reacts on temporal *changes* of signals. Therefore $K(t,\tau)$ must convert the signal [...] into an appropriate difference of signals [...] at different times.

In the limiting case of short time-intervals, both requirements 1) and 2) can be fulfilled by choosing

$$K(t,\tau) = -\delta'(t - \tau - \sigma), \qquad (7.22)$$

where δ' is the derivative of the Dirac function δ, and σ the time delay. The effect of $K(t,\tau)$ on the signal can be visualized as follows: At time $\tau = t - \sigma$ the signal can pass, and at an immediately following time instant, the signal can also pass, but with the opposite sign. Inserting (7.22) into (7.20) and using the calculus connected with δ-functions, we obtain

$$I_{12}(t) = \frac{d}{d\tau}\left[\alpha_1(x_1 - x_2)_\tau + \alpha_2(x_1 - x_2)_\tau^3\right] ; \tau = t - \sigma. \qquad (7.23)$$

The time derivative of [...] is a consequence of the δ'-function and has the effect that time-independent signals [...] lead to a vanishing interaction term I_{12} in accordance with our above postulate 2.

If the whole finger motion does not change much over the time delay, we may neglect σ compared to t in (7.23). Putting $\alpha_1 = \alpha$ and $3\alpha_2 = \beta$, (7.23) may be cast into the final form

$$I_{12} = (\dot{x}_1 - \dot{x}_2)[\alpha + \beta(x_1 - x_2)^2] \quad \text{and,} \quad I_{21} = -I_{12}. \qquad (7.24)$$

In this coupling term, the interaction between the two fingers becomes velocity-dependent. The calculation that leads from (7.16), (7.17) and (7.24) to an equation for the relative phase is based on the slowly varying amplitude approximation and the rotating wave approximation that we used in the last section. Since this calculation is somewhat lengthy, but doesn't yield any important insight, we skip it and merely quote the final result

$$\dot{\phi} = (\alpha + 2\beta r^2)\sin\phi - \beta r^2 \sin 2\phi. \tag{7.25}$$

The critical amplitude, where the change of relative phase occurs, is then given by

$$r_c^2 = \frac{-\alpha}{6\beta}, \quad \text{where} \quad \alpha < 0. \tag{7.26}$$

In the transitions generally studied in synergetics, fluctuating forces play an important role. Accordingly, a transition, say, from $\phi = \pi$ to $\phi = 0$ can be initiated only if fluctuating forces, F, are present. We thus include such forces in (7.16) and (7.17)

$$\ddot{x}_1 + f_1(x_1, \dot{x}_1) = I_{12}(x_1, x_2) + F_1(t), \tag{7.27}$$

$$\ddot{x}_2 + f_2(x_2, \dot{x}_2) = I_{21}(x_1, x_2) + F_2(t). \tag{7.28}$$

In the context of the present section it suffices to assume $F_j, j = 1, 2$ as a random small variable, which can easily be mimicked on a digital computer. The equations (7.27), (7.28) jointly with (7.24) can easily be solved on a digital computer. Figure 7.3 shows the results of the original publication by *Haken, Kelso, Bunz* (1985), in which only the Rayleigh-term was taken into account. Nevertheless, the results are in rather good agreement with the experimental data.

To summarize, it is quite clear that the main features of the experimental data are captured by our above mathematical formulation.

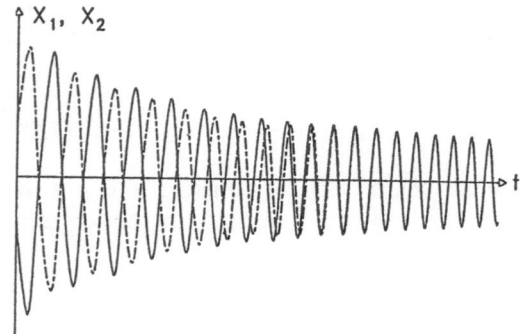

Fig. 7.3. Numerical simulation of (7.27), (7.28), (7.24) with $\epsilon_1 = 0, \epsilon_2 = 1, \omega_0^2 r_0^2 = 1, \alpha = -0.2, \beta = 0.2$. From the left to the right ω changes from $\omega = 1.17$ to $\omega = 3.05$. After *Haken, Kelso, Bunz* (1985)

7.3 Phase Transitions in Human Hand Movements During Multifrequency Tapping Tasks

In Chap. 6 and the foregoing part of Chap. 7 we presented a seemingly simple experiment on finger movement coordination at a given frequency. Nevertheless a careful experimental and theoretical study, in particular in the phase-transition region, revealed remarkable features underlying the interpretation in terms of self-organization. In this section and the following ones I want to study more complicated tasks of movement coordination. In such experiments the fingers are not tapped at the same frequency (in-phase or out-of-phase), but at different frequencies ω_1, ω_2. Hereby the task consists in keeping a fixed frequency ratio $\omega_1/\omega_2 = m/n$, where m and n are integers, and ω_1 and ω_2 are proportional to the frequency ω set by a metronome (acoustic or visual). In these experiments, interesting transitions in movement coordination occur: When the basic frequency ω is increased, the ratio m/n is changed involuntarily. As a most important result, the numbers m and n decrease during such transitions. There have been quite a variety of such experiments and theoretical approaches, in particular based on the so-called circle map. In Sect. 7.4.1 we shall present an outline of the experiments by *Beek* and *Peper* and a theoretical model which may shed some light on the mechanisms of the finger movement coupling that are not covered by previous approaches such as the circle map.

7.3.1 Experiment: Transitions in Multifrequency Tapping

Three skilled right-handed male drummers participated in the experiment. Stereo signals were generated on a micro-computer and presented through headphones. The hand movements were measured with a system which recorded the position of two light emitting diodes (LEDs) positioned at the tips of the middle fingers at a sample frequency of 122 Hz. Two signal trains with 50 ms sine-wave stimuli were generated. The difference in pitch between the two stimulus trains was small (fast train: 440 Hz [tone A]; slow train: 554 Hz [tone C#]). In all cases the initial interstimulus frequency of the fast train was 2 Hz. The frequency of the slow train was adjusted so that the frequency ratio was either 2:5 or 3:5. The time interval in which the fast train presented 5 stimuli and the slow train 2 or 3 stimuli is referred to as a *(rhythmical) cycle*. Three types of signals were presented:

- Signal Type 1 consisted of 25 rhythmical cycles at a constant rate. Signal time was 62.5 s.
- Signal Type 2 started with 10 cycles at a constant rate. Subsequently, the rate was increased by 4 per cent, so that each consecutive cycle was 4 per cent shorter than the previous cycle. This procedure was applied until the interstimulus frequency of the fast train reached 10 Hz. Thus, the total duration was 75 s.

- Signal Type 3 was constructed in the same way as Signal Type 2 with one difference: After 8 cycles (when the rate was still constant) only the first stimuli of each cycle (when both stimuli coincided) were presented. As in Signal Type 2, each new cycle period was decreased by 4 per cent, starting with the eleventh cycle. Total signal duration was thus 75 s.

The subject, wearing the headphones, sat in an upright position with the lower arms resting on a tabletop. He was instructed to tap with the hands (rotation around the wrist) on the table and to synchronize the left hand taps to the stimuli on the left channel (slow) and the right hand taps to the stimuli on the right channel (fast). In case Signal Type 3 was presented, the subject was instructed to continue tapping the prescribed rhythm even when only the simultaneous stimuli of each cycle were presented. The first taps of each cycle were to be synchronized to these stimuli so that the increase in tapping frequency was followed. The subject was instructed that in case of conflict the increasing tapping frequency should prevail over maintaining the required ratio.

A block of trials consisted of 7 signal presentations. First, Signal Type 1 was presented in a practice trial. Second, the subject practised tapping with an increasing frequency as specified by Signal Type 2. Signal Type 3 was used in the 5 experimental trials. The advantage of this signal type over Type 2 was that the presented and performed ratios did not interfere. After each trial the subject rested for 60 s. During a session both ratios were presented in an alternating fashion, three blocks of each. Halfway through the session a short break was given. A session lasted about 2 hours. Five sessions were conducted on consecutive days, leading to a total of 75 experimental trials for each ratio per subject. The experimental conditions were counterbalanced over subjects.

Following the last session, the subjects were required to tap unimanually as fast as possible. A block of three trials (20 s each) was conducted for each hand (counterbalanced). For each rhythmical cycle the mean frequency ratio was defined as $f_r/f_\ell (= f_{fast}/f_{slow})$, resulting in ratios larger than 1. Frequency locks were determined to last for at least two consecutive rhythmical cycles, during which the frequency ratios differed less than 0.1 from the average value over these cycles. During such a lock variations of ± 0.1 were allowed. In addition, the temporal pattern of the taps had to be conform to the thus obtained ratio.

The essence of the results can be described as follows:

Performance of the required ratio became unstable when movement frequency was scaled up. Importantly, the critical frequencies at which the stability of the required ratio was lost (expressed in mean frequency of the right [fast] hand: ratio 2:5: 5.95 Hz, SD: 0.79; ratio 3:5: 5.64, SD = 0.73) were circa $1 - 1.5$ Hz smaller than the maximal tapping frequencies of the right hand, as obtained in the unimanual trials. In other words, the pattern broke down as a result of coordination constraints, not as a result of constraints associated with the unit oscillators as such. In a large number of trials, the loss of stability was followed by a transition to another ratio (Fig. 7.4). In Table 7.1, the observed transitions are presented for each subject. Statistical analysis without assumptions about distribution parameters or dependencies revealed that the trials in which such a transition occurred outnumbered the

required ratio: 5/2

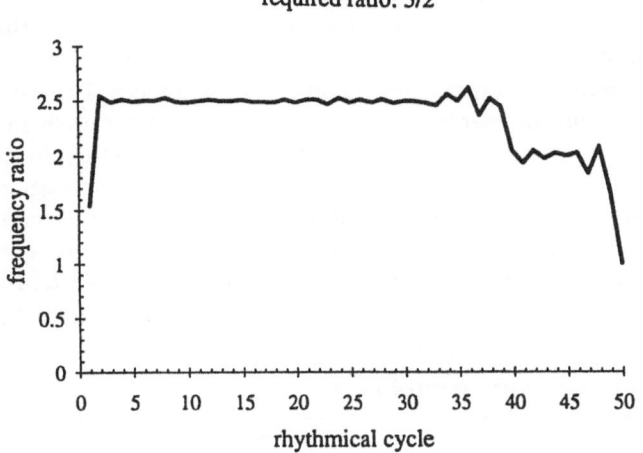

required ratio: 5/3

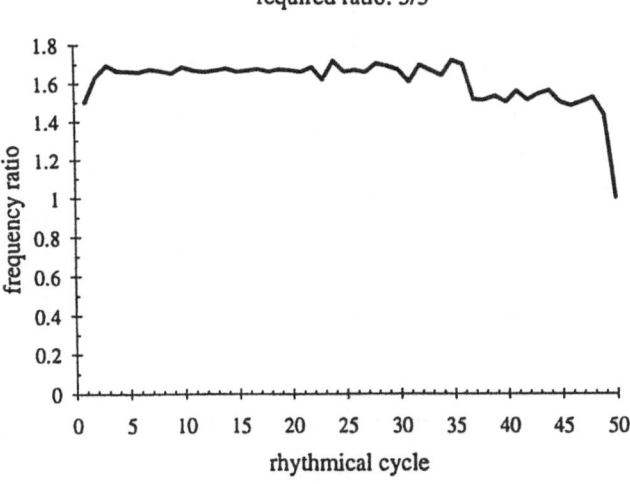

Fig. 7.4. Two typical transition routes obtained for Subject A. On the y-axis the observed mean frequency ratio ($f_{\text{right}}/f_{\text{left}}$) per rhythmical cycle is presented. Along the x-axis the tapping frequency increases over the rhythmical cycles. In the upper panel the required frequency ratio was 5:2 (2.5), showing a sudden transition to the ratio 2:1 (2.0). The lower panel shows a transition from 5:3 (1.67) to 3:2 (1.5). Both transitions conform to the unimodularity principle (*Beek, Peper* (1995))

remaining trials significantly (transition in 283 trials, no transition in 177 trials; $\chi^2 = 7.05$, $df = 1$, $p < 0.01$).

Before we describe our model, we discuss qualitatively some of the main aspects of the experimental results.

As in earlier studies on rhythmic limb movement, movement frequency functioned as a control parameter. Increase in frequency resulted in loss of stability of the performance of the required ratio. In the majority of trials transitions to other frequency ratios were observed. This corroborates the modeling of the tapping hands as a nonlinear system of coupled oscillators. Closer inspection of Table 7.1 reveals clear individual differences: For each subject preferred transitions appeared to exist. In two cases, these preferred transitions were between ratios that are not unimodularly related (Subject A: 3:5 to 3:4, Subject C: 2:5 to 2:3). These results are not easily understood from the circle map dynamics nor from an elaboration on this type of map, accounting for the dependency on the phase relation between the oscillators: the phase attractive map (*deGuzman* and *Kelso* (1991); *Kelso, deGuzman, Holroyd* (1990)). In these models the relative stability is roughly fixed (the

Table 7.1. The transition routes represented for each subject

2 : 5				**3 : 5**			
Subject A							
Transition route		# occurrences		Transition route			# occurrences
2:5	1:2	52		3:5	3:4		43
2:5		10		3:5	3:4	1:1	13
2:5	2:3	2		3:5	3:4	1:1	3:4	5
2:5	1:2	1:1	2	3:5	2:3		3
2:5	1:2	2:3	2	3:5	3:4	3:5	3:4	3
2:5	2:3	1:2	2	3:5	2:3	3:4	2
other (1 occurrence each)		5		other (1 occurrence each)			6
Subject B							
Transition route		# occurrences		Transition route			# occurrences
2:5	1:2	50		3:5			38
2:5		15		3:5	2:3		13
2:5	1:3	1:2	4	1:2			7
1:3	1:2	4		3:5	1:2		4
1:3	2:5	1:2	2	1:2	3:5	1:2	2
				other (1 occurrence each)			11
Subject C							
Transition route		# occurrences		Transition route			# occurrences
2:5		44		3:5			44
2:5	2:3	19		3:5	2:3		25
2:5	1:2	6		3:5	1:2	3:5	2
2:5	1:2	2:3	3	3:5	2:3	3:5	2
2:5	3:5	2:5	3:5	2	3:5	3:4	2:3	1
2:5	2:3	2:5	1	3:5	1:2	2:3	1

lower-order ratios h/k and h'/k' provide a larger stability region than the higher-order ratio in between $[h''/k'']$). Scaling the coupling parameter has a similar effect on the stability of all mode locks. Transitions to the closest more stable attractors will thus imply mode transitions.

The observed individual tendencies reveal that the differential stability of attracting mode locks is organized in a partly individual way. In order to understand the preferences in transition routes, a more elaborate model is required. Assuming that the unit oscillators do not change as a function of movement frequency, a more refined modeling of the coupling term is needed. In the next sections this elaboration will be performed.

7.4 A Model for Multifrequency Behavior *

The model has to account for the main features of the experimental data, which are: (i) stable performance of multifrequency ratios; (ii) loss of stability when frequency is scaled up; (iii) transitions to lower-order ratios; (iv) individual tendencies; (v) free-running solutions.

Since the mathematics of this and the following section is somewhat involved and lengthy, readers interested only in the main results can proceed directly to Sect. 7.6.

In analogy to our model of Sect. 7.2 (7.16,17), we start from the following equations for the displacements of the index fingers, x_1, x_2,

$$\ddot{x}_1 + f_1(x_1, \dot{x}_1) = I_{12} \tag{7.29}$$

and

$$\ddot{x}_2 + f_2(x_2, \dot{x}_2) = I_{21}. \tag{7.30}$$

The function f was introduced previously and is a suitable combination of the expressions of the Van der Pol and Rayleigh oscillators. Note that in contrast to Sect. 7.2, the stiffness constant a acquires two different values in the functions f_1 and f_2.

Because the tapping occurs periodically, we decompose the time dependent variables x_1, x_2 according to

$$x_j = A_j(t)\exp(i\omega_j t) + A_j^*(t)\exp(-i\omega_j t), \quad j = 1, 2 \tag{7.31}$$

where the complex amplitudes $A_j, j = 1, 2$ are decomposed into modulus r_j and phase factor according to

$$A_j(t) = r_j(t)\exp(i\phi_j(t)). \tag{7.32}$$

In the following we shall assume that r_j and ϕ_j are slowly varying functions as compared to $\exp(i\omega_j t)$, i.e., we shall use the slowly varying amplitude approximation. Furthermore, we shall occasionally use the abbreviation

$$\phi_j + \omega_j t = \chi_j(t). \tag{7.33}$$

When we insert (7.31) into the left-hand sides of (7.29), (7.30), we obtain terms of the form

$$\exp(i\omega_j t), \exp(-i\omega_j t) \tag{7.34}$$

and

$$\exp(3i\omega_j t), \exp(-3i\omega_j t). \tag{7.35}$$

According to the rotating wave approximation, the contributions due to (7.35) are smaller than those due to (7.34) and will be neglected in the following. Our main concern will be the interaction term, where we consider I_{12}. In generalization of the Haken, Kelso, Bunz model treated in Sect. 7.2, we shall assume that I_{12} may be represented as a polynomial containing terms of the form $x_1^k x_2^\ell$, i.e.

$$I_{12} = \sum_{k,\ell=0}^{K,L} c_{k\ell} x_1^k x_2^\ell \tag{7.36}$$

with time-independent coefficients $c_{k\ell}$. The sum over k and ℓ runs up to sufficiently large K and L so that the experimental data can be covered. This means in the present case $K, L = 5$. Note that $c_{00} = 0$.

In Sect. 7.2 the coupling term had to contain temporal derivatives $\dot{x}_1, \dot{x}_2$ so that adequate *phase* relationships resulted. As a closer inspection shows, the adequate generalization to the presently considered case can be achieved, when we replace I_{12} (7.36) by its time derivative. Thus our final choice of the interaction term I_{12} in (7.27) reads

$$I_{12} = \frac{d}{dt} \left(\sum_{k,\ell=0}^{K,L} c_{k\ell} x_1^k x_2^\ell \right). \tag{7.37}$$

The interaction term I_{21} has a corresponding form with coefficients $d_{k\ell}$. As long as we may consider the two hands as equivalent, we may assume $d_{k\ell} = c_{\ell k}$. We first study what the individual terms $x_1^k x_2^\ell$ of the polynomial (7.36) mean when we insert the expression (7.31), i.e., we have to study

$$x_1^k x_2^\ell = [A_1 \exp(i\omega_1 t) + A_1^* \exp(-i\omega_1 t)]^k$$
$$\times [A_2 \exp(i\omega_2 t) + A_2^* \exp(-i\omega_2 t)]^\ell. \tag{7.38}$$

With some straightforward algebra, the first bracket can be expanded in the form

$$\sum_{\kappa=0}^{k} \binom{k}{\kappa} A_1^\kappa A_1^{*k-\kappa} \exp(-i(k-2\kappa)\omega_1 t), \tag{7.39}$$

where $\binom{k}{\kappa}$ are the binomial coefficients. If k is even, all $(k-2\kappa)$ are even, and if k is odd, all $(k-2\kappa)$ are odd. Using the decomposition (7.32), each individual term under the sum in (7.39) can be written in the form

$$A_1^\kappa A_1^{*k-\kappa} \exp(-\mathrm{i}(k-2\kappa)t) = r_1^k \exp(-\mathrm{i}(k-2\kappa)\chi_1). \tag{7.40}$$

The same considerations hold, of course, for the second bracket in (7.38). All in all we are led to the idea that I_{12} is a superposition of terms of the form

$$\exp(-\mathrm{i}\underbrace{(k-2\kappa)}_{m}\omega_1 t) \cdot \exp(-\mathrm{i}\underbrace{(\ell-2\lambda)}_{n}\omega_2 t). \tag{7.41}$$

We note that

$$-k \le m \le k \tag{7.42}$$

and

$$-\ell \le n \le \ell. \tag{7.43}$$

Precisely the same considerations hold, of course, also for I_{21}.

Our basic idea may be exemplified by means of (7.29): We are looking for a resonance condition, in which the oscillation on the left-hand side, described by $\exp(\pm\mathrm{i}\omega_1 t)$, is in resonance with one of the oscillating terms on the right-hand side, which is of the general form

$$\exp(-\mathrm{i}m\omega_1 t - \mathrm{i}n\omega_2 t). \tag{7.44}$$

Note that the time-derivative that occurs in (7.37) does not influence our considerations. Equation (7.44) leads us to the resonance condition

$$\pm\omega_1 = -m\omega_1 - n\omega_2, \tag{7.45}$$

or, after a slight rearrangement,

$$(m \pm 1)\omega_1 + n\omega_2 = 0. \tag{7.46}$$

When we include the phase factors that are introduced in (7.32) and use, in addition, (7.33), we shall have to study the behavior of expressions of the form

$$\exp(\mathrm{i}[(m \pm 1)\chi_1 + n\chi_2]), \tag{7.47}$$

which result when the right-hand side of (7.29) is multiplied by $\exp(\pm\mathrm{i}\chi_1)$.

In the light of these considerations we now study the experimental results in which specific frequency ratios were observed. Let us start with the frequency ratio 2:5. This means in units of the basic frequency $\omega_1 = 2, \omega_2 = 5$. From (7.46), we obtain

$$\omega_1 = 2, \omega_2 = 5 \quad : \quad (m \pm 1)2 = -5n, \tag{7.48}$$

which has the solutions

$$m \pm 1 = 5 \quad m = \begin{cases} 6 \\ 4 \end{cases} \tag{7.49}$$

and

$$n = -2, \tag{7.50}$$

Table 7.2. Combinations of m and n which satisfy the resonance constraint $(m \pm 1)\omega_1 + n\omega_2 = 0$, determined for the frequency ratios presented in Table 7.1. The faster frequency is represented by ω_2. See text for details

$\omega_1 : \omega_2$	m	$-n$
3 : 5	4	3
	6	3
3 : 4	3	3
	5	3
2 : 5	4	2
	6	2
2 : 3	2	2
	4	2
1 : 2	1	1
	3	1
1 : 1	0	0
	2	0

where the two numbers in (7.49) result from the two different signs in (7.48). This example may suffice to show how Table 7.2 results, where we keep only low numbers of n and m. Entirely analogous considerations apply to the resonance conditions resulting for (7.30). Looking at the experimental results, there are a number of possible routes, for instance, the $n = \pm 2$ route, where $n = \pm 2$ is kept constant. The most striking feature of the experimental results reported above is the transition between the observed frequency ratios. This means that the frequency couplings must be broken and replaced by new ones. To this end, we observe that the individual exponential functions of the form (7.47) are multiplied by amplitudes according to (7.40) and correspondingly for r_2. We now assume, in agreement with former experiments discussed in Sect. 7.1, that the amplitudes drop with increasing frequency. This means that, with increasing frequency, those exponential functions of the form (7.47) that contain smaller m and n become more and more important. In other words, lower orders become more important than higher orders for increasing basic frequencies ω. When we further assume that the faster hand leads, we may consider n as a constant that remains small, and m does not grow, but rather decreases (cf. Table 7.2). In these considerations we assume that the coefficients of the polynomial that contain the exponential functions (7.47) are independent of the basic frequency ω; this is a reasonable assumption.

7.5 The Basic Locking Equations and Their Solutions *

After this general survey, we are now in a position to cast our model into an explicit form. Let us start with a more detailed study of the left-hand side of (7.29), which in the case of the mixed Van der Pol-Rayleigh oscillator, reads explicitly

$$\ddot{x}_1 + [\epsilon_1 \left(x_1^2 - r_{10}^2 \right) + \epsilon_2 \left(\dot{x}_1^2 - \omega_1^2 r_{10}^2 \right)]\dot{x}_1 + a_1 x_1^2 = I_{12}. \tag{7.51}$$

We put

$$a_1 = \omega_1^2 + \delta\omega_1(2\omega_1), \tag{7.52}$$

where $\delta\omega_1$ represents some detuning. In the following we shall consider $\epsilon_1, \epsilon_2, r_{10}$ and ω_1 as fixed constants, whereas a_1 (and correspondingly a_2) can be adjusted in a way we shall discuss later. We make the hypothesis

$$x_1 = A_1 \exp(i\omega_1 t) + \text{c.c.} \tag{7.53}$$

Using the slowly varying amplitude approximation and the rotating wave approximation, we may transform (7.51) into

$$i\omega_1 \exp(i\omega_1 t)[2\dot{A}_1 - 2i\delta\omega_1 A_1 + (\epsilon_1 (\mid A_1 \mid^2 - x_{01}^2)$$
$$+ \epsilon_2 \omega_1^2 (3 \mid A_1 \mid^2 - x_{01}^2))A_1] + \text{c.c.} = I_{12}. \tag{7.54}$$

At this point, our fundamental resonance condition (7.45) comes into play: on the left- and right-hand side we compare the coefficients of $\exp(i\omega_1 t)$ and $\exp(-i\omega_1 t)$. Furthermore we put

$$A_1(t) = r_1(t) \exp\left(i\phi_1(t)\right). \tag{7.55}$$

By means of (7.55) and

$$\chi_1 = \omega_1 t + \phi_1, \tag{7.56}$$

we obtain, after dividing the coefficient of $\exp\left(i\omega_1 t\right)$ in (7.54) by $2i\omega_1$,

$$\exp(i\chi_1) \left(\dot{r}_1 + i\dot{\phi}_1 r_1 - i\delta\omega_1 r_1 + N_1(r_1)\right) = R_1, \tag{7.57}$$

where

$$N_1(r_1) = \frac{1}{2} \left\{\epsilon_1 \left(\mid r_1 \mid^2 - x_{01}^2\right) + \epsilon_2 \omega_1^2 \left(3 \mid r_1 \mid^2 - x_{01}^2\right)\right\} r_1. \tag{7.58}$$

The right-hand side denoted by R_1 stems from the coupling to the other variable, x_2. Multiplying (7.57) by $\exp(-i\chi_1)$ and taking the real part results in

$$\dot{r}_1 + N_1(r_1) = \text{Re} \left\{\exp(-i\chi_1)R_1\right\}. \tag{7.59}$$

The form of $N_1(r_1)$ guarantees a stable solution r_1 that is changed only slightly by the right-hand side of (7.59) and that decreases with increasing frequency ω_1. By taking the imaginary part, we obtain

$$\dot{\phi}_1 = \frac{1}{r_1}\text{Im} \left\{\exp(-i\chi_1)R_1\right\} + \delta\omega_1. \tag{7.60}$$

To demonstrate the model, we choose to examine how it accounts for the cascade 5:2, 3:2, 2:1. As can be seen in Table 7.1, this particular cascade was observed twice for Subject A. The first transition (from 2:5 to 2:3) turned out to be a preferred transition for Subject C. We now write down (7.60) explicitly with its right-hand side terms that are responsible for this particular transition route

$$\dot{\phi}_1 = -\delta\omega_1 + ar_1^3 r_2^2 \sin\left(5\chi_1 - 2\chi_2\right) + br_1 r_2^2 \sin\left(3\chi_1 - 2\chi_2\right)$$
$$+ cr_2 \sin\left(2\chi_1 - \chi_2\right). \tag{7.61}$$

In complete analogy to the phase equation for the first hand, we may derive an equation for the second hand, which reads explicitly

$$\dot{\phi}_2 = -\delta\omega_2 + a'r_1^5 \sin\left(5\chi_1 - 2\chi_2\right) + b'r_1^3 \sin\left(3\chi_1 - 2\chi_2\right)$$
$$+ c'\left(r_1^2/r_2\right) \sin\left(2\chi_1 - \chi_2\right). \tag{7.62}$$

In (7.61) and (7.62) we included only the lowest powers of r_1 and r_2 that are possible. Note that the coefficients a', b' and c' may be different from those of (7.61) and that particularly the dependence on the mode amplitudes is different. In the following we assume that the coefficients a, b, c, a', b', c' are independent of the basic frequency ω. [This can actually be derived explicitly: The original coefficients $c_{k\ell}, d_{k\ell}$ were assumed to be independent of ω. When proceeding from (7.54) to (7.57) we divided the interaction term by $2i\omega_1$, which is proportional to ω. But, on the other hand, the time-derivative in (7.37) produces a factor $-i(m\omega_1 + n\omega_2)$, which is proportional to ω. In this way, in the final result, ω cancels. At the same time we note that the factor i also cancels. Thus the prefactors of the exponential functions are real and we may replace the imaginary part of the coupling by sine-functions.]

Equations (7.61) and (7.62) are coupled equations for the phases ϕ_1 and ϕ_2. As we shall demonstrate below, these equations contain the key to our understanding of the observed transition in human hand movements. When the basic frequency ω is increased, the amplitudes r_1, r_2 decrease. As a consequence, the coefficients with higher powers of r_1, r_2 decrease more quickly than those with lower powers. This implies that the coefficient of, for instance, $\sin(5\chi_1 - 2\chi_2)$ decreases more rapidly than that of $\sin(3\chi_1 - 2\chi_2)$. As we shall see, this causes a switch in behavior. It is interesting to note that this decrease of r_1, r_2 also lies at the root of the phase transitions in finger movements in our model of Chap. 6.

In order to bring out the essentials we drop the third terms on the corresponding right-hand sides of (7.61) and (7.62) so that in the following we shall explicitly be concerned with the 2:5 $\rightarrow$ 2:3 transition. The resulting equations can be further transformed. We introduce the variables Δ_1, Δ_2, which we define as

$$\Delta_1 = 5\chi_1 - 2\chi_2, \tag{7.63}$$

and

$$\Delta_2 = 3\chi_1 - 2\chi_2 \tag{7.64}$$

where (cf. (7.56))

$$\chi_j = \phi_j + \omega_j t, \quad j = 1, 2. \tag{7.65}$$

Conversely, χ_1 and χ_2 may be expressed in terms of these new variables as

$$\frac{\Delta_1 - \Delta_2}{2} = \chi_1 \tag{7.66}$$

and

$$\frac{3\Delta_1 - 5\Delta_2}{4} = \chi_2. \tag{7.67}$$

We introduce (7.65), (7.63), and (7.64) into (7.61) and (7.62). Multiplying (7.61) by 5 and multiplying (7.62) by 2 and subtracting the resulting equations from each other, we obtain an equation of the form

$$\dot{\Delta}_1 = a_1 - B_1 \sin \Delta_1 - C_1 \sin \Delta_2. \tag{7.68}$$

In a similar fashion, we may form an equation for Δ_2, and obtain

$$\dot{\Delta}_2 = a_2 - B_2 \sin \Delta_1 - C_2 \sin \Delta_2. \tag{7.69}$$

We have introduced the abbreviations $a_j = \Omega_j + \eta_j, j = 1, 2$, where

$$\Omega_1 = 5\omega_1 - 2\omega_2, \quad \eta_1 = -5\delta\omega_1 + 2\delta\omega_2 \tag{7.70}$$

and

$$\Omega_2 = 3\omega_1 - 2\omega_2, \quad \eta_2 = -3\delta\omega_1 + 2\delta\omega_2. \tag{7.71}$$

Note that the coefficients $B_j, C_j, j = 1, 2$ contain powers of r_1, r_2 according to (7.61), (7.62). We first consider the case in which the frequency ratio 2:5 is achieved and thus assume

$$\Omega_1 \equiv 5\omega_1 - 2\omega_2 = 0, \quad \eta_1 \equiv -5\delta\omega_1 + 2\delta\omega_2 = 0 \quad \text{or small.} \tag{7.72}$$

Then (7.68) and (7.69) reduce to

$$\dot{\Delta}_1 = \eta_1 - B_1 \sin \Delta_1 - C_1 \sin \Delta_2, \tag{7.73}$$

and

$$\dot{\Delta}_2 = \Omega_2 + \eta_2 - B_2 \sin \Delta_1 - C_2 \sin \Delta_2. \tag{7.74}$$

We first consider the case

$$| B_1 | > | C_1 |, \tag{7.75}$$

and

$$| \Omega_2 | > | B_2 | > | C_2 | \tag{7.76}$$

and put $\eta_1 = 0$. As we shall see below, the conditions (7.75) and (7.76) cause quite different behavior of Δ_1 and Δ_2. While Δ_1 is practically frequency locked, i.e. $\Delta_1 \equiv 5\chi_1 - 2\chi_2 \approx \text{const.}$, Δ_2 increases in time essentially proportional to Ω_2. This is precisely the behavior we expect in the case of 2:5 frequency locking. On the other hand, when condition (7.75) is violated, i.e. $| C_1 | > | B_1 |$, a new behavior results: frequency locking is lost and we are dealing with a free-running case. Under these conditions, however, frequency locking with a ratio of 2:3 can be attained.

Let us treat these cases in more detail. We claim that if the conditions (7.75) and (7.76) are fulfilled, the solutions of (7.73), (7.74) have the following properties:

$$\Delta_1 = \text{constant} + \text{small oscillations at } \Omega_2 \text{ and higher harmonics,} \qquad (7.77)$$

$$\Delta_2 = \Omega_2 t + \text{small oscillations at } \Omega_2 \text{ and higher harmonics.} \qquad (7.78)$$

To prove our assertion, we again put $\eta_1 = 0$ and

$$\Delta_1 = \Delta_0 + \delta_1(t), \qquad (7.79)$$

where Δ_0 is the solution of

$$- B_1 \sin \Delta_0 - \overline{C_1 \sin \Delta_2} = 0 \qquad (7.80)$$

with the bar denoting time-average. Because of (7.75), this relation can always be fulfilled by a real, time-independent constant Δ_0. We insert (7.79) into (7.73) and obtain, using (7.80) and

$$\sin \Delta_1 = \sin \Delta_0 \cos \delta_1 + \cos \Delta_0 \sin \delta_1$$

for small enough δ_1:

$$\dot{\delta}_1 = -B_1 \cos \Delta_0 \delta_1 - C_1(\sin \Delta_2 - \overline{\sin \Delta_2}). \qquad (7.81)$$

Apart from decaying transients, the solution of this equation is a purely oscillating function of time with the same frequency dependence as the term in brackets. Jointly with (7.79) and anticipating (7.78) in a self-consistent manner, this proves assertion (7.77). To prove our assertion (7.81), we insert the hypothesis

$$\Delta_2 = \Omega_2 t + \delta_2(t) \qquad (7.82)$$

into (7.74) and obtain, using

$$\sin \Delta_2 = \sin \Omega_2 t \cos \delta_2 + \cos \Omega_2 t \sin \delta_2,$$

for small enough δ_1, δ_2 (neglecting terms of higher order of δ_1, δ_2 and $C_2\delta_2$):

$$\dot{\delta}_2 = \eta_2 - B_2 \sin \Delta_0 - B_2 \cos \Delta_0 \delta_1 - C_2 \sin \Omega_2 t. \qquad (7.83)$$

If we assume $| \Omega_2 | >> | B_2 |$, $B_2 \sin \Delta_0$ is a small quantity compared to Ω_2. The subject may thus slightly change the stiffness that enters through η_2 into (7.83) to compensate $B_2 \sin \Delta_0$. Under this condition the steady state solution, $\delta_2(t)$, of (7.83) consists of pure oscillations caused by δ_1 and $\sin \Omega_2 t$, which both oscillate at Ω_2 and higher harmonics.

In conclusion, we may state that under the condition (7.75) and (7.76) the system is able to maintain the ratio 2:5. This result is fully substantiated by the numerical solution of (7.68), (7.69) presented in Fig. 7.5.

We now assume that with increasing frequency the amplitudes r_1, r_2 drop and that, eventually,

$$| B_1 | < | C_1 | \overline{\sin \Delta_2}. \qquad (7.84)$$

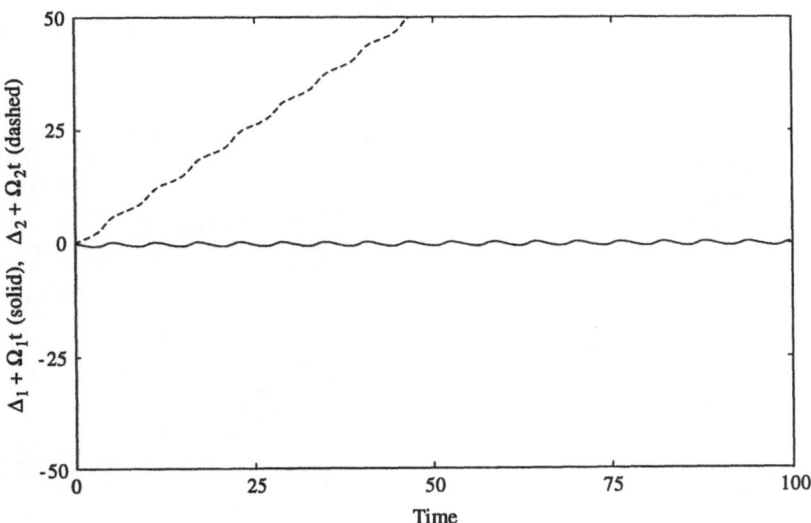

Fig. 7.5. The results of the integration of (7.68), (7.69) with the initial condition $\Delta_i(0) = 0$. The solid curve represents $\Delta_1 + \Omega_1 t$, the dashed curve $\Delta_2 + \Omega_2 t$. Note that $\Delta_1 + \Omega_1 t$ is *stable*. Parameters:

	η_i	B_i	C_i	Ω_i
i = 1	0	1	0.75	0
i = 2	0	1	0.9	1.1

Roughly speaking, in this case the term containing B_1 can no longer compensate the last term in (7.73), if η_1 is small. By adding and subtracting $C_1\overline{\sin\Delta_2}$, we write (7.73) in the form

$$\dot{\Delta}_1 = \eta_1 - C_1\overline{\sin\Delta_2} - B_1\sin\Delta_1 - C_1(\sin\Delta_2 - \overline{\sin\Delta_2}).$$

Its solution reads

$$\Delta_1 = (\eta_1 - C_1\overline{\sin\Delta_2})t + \text{small oscillations.} \tag{7.85}$$

This form of solution can be readily verified by considering the limiting case $B_1 \to 0$.

The general result (7.85) is fully substantiated by the numerical solution of (7.71), (7.72), as illustrated in Fig. 7.6. When we remember that according to its definition,

$$\Delta_1 = 5\omega_1 - 2\omega_2 + 5\phi_1 - 2\phi_2$$

the result (7.85) implies that frequency locking has been lost, because in the locking case ϕ_1 and ϕ_2 are time independent or, at least, do not increase linearly with time. Because $\Omega_1 = 0$, a rather large change in η_1 (compared to Ω_1) would be required to maintain the 2:5 frequency locking. If, however, η_1 can be changed only slightly, the mode-locked state is lost. Changing the

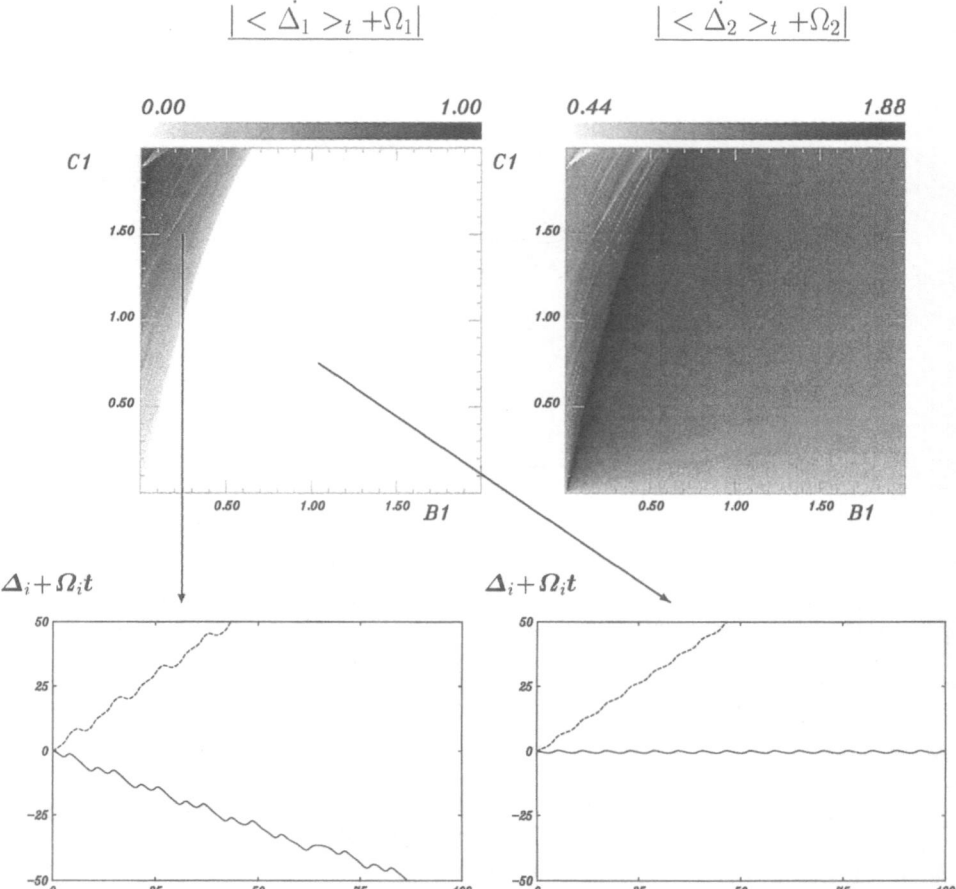

Fig. 7.6. In general the condition of frequency locking or *stability* can be expressed by $\Delta_i + \Omega_i t = \text{constant} + \text{small oscillations}$ or in an analogous fashion by $\langle \dot{\Delta}_i \rangle_t + \Omega_i \rightarrow 0$. Thus, the parameter range of *steady* states can be visualized by computing the time-averaged derivatives of Δ_i for different parameters a_i, B_i, C_i, or C_i, or Δ_i. The resulting stability diagrams are coded so that large values of $\langle \dot{\Delta}_i \rangle_t + \Omega_i$ correspond to darker displays. *Upper part*: *Stability-diagrams* show the absolute values of the time-averaged derivatives of $\Delta_i + \Omega_i t$. The white area reveals the range of frequency locking for $\Delta_i + \Omega_1 t$, while the behavior of $\Delta_2 + \Omega_2 t$ always remains *unstable* ($\propto t + \text{small oscillations}$). The precise correspondence with the grey areas is plotted on top. *Left*: i = 1; *right*: i = 2. *Lower part*: *Evolutions* of $\Delta_i + \Omega_i t$ (i = 1: *solid line*; i = 2: *dashed line*) are shown. The parameters B_1 and C_1 are chosen such that $\Delta_1 + \Omega_1 t$ becomes either *stable* (*right panel*) or *unstable* (*left panel*). While in the former case frequency locking occurs, in the latter case it is lost. Parameters:

	η_i	B_i	C_i	Ω_i
i = 1 (*left*)	0	0.15	1.5	0
i = 1 (*right*)	0	1	0.75	0
i = 2	0	1	0.9	1.1

roles of Ω_1 and Ω_2 then results in the analogue of the previously considered case, but now locking at the frequency ratio 2:3. In other words, the subject may realize a new frequency-locked state by means of a small detuning in the frequencies of the hands.

7.6 Summary of the Main Theoretical Results

In Sects. 7.4 and 7.5 we developed a theoretical model to account for the main features of the empirical data: (i) frequency locking; (ii) loss of stability, when movement frequency was scaled up; (iii) transitions to one or more lower-order ratios in the majority of trials; (iv) individual tendencies in these transition routes; and (v) free-running solutions in a number of trials. In accordance with Chap. 6 and parts of Chap. 7, the displacement x_1, x_2 of the finger tips was modeled by means of mixed Van der Pol–Rayleigh oscillators. Central to our approach is the form of the coupling between the two finger oscillators. The coupling term was formulated as a polynomial in x_1, x_2 of increasing power. Decomposing $x_j, (j = 1$ or $2)$ into a real amplitude r_j, a frequency factor $\exp(i\omega_j t)$ and a phase-factor $\exp(i\phi_j)$, we obtained coupling terms that depend on exponential functions containing various frequency combinations of the form $m\omega_1 - n\omega_2$, with m and n integers. The prefactors of the exponential function contain constant coefficients $a, b, \ldots$ and powers of r_1 and r_2. The higher m and n, the higher the powers of r_1 and r_2. We then studied these cases, where the frequency combination $m\omega_1 - n\omega_2$ resonated with the individual oscillation frequencies ω_1 and ω_2 of the finger tip movement. Which of these resonance conditions can be fulfilled depends on ω_1, ω_2 and, in particular, on the amplitudes r_1 and r_2. Because increase in movement frequency ω is associated with a drop in amplitudes r_1 and r_2, the relative weights of the coupling terms depend on this control parameter, ω. If movement frequency is increased, lower-order terms become more important than higher-order terms. If stability of the required (higher-order) ratio is lost, the system may be detuned to attain a lower-order frequency lock.

The working of the model was exemplified for the transition route 2:5, 2:3, 2:2. In the frame of the present approach the occurrence of this particular cascade of transitions may be interpreted as follows: By learning or practice the subject becomes able to set the coefficients a, b and a', b', in such a way that a stable locking at 5:2 can be achieved for the basic driving frequency ω, according to our model.

However, when this frequency is increased, the amplitudes r_1, r_2 drop as we know from the beginning of this chapter. As a consequence, in the coupling the relative weight of those frequency combinations $m\omega_1 - n\omega_2$ increases which are associated with powers of r_1, r_2 with *lower* m and/or n. As a result, the frequency locking condition for the ratio 5:2 may be violated and instead of frequency locking we obtain a free-running solution or the system (i.e. the subject's brain) has to search for a new compatible solution. Then under only slight changes of the parameters, a new frequency locking condition can be fulfilled and the transition to the frequency ratio 3:2 is achieved. Our approach can be continued if we take into account a further coupling term. In this way the transition 3:2 to 2:1 can be obtained along similar lines, as indicated above.

To generalize this survey on the working of our model, we emphasize that the coupling function consists of a large number of coupling terms, not just the ones that specify the empirically observed frequency locks. The relative weights of these terms depend on intrinsic dynamics, learning, and intention. If we visualize the resulting generalized potential landscape as a function of the frequency ratio between the hands, the minima indicate which ratios may be attained. When the driving frequency ω is changed, some minima may disappear and transitions between frequency lockings may occur. However, in the presence of several minima the transitions are not uniquely determined and depend in part on chance events in the system. When the required ratio loses stability, the system may achieve a new frequency lock by detuning the stiffness parameter. In general, this detuning will be small. This implies that the newly established ratio will be either situated near the required ratio (on the unit interval) or it may be a ratio with large weights (e.g. as a result of practice). In either case only a little detuning is required to be attracted to the new ratio. This may explain the strong individual tendencies in the empirically observed transition routes. Asymmetries between the fast hand and slow hand (cf. *Summers, Ford, Todd* (1993); *Summers, Kennedy* (1992); *Peper* et al., in press) can be accounted for by allowing larger detuning for the slow hand than for the fast hand.

In closing, it is useful to compare the present approach with that of Sect. 7.1, which modeled the transition between the in-phase and the antiphase motion in a 1:1 frequency coordination task. In our approach, in contrast, we were concerned with changes between frequency ratios. While the transition in the case of Sect. 7.1 was caused by a potential function for the relative phases (where one minimum vanished under increasing frequency), the present case is more complicated. In line with the model of Sect. 7.1 we observed that coupling terms between the two hands play a decisive role both in establishing the frequency coupling and its loss. But, in contrast to that former case, additional requirements were needed. First of all, the requirement that ω_1 and ω_2 are integer multiples of ω must be fulfilled. *Kelso* et al. (1992) presented results which showed adequate brain functioning in this respect (cf. Chap. 15). Their MEG data showed that while tapping to an acoustic signal not only the basic frequency of that signal but also higher harmonics were generated (cf. Chap. 15). These higher harmonics may be used to fix the eigenfrequencies of the limb movements by means of the stiffness factors a_1 and a_2. This mechanism would guarantee a certain automatism for the transition from one frequency ratio to another. In addition, the subjects seemed to avoid free-running solutions. This implies an additional control mechanism which checks whether frequency locking is lost and scans the new possibilities for coordinated behavior. In other words, the mechanism of transition between different coordination modes is not as stringent as in the case of the finger movement experiments by *Kelso* and his co-workers (e.g. *Kelso* (1994), *Kelso* et al. (1986)). This is corroborated by the experimental results quoted in Sect. 7.3, which indeed showed free-running solutions as well as variation and individual tendencies in transition routes.

7.7 Summary and Outlook

Looking back at this chapter we may state that a coherent picture of oscillatory finger movements is emerging both in relation to the movement of a single (index) finger and to coupled movements. In Sect. 7.3 we developed a model that accounts for rather complicated phenomena whose explanation goes beyond previous approaches, in particular by the circle map. Preliminary studies of mine indicate that the theoretical approach developed in Sect. 7.3 can also be applied to explain the theoretical results that are phenomenologically described by the Farey tree, so that our model may become an interesting alternative to that of the circle map. Continuing the discussion at the end of Chap. 6, we may expect that the model of Sect. 7.3 will also be applicable to other coordination tasks.

8. Learning

In this chapter we will discuss various kinds of learning and relate them to the order parameter concept.

8.1 How Learning Changes Order Parameter Landscapes

The finger movement paradigm that we discussed in the preceding sections allows us to shed new light on mechanisms of learning. Relevant experiments were performed by *Yamanishi, Kawamoto, Suzuki* (1980), *Tuller* and *Kelso* (1989), and *Zanone* and *Kelso* (1992). Let us describe the experiment by *Yamanishi* et al. first. Subjects had to follow two visual metronomes, one for each finger, by their finger movement, where the frequency was fixed but the relative phase between metronomes was varied. Subjects were asked to produce several such phasing patterns until a level–of–performance criterion was attained. Then the subjects were required to reproduce a memorized relative phase. In the Tuller/Kelso experiment the subjects had to follow the metronomes all the time.

Let us describe the Zanone/Kelso experiment in more detail. The learning procedure consisted of five consecutive daily sessions that was followed one week later by a test session. The visual metronome consisted of two light-emitting diodes (LED) placed 8 cm apart at gaze height. The hands were slipped into a bimanual finger apparatus that measured the flexion–extension movements of the index fingers. The task was to flex each finger in temporal coincidence with the onset of the ipsilateral LED. The variable to be measured was the point estimate of the relative phase actually produced. The pattern to be learnt consisted of finger movements with a relative phase of 90°. We shall not dwell on the details of the careful measurements, but simply present some of the most salient features which will lead us to a basic conclusion about learning. To check its effect, the finger movement was scanned over prescribed relative phases in the whole range from 0° to 180°, whereby the phases were increased stepwise by 15°. Figure 8.1 shows the resulting curves for three subjects. For example, let us consider the uppermost part of this figure. To understand it, we define the error $\phi - \psi_{\mathrm{env}}$ by the actually performed phase minus the prescribed phase. The solid curve shows the error versus the

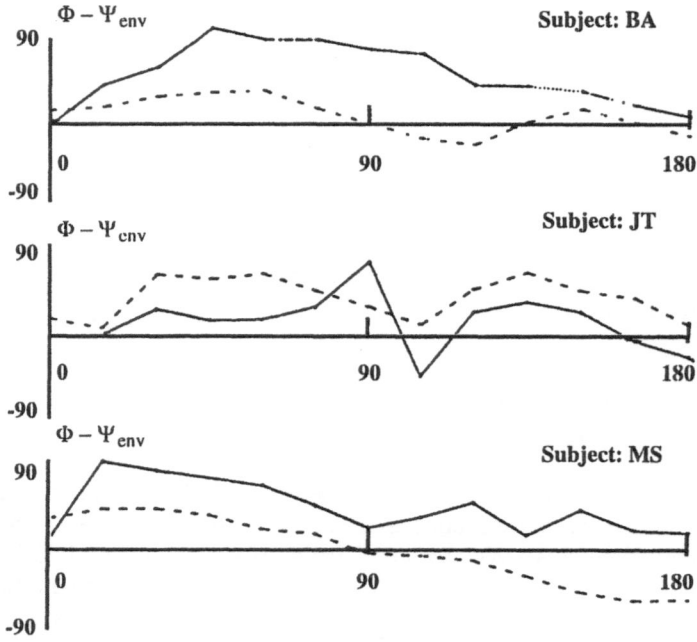

Fig. 8.1. In each of the three diagrams, referring to test persons BA, JT, and MS, the scanning error ($\phi - \psi_{env}$) is plotted versus ψ_{env} before learning (*solid curve*) and after learning (*dashed curve*). After *Schöner, Zanone,* and *Kelso* (1992)

prescribed relative phase on day 1 before learning took place, the dashed curve on day 5 after learning took place. Quite evidently, the error has been minimized for 90° in all three cases. A more detailed insight into the *statistical nature* of these results is revealed by Fig. 8.2. In it the relative frequency of an achieved relative phase is plotted versus the prescribed relative phase. Each individual curve refers to learning after 1 to 5 days, as indicated on the lower part of the ordinate. Note the correspondingly shifted zero-level of the distribution function. After one day there is a two-peaked frequency distribution, which after progressive learning is transformed into a single peak curve.

An experimentally measured relative frequency can be immediately linked with a probability distribution of the kind we have been discussing in Sect. 6.4. There we saw how a probability distribution resulted from two sources, namely a deterministic driving force connected with a potential landscape and a fluctuating force. Let us assume that the fluctuating force remains constant even in the presence of learning. Then the result of Fig. 8.2 strongly suggests that the underlying potential landscape has been changed in the course of learning. In order to construct such a landscape, we consider (6.28) with a new potential function V_ψ

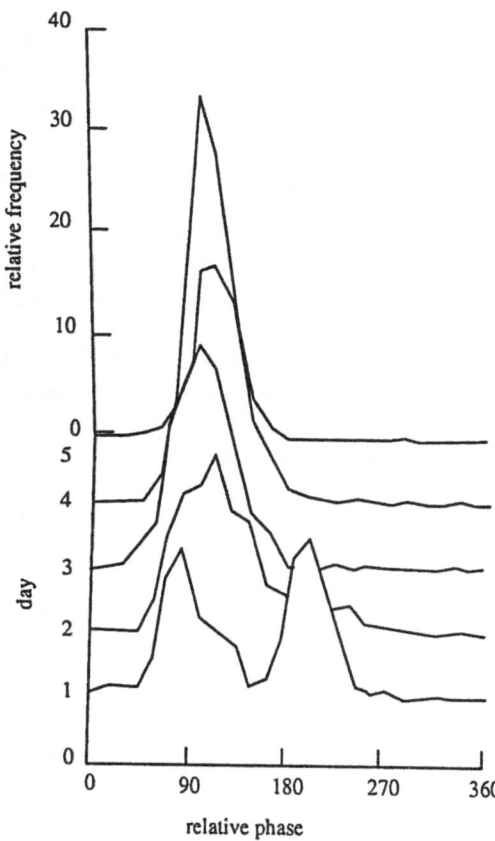

Fig. 8.2. The performance (in this case of Subject JT) can be represented by means of a histogram, i.e., by plotting the relative frequency of the occurrence of the relative phase versus phase. Note the transition from the bimodal to the monomodal distribution. (Redrawn after *Schöner*, et al. (1992) This change implies a corresponding change in the potential landscape from two minima to having a single minimum

$$\dot{\phi} = -\frac{\partial V_\psi}{\partial \phi} + F(t). \tag{8.1}$$

The index ψ of V_ψ indicates that the phase ψ has to be learnt. We decompose V_ψ into two parts according to

$$V_\psi = V_{\text{intr.}} + V_{\text{env}} \tag{8.2}$$

or

$$V_\psi = V_{\text{intr.}} + V_{\text{mem}}. \tag{8.3}$$

In it the first part is the well-known potential function that we denote as intrinsic potential:

$$V_{\text{intr.}} = -a\cos\phi - b\cos 2\phi. \tag{8.4}$$

The additional parts V_{env} and V_{mem} correspond to the change of the potential by means of the *environment* as in the Tuller/Kelso experiment or to the effect of *memory* as in the experiment of *Yamanishi* et al. We may think of a third similar term if *intention* is involved. A new potential term must take care of the fact that ψ is to become the preferred phase after learning, i.e. this new potential must have a minimum at $\phi = \psi$. We may consider several forms, the simplest compatible with the requirement of a minimum at $\phi - \psi$ and a periodicity constraint being

$$V_{\text{mem}}^{\text{env}} = -c\cos((\phi - \psi)/2) \tag{8.5}$$

as suggested by *Schöner* and *Kelso* (1988). If we superimpose (8.5) on (8.4) with increasing value of $c > 0$, we indeed find a potential curve that has as its stationary solution the probability distribution (Fig. 8.3) corresponding to the relative frequency distribution of Fig. 8.2.

There is actually a still more direct way to deduce the potential curve from Fig. 8.2. To this end we start from the explicit form of the stationary solution of the one-dimensional Fokker–Planck equation (6.31) of Sect. 6.4. Solving this expression for $V_\psi(\phi)$, we obtain the desired potential

$$V_\psi(\phi) = \frac{Q}{2}\left(\ln N - \ln P_\psi(\phi)\right). \tag{8.6}$$

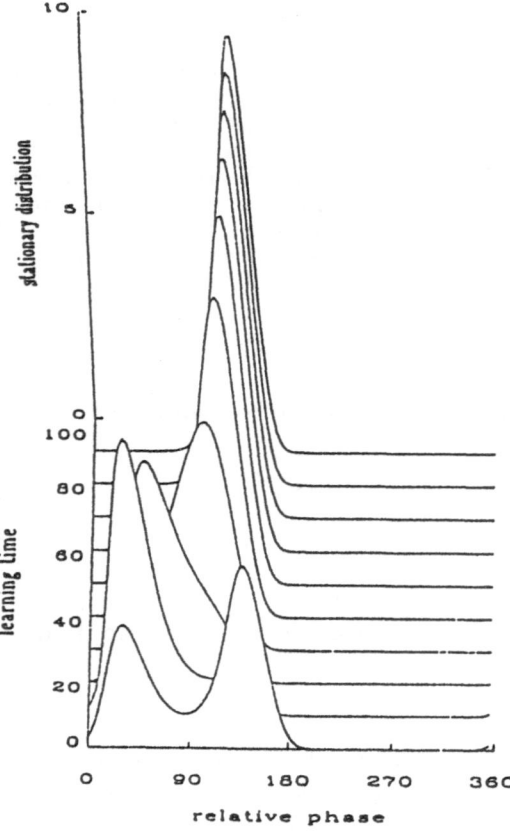

Fig. 8.3. The stationary probability distribution, P_{stat} (units in percent), of relative phase, ϕ, is plotted as a function of learning time. The distribution is bimodal initially, reflecting the underlying bistability of the dynamics. At the first instability, the distribution changes to monomodal. The second instability can be observed as an abrupt shift of the peak position accompanied by a broadening of the distribution. After *Schöner* et al. (1992)

Q is the strength of the fluctuating forces which can be determined from other experiments as explained in Sect. 6.4. The first term in the bracket on the right-hand side of (8.6) stems from the normalization and represents just a constant shift of the potential which is unimportant. The last term establishes the important relationship between the potential $V_\psi(\phi)$ and the relative frequency (or probability) which is denoted here by $P_\psi(\phi)$. ψ is kept constant at $90°$ in the present case. Quite clearly, this is a simple means to check the accuracy of the decomposition (8.2) with (8.4) and (8.5) against experimental data.

These experiments and their theoretical interpretation by *Schöner* and *Kelso* shed new light on learning. Quite evidently, the intrinsic dynamics that may have been (genetically) inherited or had been learnt earlier influences the learning dynamics. Points that were attractors previously show their influence on the learning of new attractor states. This may have important consequences, for instance, for training methods in sports. For example, when people learn skiing, usually they first learn the snow plough and only later on swinging. But as many skiers' experience shows, when difficult situations arise, they still resort to the formerly learnt primitive method of snow plough rather than to apply the still more effective method of swinging. Quite evidently, in terms of the interpretation we just came across, one may say that the old attractors are still there and may be even more stable than the newly learnt ones.

8.2 How Learning Changes the Number of Order Parameters

There may be several mechanisms by which movement patterns described by order parameters can change. In our discussion above the *dynamics of the order parameters* was changed in that the potential landscape underwent a transition. There may be other mechanisms possible, however, that to our knowledge have not yet been discussed, namely *new order parameters* may emerge by means of the cooperation of old order parameters. Let us again borrow an example from physics, in particular, fluid dynamics. Under certain circumstances, a fluid heated from below forms rolls that may be oriented in three different directions. Under specific conditions of heating and surface tensions, these three modes of behavior *cooperate* to form hexagons which are governed by a *single* new order parameter.

In Chap. 12 we shall analyse how people learn to drive a pedalo. We shall demonstrate that, as learning proceeds, fewer and fewer degrees of freedom dominate the movement pattern, or, in other words, a smaller number of order parameters occurs. Eventually, the movement pattern is governed by a single order parameter.

8.3 How Learning Gives Rise to New Order Parameters

Finally, we may discuss learning as a process in which new order parameters emerge from microscopic parts. An example is provided by the synergetic computer which we shall discuss later (Sect. 16.1). This latter example will provide us with a simple means to study learning at the level of neurons. In this case, learning changes control parameters at the microscopic level. These control parameters are the synaptic strengths and the learning rule is identical with, or, at least related to, Hebb's rule. As a consequence of the resulting new dynamics, entirely new order parameters result, representing new patterns that can now be recognized.

9. Animal Gaits and Their Transitions

9.1 Introductory Remarks

The gaits of animals, for instance of horses, provide us with beautiful examples of behavioral patterns and their changes. In this chapter we want to focus our attention on the locomotory gaits in quadrupeds. Its empirical study has a long history. More recently, theoretical studies have also started; in Sects. 9.4 – 9.7 we shall essentially follow *Schöner, Yiang* and *Kelso* (1990), who based their approach on synergetics. In the context of this book it seems obvious to try a modeling of gaits and their transitions similarly to our treatment of finger movements in Chap. 6. Quite clearly, the new problem is more complex because instead of the previous two variables of the phases of the two index fingers or the one relative phase, now we have to deal with the four phases of the animal's limbs. This implies that the corresponding equations must contain more terms and there will be more adjustable constants. Thus we need an additional guideline to establish the equations for the phases and also to classify the evolving movement patterns. An elegant approach is provided by group theory, an important branch of mathematics. Since we do not assume that the reader is familiar with group theory, we will provide him or her with the necessary information. As the reader will note in a minute, group theory is pure fun because it can be connected with a pictorial representation of symmetries. The plan of the present chapter will be as follows:

We first introduce basic and relevant concepts of group theory, then we shall convince ourselves that animal gaits have a strong connection with symmetries. After that we shall establish the equations of motion of the phases and discuss solutions. Finally, we shall be concerned with the question of transitions.

9.2 Symmetries and Groups

Let us consider the square of Fig. 9.1, whose corners are numbered 1,2,3,4. We first introduce the notion of symmetry operations. We may mirror the square with respect to the vertical axis as shown in Fig. 9.2, upper left. In this way, the corners are transformed with a new numbering as seen in Fig. 9.2,

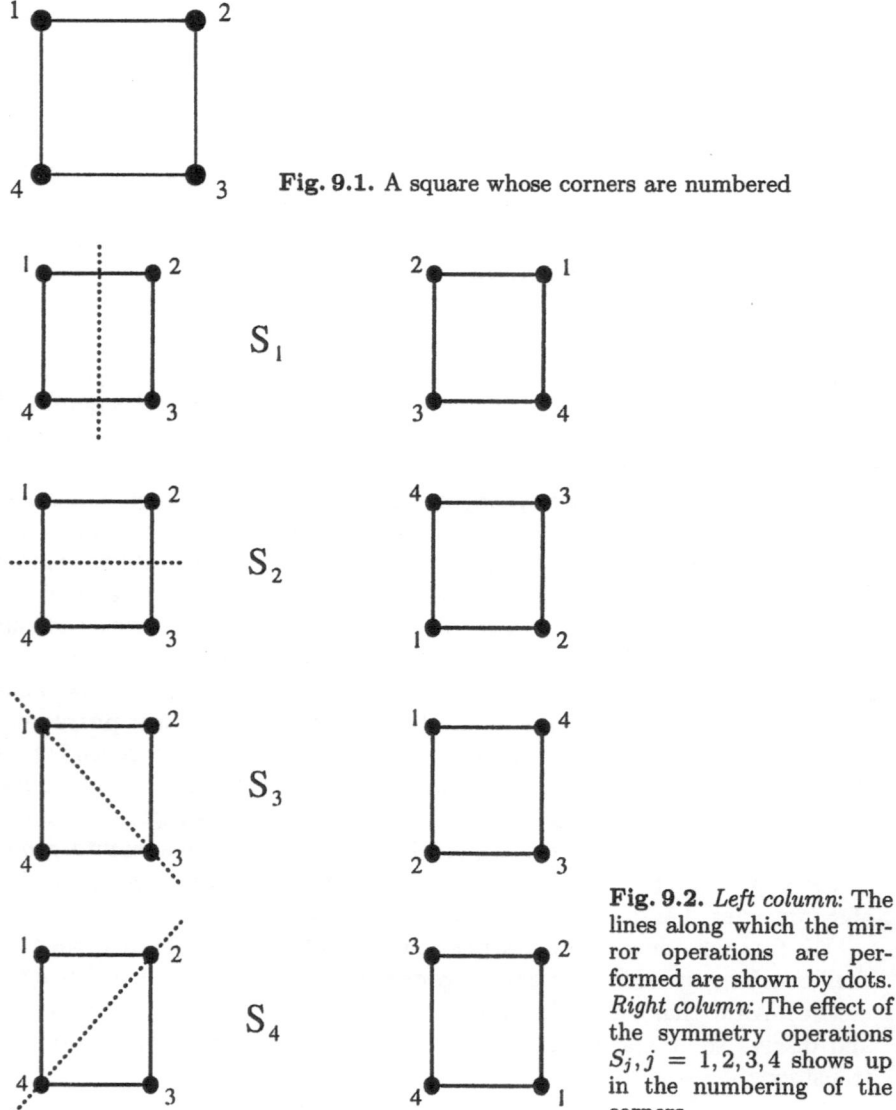

Fig. 9.1. A square whose corners are numbered

Fig. 9.2. *Left column*: The lines along which the mirror operations are performed are shown by dots. *Right column*: The effect of the symmetry operations $S_j, j = 1, 2, 3, 4$ shows up in the numbering of the corners

upper right. We shall denote this reflexion by S_1. There are three further reflexions possible, namely reflexion in the horizontal axis, in the diagonal from 1 to 3, and in the diagonal from 4 to 2 (cf. Fig. 9.2). In addition, we may have a rotation of the square. We shall denote the rotation by R (Fig. 9.3). If we rotate the square to the right instead of the left, we compensate for the original rotation. We call this operation the inverse and denote it by R^{-1}. Quite evidently, we may rotate the square not only once, but two or three times. The results are shown in Fig. 9.3. When we apply the rotation

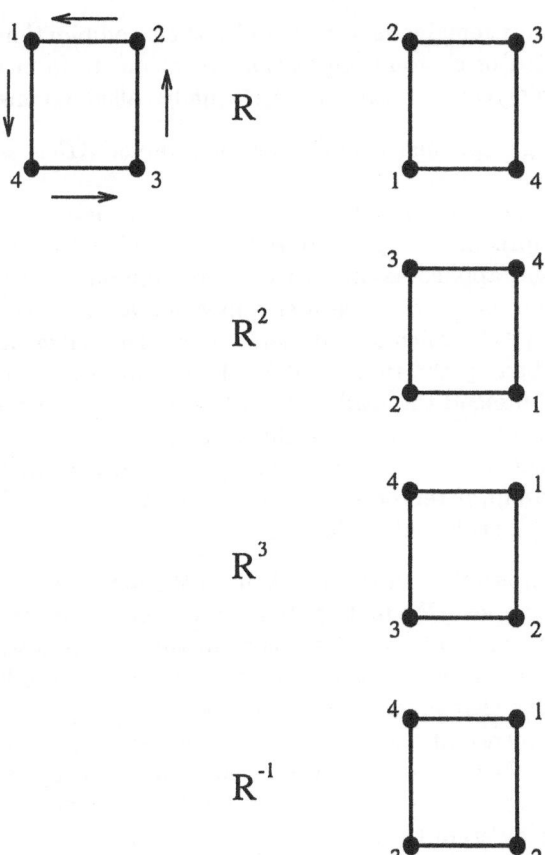

Fig. 9.3. *Upper left corner:* Application of a rotation. *Upper right corner:* Effect of that rotation on the numbering of the square. If rotation is applied two or three times, the numbering is as shown on the right-hand side of this figure. *Lowest diagram:* The inverse of the rotation

four times, we arrive at the original square of Fig. 9.3, upper left, so that the operation R^4 is equal to the identity operation that leaves the square unchanged in its orientation.

Except for the numbering, nothing has changed after the symmetry operation. This leads us directly to the definition of symmetry. We call a configuration, in our case the square, symmetric with respect to a symmetry operation, if after that operation the final state cannot be distinguished from the initial state. Note that there is a conceptual difficulty: on the one hand we have distinguished the different corners by numbers, on the other hand, we eventually disregard this numbering. Thus when we want to perform symmetry operations, we have first to mark the relevant features, in our case the corners, then perform the operation and, then finaly forget this marking. Instead of saying that a configuration is *symmetric* with respect to an operation, we may also say it is *invariant* under that operation. We can combine operations by performing several of them, one after the other. For instance, we may apply two operations. Because the square remains invariant under each of these operations, it also remains invariant under their combination. This leads us to the definition of a group. Operations or elements form a group, if they obey the following conditions, which are usually called axioms:

1) If A, B are elements of a group, their combination or product AB is an element of the same group.
2) The so-called associative law is valid $(AB)C = A(BC)$. The right-hand side means that we first form BC to get a new element. After that element has been formed and applied as an operation, the operation A is applied. The left-hand side means that we first apply operation C and then apply the operation (AB) which is considered as a single element. In these two ways, described by the right- and left-hand side, the same result is achieved. We recommend the reader to verify this statement for a few examples using symmetry operations on the square.
3) There exists the identity element E which has the property that combined with any element of the group, it reproduces the element, $AE = A$.
4) There exists an inverse A^{-1}, such that $A^{-1}A = E$.

Let us consider a few examples that shed some light on the nature of the multiplication rules introduced above. We first apply the rotation R and then reflexion S_1. The result is shown in Fig. 9.4. Note that in this notation we read from right to left, i.e. the operation that must be performed first stands to the right of those operations that must be performed subsequently. Now let us consider the reverse sequence of operations, namely we apply S_1 first and then R afterwards. As may be seen from a comparison between the upper and lower parts of Fig. 9.4, the resulting numbering of the squares is different, or written as a formula, we may state that $S_1 R \neq R S_1$. The multiplication

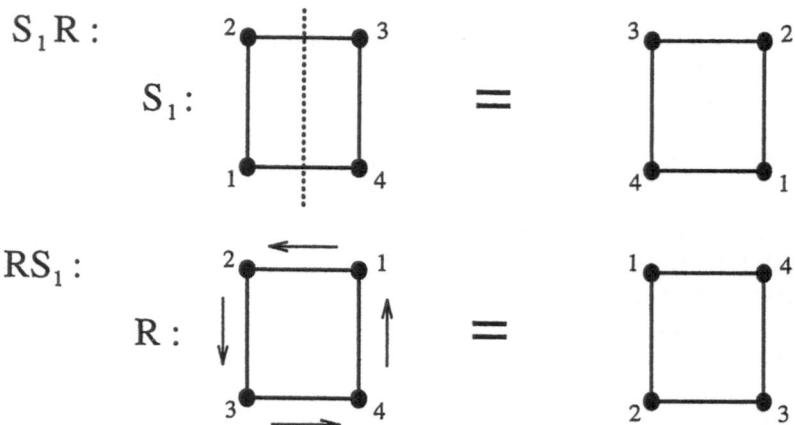

Fig. 9.4. *Upper part:* After rotation has been performed on the original square, a reflexion S_1 is applied. *Lower part:* First a reflexion S_1 is applied and then the rotation. Note that the corresponding numbering in the resulting upper and lower parts is different

rule of groups may differ from the one we are used to for ordinary numbers, in which the commutative law $AB = BA$ holds. In other words, groups may be noncommutative. An important notation is that of *subgroups*. An example is provided by the two elements E, S_1. When we perform, for instance, the reflexion S_1 twice, we obtain, of course, the original state, or written in a formula, $S_1^2 = E$, or in other words, S_1 is at the same time its own inverse, $S_1^{-1} = S_1$. All the elements of this subgroup are thus given by E, S_1 and form, quite evidently, a group. Another example is provided by the rotations including the identity E, R, R^2, R^3. Because $R^4 = E$, we have the definition of the inverse of R in the form $R^{-1} = R^3$. Closely connected with the notion of subgroups is that of *broken symmetry*. Consider Fig. 9.5, where the square is, of course, invariant under the operations S_1 and S_2. Now change the square materially, for instance by making the corners 1 and 2 distinct from the corners 3 and 4, e.g. by attaching spheres to the corners 1 and 2 and 3 and 4, where the spheres 1 and 2 are the same but different from the spheres 3 and 4 that are the same as each other. Then quite clearly this configuration is still invariant under S_1, but not under S_2. Quite evidently, the square is now only invariant under the group E, S_1, which is a subgroup of the group G. Going from the situation of the upper part of Fig. 9.5 to that of the lower part, we witness a broken symmetry.

The symmetry operations of reflexions and rotations can also be described by concentrating on the changes of the numbers 1,2,3,4 under these opera-

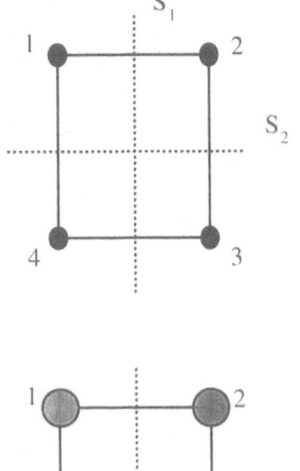

Fig. 9.5. *Upper part:* If all corners are equivalent, both the symmetry operations S_1 and S_2 can be applied. *Lower part:* If the two upper corners of the square are made different from the lower two corners, only the symmetry operation S_1 can be applied

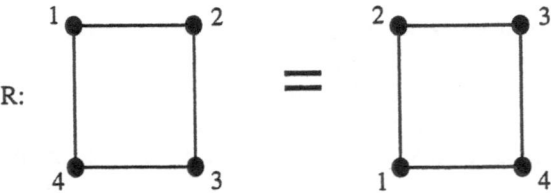

R:

Fig. 9.6. The rotation of the square on the left-hand side induces a permutation of the number of the corners

tions. For instance, when we rotate the square according to the operation R, (Fig. 9.6), we obtain a renumbering of the corners according to the law $1 \rightarrow 2, 2 \rightarrow 3, 3 \rightarrow 4, 4 \rightarrow 1$. In a briefer notation, we may write this as $1 \rightarrow 2 \rightarrow 3 \rightarrow 4 \rightarrow 1$. A still shorter version is given by $(2, 3, 4, 1)$, where, as we imagine, the first place 1 is replaced by 2, then 2 goes over into 3, and so on. In other words, the original sequence 1,2,3,4 is permuted into the sequence 2,3,4,1. Quite evidently, all symmetry operations can be represented as permutations, as may be exemplified by the square. Because the symmetry operations form a group, their representations by means of permutations form a group as well. So we are led to the definition of a *permutation group*.

9.3 An Empirical Study of Quadruped Gaits

In order to be able to model gaits, we first take a look at their pictorial representation. An example is provided by Fig. 9.7 which shows the different stages of the movement pattern of a trotting horse. Reading the picture from left to right, we first observe that the horse has no ground contact, its right front- and hind-leg are stretched outwards, the legs on the left-hand side stretched inwards. Then the right front-leg and left hind-leg hit the ground and stay there. After that the horse again takes off and the left front- and hind-leg are stretched outwards, while the right legs are stretched inwards. Then the left front-leg and right hind-leg hit the ground and stay there for a while. After these four stages, the cycle starts again. These different stages

Fig. 9.7. Trot of a horse; from *Gambaryan* (1974). Figures 9.9, 9.10, 9.12-15 are also taken from this book

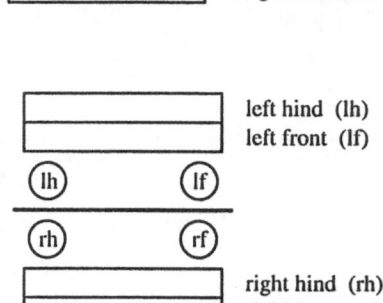

ground contact
no ground contact

left hind (lh)
left front (lf)

right hind (rh)
right front (rf)

Fig. 9.8. *Upper part*: Visualization of ground contact (*shaded*) and of no ground contact (*unshaded*). *Lower part*: How the gait patterns are encoded in the individual figures. The length of the bars indicate the duration in time; the solid circles show which feet have ground contact

are encoded in the lower part of the figure. To explain it, consider Fig. 9.8. The shaded area means ground contact of the corresponding leg, the white area means that the corresponding leg has no ground contact. The middle bars indicate the body axis, the dots in the subsequent figures indicate which foot has ground contact. At this stage we may introduce the notion of phase, where we shall be guided by the analogy with the finger movements. We immediately observe that there is a pronounced phase relation between the different legs of the trotting horse. We note that the right front-leg and the left hind-leg always move in parallel and that the same holds for the left front-leg and the right hind-leg. Each of these pairs has the same phase. On the other hand, when we compare the two front legs, they evidently move in the opposite direction, i.e., they move in an antiparallel fashion, or, expressed yet another way, their phase difference is π. This example shows strikingly that phase again appears as a relevant macroscopic variable and furthermore that symmetries are involved.

In the following, we shall consider a number of important examples of gaits and we shall plot their empirically observed phase angles. There are a variety of different definitions of the phase angle in use and here we shall consider two definitions that are of particular relevance for our study. When we want to attribute a number to the phases, we may define the relative phase ϕ_i of foot i as the fraction of the gait cycle between ground contact of a reference foot, for instance the left forelimbs in quadrupeds, and ground contact of foot i. This will lead us to the numbers that occur in Figs. 9.11 and 9.16 below. On the other hand, when we wish to consider the whole dynamics of the movement of the limbs, we have to consider ϕ_i as a function of time. We shall come back to this definition later.

For the time being, let us consider examples that are most relevant to our present analysis. The simplest example of a movement is the pronk. All four legs move together and in phase. This gait is sometimes used by young animals and gazelles. The next simple movement pattern is that of a bound or an idealized gallop (Fig. 9.9). The front legs move together and in phase,

Fig. 9.9. Bound of long-tailed Siberian souslik

Fig. 9.10. Pace of a camel

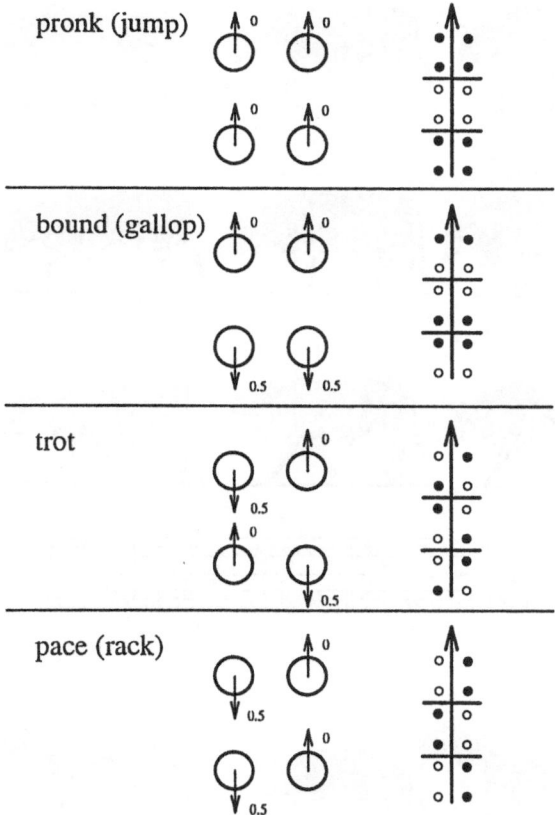

Fig. 9.11. Visualization of the individual phase angles in units of 2π corresponding to various gaits. In the column on the right-hand side, the ground contacts (*solid circles*) and no ground contacts (*open circles*) are indicated. Time runs from bottom to top

the back legs move together half a period out of phase with the front pair. We already came across the trot (Fig. 9.7), where the diagonally opposed legs move together and in phase. The right front- and left hind-legs move together half a period out of phase with the other pair. In the pace (Fig. 9.10), the left legs move together and in phase, and the right legs move together half a period out of phase with the left legs. Figure 9.11 gives a summary of these four gaits. Note that the numbers are in units 2π so that 0.5 corresponds to a phase angle of π.

Let us now consider a somewhat more complicated set of movement patterns, namely the walk (Fig. 9.12), canter (Fig. 9.13), transverse gallop (Fig. 9.14) and rotary gallop (Fig. 9.15). The phase relations of the limbs are summarized in Fig. 9.16. We note that the phase relations are no longer quite so clearcut as in Fig. 9.11. Now phases such as 0.8, 0.1, 0.6 appear instead of

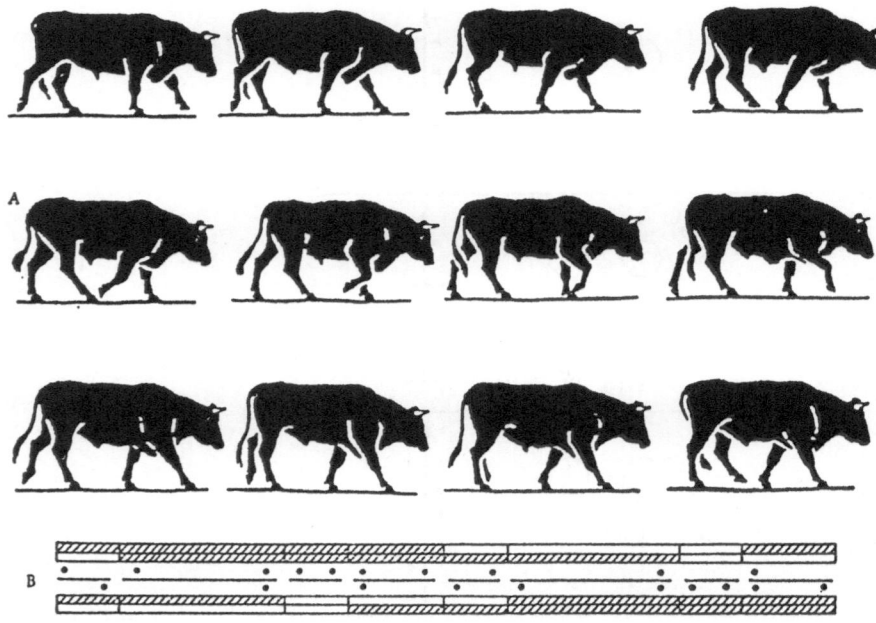

Fig. 9.12. Walk of oxes

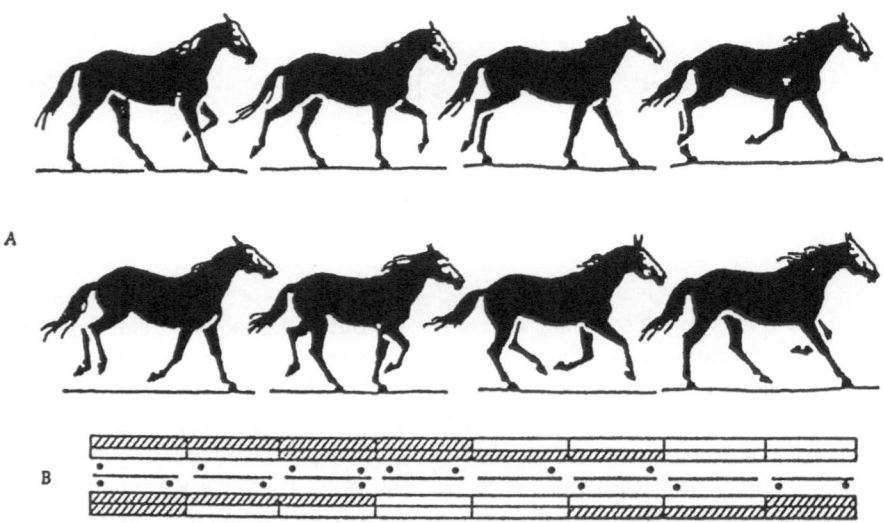

Fig. 9.13. Canter of horses

phases 0 and 0.5. In order to bring out the essentials, we consider the ideal-ized cases that are depicted in Figs. 9.17 and 9.18 instead of Figs. 9.11 and 9.16, respectively. At any rate, both in the empirically observed cases and the

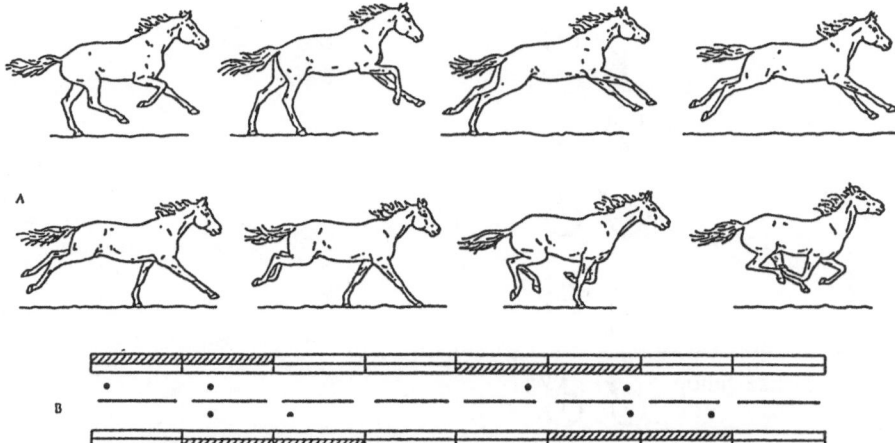

Fig. 9.14. Transverse gallop of horses

Fig. 9.15. Rotary gallop of cheetah

idealized cases, we are dealing with pronounced symmetries. How can such symmetries result from the underlying dynamics, or mathematically speaking, from the equations of motion?

9.4 Phase Dynamics and Symmetries

To arrive at a definition of the phases that takes into account their time dependence, we start from the vertical elevations of the end effectors, which we distinguish by an index k. Denoting this elevation by x_k and the corresponding velocity by v_k, we may describe the motion of the endeffector k by means of a phase plane representation (cf. Sect. 5.2 and Figs. 5.9 – 5.11) in the form

$$x_k(t) = r_k(t) \cos \phi_k(t) \tag{9.1}$$

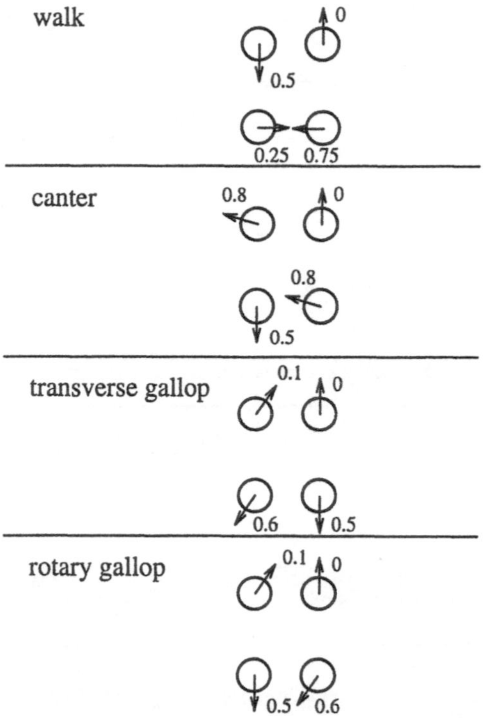

walk
0
0.5
0.25 0.75
canter
0.8
0
0.8
0.5
transverse gallop
0.1
0
0.6
0.5
rotary gallop
0.1
0
0.5
0.6

Fig. 9.16. Phase angles in units of 2π for different gaits

pronk
bound
trot
pace

Fig. 9.17. A schematic representation of the phase angles of different gaits. (Redrawn after *Schöner, Yiang, Kelso* (1990))

and

$$v_k(t) = r_k(t) \sin \phi_k(t), \quad k = 1, 2, 3, 4. \tag{9.2}$$

In the following we shall not consider the time dependence of the amplitudes r_k, but focus our attention instead on the time dependence of the phases ϕ_k.

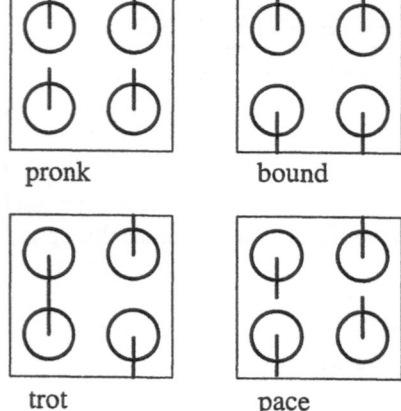

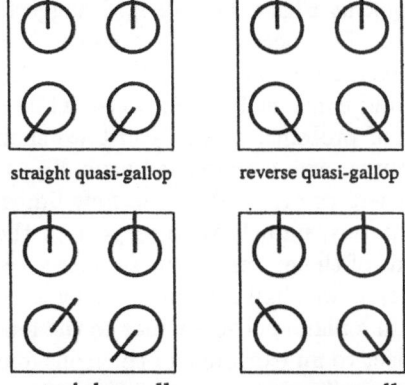

straight quasi-gallop   reverse quasi-gallop

straight walk   reverse walk

Fig. 9.18. Simplified schematic representation of phase angles of different gaits. (Redrawn after *Schöner* et al. (1990)

We shall distinguish between the individual limbs by corresponding indices, according to

$$\text{right(r), left(ℓ), front(f), and hind(h) limbs.} \tag{9.3}$$

Then the phase angles of the limbs "right front", etc., will be denoted by

$$\phi_{\rm rf}, \quad \phi_{\ell \rm f}, \quad \phi_{\rm rh}, \quad \phi_{\ell \rm h}. \tag{9.4}$$

It will be convenient to have an abbreviation in the form

$$\phi_{\rm ij}, \quad i = r, \ell; \quad j = f, h, \tag{9.5}$$

where the first index i refers to right/left, and the second index j to front/hind. Because cosine and sine are periodic functions, we need to consider the phase angles only in the region

$$-\pi, ... + \pi, \quad \text{modulo} \quad 2\pi. \tag{9.6}$$

Occasionally it will be convenient to combine the four phases into a phase vector according to

$$\phi = (\phi_{\rm rf}, \phi_{\rm rh}, \phi_{\ell \rm f}, \phi_{\ell \rm h}). \tag{9.7}$$

Since the displacement of all phases by the same constant just means a shift in time and has nothing to do with the relative phases, we may consider ϕ and $\phi + c\mathbf{1}$ as identical, where

$$\phi, \phi + c\mathbf{1} \quad \text{identical} \tag{9.8}$$

and

$$\mathbf{1} = (1, 1, 1, 1) \tag{9.9}$$

holds. From a more formal point of view, we can lump the vectors (9.8) into a vector space

$$\boldsymbol{P}_4 = \{\phi \equiv \phi + c\mathbf{1} \quad \text{for all real} \quad c\}. \tag{9.10}$$

Where does symmetry come in? A look at Figs. 9.17 and 9.18 suggests that symmetries must play a prominent role. One occasionally hears the joke about how a physicist or a chemist would characterize a horse. It is said that, to a lowest approximation, a physicist would consider it as a sphere, while a chemist would consider a horse as a bag of proteins. In this spirit we shall characterize the horse or any other quadruped as a simple geometric object, for our purposes as a square. Its four corners correspond to the four limbs. While we recognize one symmetry immediately, namely with respect to the vertical plane through the longitudinal axis of the animal, some further symmetries are of a more artificial nature, but, as we shall see, quite reasonable. Such a symmetry is a reflexion symmetry in a plane perpendicular to the longitudinal axis of the animal, i.e. with respect to an exchange of the front and hind girdle. While front and hind legs are usually anatomically different, this need not be so with respect to the neural substrate which serves to trigger and correlate the motions of the legs. Going one step further, we may allow symmetry operations that are effected by all possible permutations of the four corners of the square, or now of all the four legs or their corresponding phase variables ϕ_{ij}. Finally, when we only consider the kinematics, we may imagine that the time runs backwards, or in other words, that the system is invariant under a simultaneous change of the phase angles from ϕ into $-\phi$. This then defines the largest symmetry group that we call A. It comprises all the permutations of the phase variables ϕ_{ij} and the inversion $-E : \phi \rightarrow -\phi$. Note that this group also contains rotations.

Let us study which of the patterns of Figs. 9.17 or 9.18 are invariant under this symmetry group. Quite obviously this group leaves the square invariant if the corners of the square are indistinguishable from one another. The only pattern that has this property is that of the pronk (Fig. 9.17). In order to capture the symmetries of the other configurations, let us consider a subgroup of A, called B, that describes the symmetries of quadrupeds in a more realistic fashion. Here we admit the following basic group operations: First, the reflexion in the longitudinal vertical plane, or, in other words, the exchange of right and left. We shall denote this operation by $O_{r\ell}$. Similarly we use the symmetry with respect to the vertical plane perpendicular to the longitudinal axis, or, in other words, the exchange of front and hind limbs denoted by O_{fh}. Finally we include the reversal of the phase angles, i.e. the operation $-E$.

In summary, the symmetry group B is based on the following elements

$$B = \{E, O_{r\ell}, O_{fh}, -E\}. \tag{9.11}$$

We can show that all the patterns of Fig. 9.17 are invariant under this symmetry group. To this end, let us consider the example of a gallop (bound). Quite evidently, the pattern shown in Fig. 9.19 remains the same when we reflect it in the longitudinal axis, i.e. this pattern is invariant under the operation $O_{r\ell}$. If we apply the operation O_{fh}, i.e. exchange front and rear, we arrive at Fig. 9.20. However, we may rotate the four pointers (or phases) by π, which

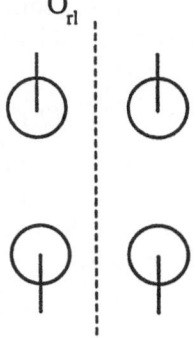

Fig. 9.19. Application of the symmetry operation $O_{r\ell}$ to the pattern of a bound.

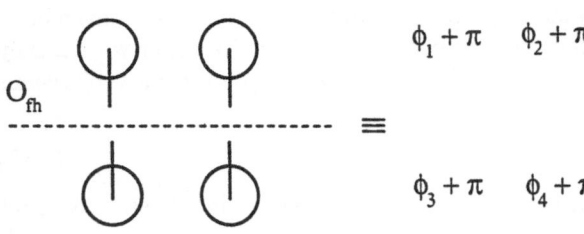

Fig. 9.20. Application of the operation O_{fh} to Fig. 9.19 induces the new pattern shown here, in which each phase angle appears to a rotated by π as compared to the pattern of Fig. 9.19

leads us back to the original figure. As we know, according to (9.8), this simultaneous rotation is possible and just corresponds to a shift in time, but not to a change of relative phases. Finally, when we apply $-E$ (Fig. 9.21), we find the same situation as in Fig. 9.20, which, as we have just seen, is equivalent to that of the original Fig. 9.19.

Let us consider the trot as a second example. The right/left symmetry is not obvious, but we note that after the right- and left-hand sides have been exchanged, we may shift the phases simultaneously by π and come back to the original figure. The symmetry with respect to the exchange of front and rear leads to a figure in which all phases are simultaneously changed by π and are thus again equivalent to the original. Similarly one may show that the trot is invariant under the operation $-E$. When one tries to apply the

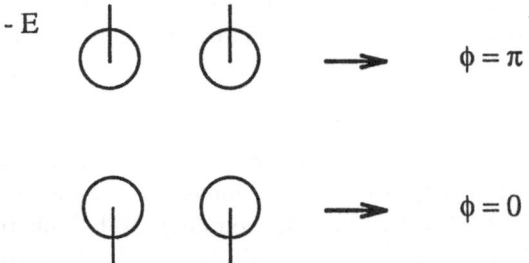

Fig. 9.21. Application of the operation $-E$ to the pattern of a bound. The phase angle in the upper row is shifted to $\phi = \pi$ and that of the lower row to $\phi = 0$. The pattern can be restored by a simultaneous shift of all ϕ_j by π

symmetry group B to the patterns of Fig. 9.18, one quickly realizes that these patterns are not invariant. To achieve invariance here, we have to consider a still smaller group, namely

$$C = \{E, O_{r\ell}, -EO_{fh}\}.\tag{9.12}$$

This group contains the exchange of the right- and left-hand sides as a symmetry element, and the exchange of the front and hind limbs provided this exchange is accompanied by time-reversal. We leave the proof that the patterns of Fig. 9.18 are invariant under C as an exercise for the reader.

9.5 Equations of Phase Dynamics

Let us be guided by the same hypothesis as in the case of finger movements, namely that the phases may serve as order parameters. In other words, they are assumed to obey equations that describe their temporal change by means of

$$\dot{\phi} = K(\phi).\tag{9.13}$$

Since we know that a simultaneous shift of all the phases by the same amount does not change the dynamics, we have to require

$$K(\phi) = K(\phi + cI).\tag{9.14}$$

In complete analogy to the finger movement case, we may conclude that K depends only on relative phases. For instance, when we consider the individual components of (9.14), say, the components with index rf, the right-hand side of (9.13) takes a particular form such that we may write

$$\dot{\phi}_{rf} = K_{rf}(\phi_{rf} - \phi_{\ell f}, \phi_{rf} - \phi_{rh}, \phi_{rf} - \phi_{\ell h}).\tag{9.15}$$

Here K_{rf} on the right-hand side depends only on relative phases, where ϕ_{rf} is taken as reference phase. Corresponding equations can, of course, be formulated for $\phi_{rh}, \phi_{\ell h}$, and $\phi_{\ell f}$. The right-hand side of (9.15) is still far too general and must be specified further. Again the analogy with the finger movement example helps. We shall assume that direct interaction between the limbs is pairwise, i.e. that the movement of the right fore limb is either directly connected with that of the left fore limb or the right hind limb, or diagonally with the left hind limb. In other words, we will use a decomposition of K_{rf} in the form

$$K_{rf} = K_{rf}^H(\phi_{rf} - \phi_{\ell f}) + K_{rf}^I(\phi_{rf} - \phi_{rh}) + K_{rf}^D(\phi_{rf} - \phi_{\ell h}).\tag{9.16}$$

The indices H, I, D are chosen to indicate H : homologous contra-lateral, I: ipsi-lateral, D: diagonal contra-lateral. To make our analysis as simple and transparent as possible, we shall neglect the diagonal coupling. To be able to write down an explicit form of the terms K on the right-hand side of (9.16), it might be tempting to directly resort to the corresponding terms in the

finger movement paradigm (Sect. 6.2). In this way, however, we would lose
the important benefits of symmetry considerations. To outline the basic idea,
we proceed as follows: Each of the functions on the right-hand side of (9.16)
is periodic in its argument. If we denote this argument, i.e. the relative phase,
by ϕ, we may decompose

$$K_{\mathrm{rf}}(\phi) \tag{9.17}$$

into a superposition of the functions

$$\cos n\phi, \sin n\phi, \quad n = 0, 1, 2, \ldots. \tag{9.18}$$

By demanding that the phase equations are invariant under the group B,
these can be considerably simplified. In order not to overload our presentation, we illustrate this procedure by means of two examples. Then we shall
immediately write down the full equations that result. We first consider as
an example the application of the group operation $-E$. Consider the simple
case of the following equation

$$\dot{\phi} = A_0 + A_1 \sin \phi + B_1 \cos \phi. \tag{9.19}$$

Let us apply to it the operation

$$-E : \phi \rightarrow -\phi. \tag{9.20}$$

Then (9.19) transforms into

$$-\dot{\phi} = A_0 + A_1 \sin(-\phi) + B_1 \cos(-\phi). \tag{9.21}$$

Using the symmetry of the trigonometric functions

$$\left. \begin{array}{l} \sin(-\phi) = -\sin \phi \\ \cos(-\phi) = \cos \phi \end{array} \right\} \tag{9.22}$$

and adding the equations (9.19) and (9.21), we immediately arrive at

$$0 = A_0 + B_1 \cos \phi, \tag{9.23}$$

which must hold for all values of ϕ, from $-\pi \ldots +\pi$. We immediately find that

$$A_0 = 0, B_1 = 0 \tag{9.24}$$

must hold. This example teaches us that, by means of invariance considerations, the equations for the order parameters ϕ can be considerably simplified.

As a second example we study the group operation $O_{r\ell}$, i.e. the exchange
between right and left. As we have just seen, it will be sufficient to include
sine-functions only. We write down the corresponding equations for ϕ_{rf} and
$\phi_{\ell\mathrm{f}}$ and keep only the lowest order sine-functions. The equations to be considered then read

$$\dot{\phi}_{\mathrm{rf}} = A \sin(\phi_{\mathrm{rf}} - \phi_{\ell\mathrm{f}}), \tag{9.25}$$

$$\dot{\phi}_{\ell\mathrm{f}} = B \sin(\phi_{\ell\mathrm{f}} - \phi_{\mathrm{rf}}). \tag{9.26}$$

Application of the group operation $O_{r\ell}$ means that we have to exchange the indices r and ℓ everywhere. In this way, (9.25) is transformed into

$$\dot{\phi}_{\ell f} = A \sin(\phi_{\ell f} - \phi_{rf}) \tag{9.27}$$

and (9.26) into

$$\dot{\phi}_{rf} = B \sin(\phi_{rf} - \phi_{\ell f}). \tag{9.28}$$

Invariance means that the equations remain the same, i.e. the equation for $\phi_{\ell f}$ (9.27) must coincide with the former one for $\phi_{\ell f}$, namely (9.26), and the same must hold for ϕ_{rf}, eqs. (9.28) and (9.25). The direct comparison between (9.27) and (9.26) and similarly between (9.28) and (9.25) yields the condition

$$A = B. \tag{9.29}$$

Thus we see that the invariance requirement for the equations under the operation $O_{r\ell}$ leads to a relation between the coefficients in the corresponding equations. The reader should be warned of a possible trap that arises in the following way: Instead of writing coefficients A and B, we could equally well have denoted the coefficients differently, namely

$$A = A_{rf} \tag{9.30}$$

and

$$B = A_{\ell f}, \tag{9.31}$$

in order to bring out the fact that the first coefficient belongs to an equation for ϕ_{rf} and the second coefficient to one for $\phi_{\ell f}$. When we now apply the group operation, we must not apply the exchange operation to the indices r, ℓ of A in (9.30) and (9.31) but only to the variables ϕ.

Now let us use these ideas to simplify (9.15) with the right-hand side of (9.16) and the corresponding expressions for $\phi_{rh}, \phi_{\ell h}, \phi_{\ell f}$. Imagine that we expand the right-hand sides into superpositions of functions of the form (9.18) with as yet arbitrary coefficients $A_1, B_1, \ldots$. Now require that the equations are unchanged, i.e. remain invariant, under the operations of the group B. Then an astonishingly simple result appears after some elementary analysis. To bring out the beauty of the resulting equations, we introduce the "hat-notation"

$$\hat{i} : \text{if} \begin{cases} i = r, \\ i = \ell, \end{cases} \text{then} \quad \begin{aligned} \hat{i} &= \ell \\ \hat{i} &= r \end{aligned} \tag{9.32}$$

and

$$\hat{j} : \text{if} \begin{cases} j = f, \\ j = h, \end{cases} \text{then} \quad \begin{aligned} \hat{j} &= h \\ \hat{j} &= f \end{aligned}. \tag{9.33}$$

The hat above i means that one has to take the counterpart, for instance, if i means right, $\hat{i}$ means left. A similar convention applies to j with respect to front and hind. The resulting equations read

$$\dot{\phi}_{ij} = A_1 \sin\left[\phi_{ij} - \phi_{\hat{i}j}\right] + A_2 \sin\left[2\left(\phi_{ij} - \phi_{\hat{i}j}\right)\right]$$
$$+ C_1 \sin\left[\phi_{ij} - \phi_{i\hat{j}}\right] + C_2 \sin\left[2\left(\phi_{ij} - \phi_{i\hat{j}}\right)\right] \equiv S_{ij}, \tag{9.34}$$

where S_{ij} serves merely as an abbreviation for the preceding part of the equation and will be needed only later. Actually symmetry arguments would not rule out that higher order terms of the form

$$\sin\left[n\left(\phi_{ij} - \phi_{\hat{i}j}\right)\right], \quad n = 3, 4, \dots \tag{9.35}$$

appear. It is here that we once more employ the analogy with the finger movement paradigm, where it turned out that the first term proportional to $\sin\phi$ is not sufficient to describe bistability or phase transitions, but that a second term is needed. Higher order terms would make things more complicated and are not needed in our analysis.

We wish to convince ourselves that the equations (9.34) have the required invariance properties. Let us consider as an example the operation $O_{r\ell}$, which implies that we have to exchange i with $\hat{\text{i}}$. Applying this operation to (9.34) leads us to

$$\dot{\phi}_{\hat{i}j} = A_1 \sin\left[\left(\phi_{\hat{i}j} - \phi_{ij}\right)\right] + A_2 \sin\left[2\left(\phi_{\hat{i}j} - \phi_{ij}\right)\right]$$
$$+ C_1 \sin\left[\phi_{\hat{i}j} - \phi_{\hat{i}\hat{j}}\right] + C_2 \sin\left[2\left(\phi_{\hat{i}j} - \phi_{\hat{i}\hat{j}}\right)\right]. \tag{9.36}$$

When we write down the sets of equations (9.34) and (9.36) by specifying the indices ij according to rh, ℓh, rf, ℓf, we immediately realize that the set of equations (9.36) remains the same as (9.34); only the sequence of equations has been changed.

9.6 Stationary Solutions

Here we require that ϕ does not change in the course of time, i.e.,

$$\dot{\phi}_{ij} = 0. \tag{9.37}$$

Furthermore, since we are concerned with relative phases only, we may choose one of them arbitrarily and put

$$\phi_{rf} = 0. \tag{9.38}$$

Let us consider one equation of (9.34) , namely the one with the index rf. This equation then reads

$$A_1 \sin\phi_{\ell f} + A_2 \sin(2\phi_{\ell f}) + C_1 \sin\phi_{rh} + C_2 \sin(2\phi_{rh}) = 0. \tag{9.39}$$

When we assume a left/right symmetry of the front girdle, then we may either put

$$\phi_{\ell f} = 0, \tag{9.40}$$

which jointly with (9.38) describes a symmetric pattern or

$$\phi_{\ell f} = \pm \pi \qquad (9.41)$$

which jointly with (9.38) describes an antisymmetric pattern. Using (9.40) and (9.41) in (9.39), the terms containing A_1, A_2 become zero and we are left with

$$C_1 \sin \phi_{rh} + C_2 \sin (2\phi_{rh}) = 0. \qquad (9.42)$$

Using the trigonometric relation

$$\sin 2\phi = 2 \sin \phi \cos \phi, \qquad (9.43)$$

we may cast (9.42) into the form

$$(C_1 + 2C_2 \cos \phi_{rh}) \sin \phi_{rh} = 0. \qquad (9.44)$$

This equation admits four different solutions: If the sine-function vanishes, we obtain

$$\phi_{rh} = 0 \qquad (9.45)$$

or

$$\phi_{rh} = \pm \pi, \qquad (9.46)$$

or, if the bracket vanishes, we find additional patterns described by

$$\phi_{rh}^0 = \pm \arccos \left(-\frac{C_1}{2C_2} \right), \quad \phi_{\ell h}^0 = \phi_{rh}^0, \quad \phi_{\ell f}^0 = 0 \qquad (9.47)$$

and

$$\phi_{rh}^0 \quad \text{same}, \quad \text{but} \quad \phi_{\ell h}^0 = \phi_{rh}^0 - \pi, \quad \phi_{\ell f}^0 = \pi. \qquad (9.48)$$

Here we have supplemented the results stemming from (9.44) by those solutions of the remaining equations (9.34).

Let us discuss (9.47) and (9.48) in more detail. If we consider (9.38) and (9.47) we note that the front limbs have both phases equal to zero, while the left and right hind limbs have the same phase which is, however, different from zero and, say, π. This configuration corresponds to the quasi-gallops shown in Fig. 9.18. On the other hand, using (9.38) and (9.48) again, we note that the phases of the front limbs differ by a phase angle π as do those of the hind limbs, but the phase relation between the front and hind limbs is more complicated than zero or π, being given instead by the arccos function. The type described by (9.48) corresponds to the walks. So we note two kinds of solutions, namely straight and reverse walks and straight and reverse quasi-gallops, all shown in Fig. 9.18. When the argument of arccos approaches plus or minus 1, the clear relationships of that figure merge, namely for $C_1/2C_2 \to 1$, the straight and reverse quasi-gallop merge into the idealized gallop (bound) and the straight and reverse walk merge into trot. On the other hand, for $C_1/2C_2 \to -1$, the straight and reverse quasi-gallop merge into pronk, while the straight and reverse walk merge into pace. We mention that another set of solutions is obtained by rotating the system by 90^0 and that this set of solutions also occurs in nature. Furthermore less symmetric solutions can be found by a numerical solution of these equations. It is interesting to note that the motion of the phases can be visualized in complete analogy to the finger movement case by means of the motion of a ball in a hilly landscape. In analogy to the finger movement case, it becomes possible in this way to study the stability of the various configurations, if the parameters in the equations are changed. Such a discussion requires a

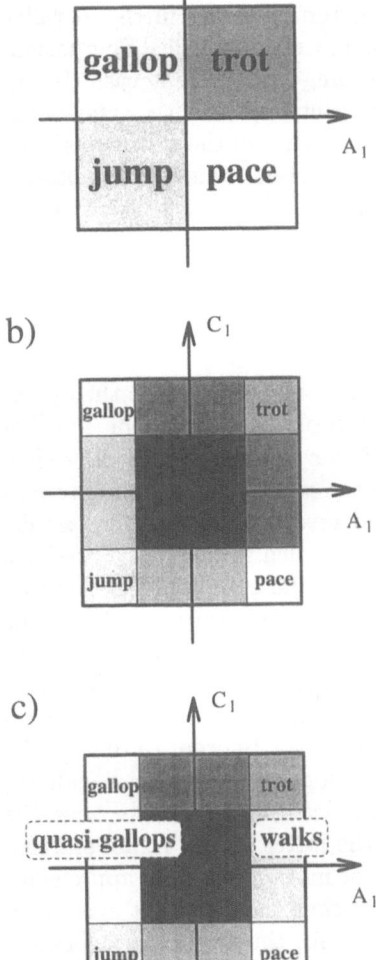

Fig. 9.22. Sections of the phase diagram in the A_1, C_1 parameter plane show regimes where the different solutions are stable for various values of (a) A_2 and C_2: $A_2 = C_2 = 0$; (b) $A_2 = -0.3\text{Hz}, C_2 = -0.2\text{Hz}$; (c) $A_2 = -0.3\text{Hz}, C_2 = 0.2\text{Hz}$. (Redrawn after *Schöner* et al. (1990))

high dimensional space. For graphical representations it is preferable to keep some parameters fixed and change only two particular parameters. Examples are shown in Fig. 9.22 which presents phase diagrams in the (A_1, C_1) plane. The simplest case occurs when $A_2 = C_2 = 0$. As shown in Fig. 9.22a only the four basic patterns exist and at each parameter point only one of these patterns is stable. If both $A_2 < 0$ and $C_2 < 0$ the stability regions of the basic patterns overlap (Fig. 9.22b). In two linear stripes of width $4\,|C_2|$ and $4\,|A_2|$ along the A_1 and C_1 axis the system is bistable. Where these bistable regions overlap, all four basic patterns are stable. Exactly on the A_1 and C_1 axis other patterns are stable. Patterns of lower symmetry first appear when $A_2 < 0$ and $C_2 > 0$. In a stripe of width $4C_2$ along the A_1 axis, the four idealized patterns become unstable and quasi-gallops and walks appear (Fig. 9.22c). As the reader may note, there is a rich variety of existence and coexistence of movement patterns, but we shall refrain from a further detailed discussion of them. When stability is lost, we may expect critical fluctuations and critical slowing down in analogy to the finger movement case. In the finger movement example, there was only one control parameter, namely the frequency of the movement. Similarly, it appears that in the quadruped case there are only one or a few essential control parameters. This is substantiated by experiments on horses in treadmills, and on decerebrate cats which we will describe at the end of this section.

9.7 Gait Dynamics of Lower Symmetry

In Sect. 9.5 we derived phase equations that were invariant under the operations of the group B. We found solutions with broken symmetry, or, in other words, with lower symmetry than that of B, for instance, the walk. When we apply certain group operations of B on the walk pattern, we may find other solutions, for instance, the straight and reverse walk. In other words, the solutions are degenerate. In order to remove such degeneracies, we have to reduce the symmetry of the dynamics, i.e., of the equations. In this way we find that the interplay of symmetry and degeneracy throws light on the origin of functions in biological systems. As an example we will show how to differentiate between the straight and the reverse walk by removing the front/hind symmetry. In many species straight and reverse walk are functionally different and only one form occurs frequently. In order to remove the degeneracy of straight and reverse walk and analogously so for quasi-gallops, we will reduce the symmetry group of the dynamics from B to C, because the walks and quasi-gallops are invariant under the latter group. Again starting from a general formulation of the equations of motion, we may apply symmetry conditions to these equations. We note that the right/left symmetry is kept, but not the front/hind symmetry. Instead, the symmetry operation $-EO_{\text{fh}}$ may be applied. Because of the lack of the front/hind symmetry, we have to distinguish between equations for the front limbs and those for the hind limbs. Recalling the definition of S_{ij} from equations (9.34), we may cast the new equations into the form:

front limbs, $i = r, \ell$

$$\dot{\phi}_{if} = S_{if} + A_0 + D_1 \cos\left[(\phi_{if} - \phi_{ih})\right] + D_2 \cos\left[2\left(\phi_{if} - \phi_{ih}\right)\right] \tag{9.49}$$

and hind limbs, $i = r, \ell$

$$\dot{\phi}_{ih} = S_{ih} - A_0 - D_1 \cos\left[\phi_{if} - \phi_{ih}\right] - D_2 \cos\left[2\left(\phi_{if} - \phi_{ih}\right)\right]. \tag{9.50}$$

Quite clearly, the loss of the front/hind symmetry allows for additional terms compared to (9.34). These equations can be solved numerically fairly easily, but we shall not dwell on the detailed results. We simply mention a few features of particular interest. When, for instance, the lower symmetry term is present with $D_1 = 0.2\text{Hz} > 0$, the four basic patterns are replaced by walks and quasi–gallops everywhere and are recovered only in the limits $C_1 \to \pm\infty$. Note that only straight patterns exist. The symmetry–lowering term induces the deviation from the basic patterns in a definite direction. Reversing the sign of D_1 would produce the opposite effect, where trot and pace were bistable for $D_1 = 0$. We now find two walks, one closer to a trot, the other closer to a pace and analogously for the gallop and pronk. Again we shall not dwell on the details of the very rich structure of the phase diagrams of the dynamics; rather, we wish to point out a new aspect that appears when we analyse the simplest symmetry–lowering term, namely the constant A_0. The most interesting aspect now is that, in a strip along the A_1 axis, no stable stationary patterns exist any more. The constant A_0 represents the difference in the natural frequencies of the front and hind girdle. If such frequencies are sufficiently different, i.e. if $\Delta\omega = \omega_2 - \omega_1$ is large enough, frequency and phase locking are lost and the front and hind girdle may adopt slightly different movement frequencies. The situation is similar to that discussed in Fig. 9.23. In addition, the dynamic coupling functions of the different components give some structure to the relative phases. The transient of the phases slows down at those patterns for which the slope of the curves of Fig. 9.23 is minimal. This attraction to preferred phase positions has been observed experimentally in several species including quadrupeds by *von Holst* (1939). He coined the word *relative coordination* for this phenomenon and understood it in sufficient detail to be able to build fluid-mechanical oscillator models that reproduced this phenomenon. His time and speed tables can be interpreted as return maps of discretely sampled relative phases that exhibit the attraction of relative phase to certain preferred values. In such a case front and hind girdle are no longer phase locked while within a girdle the system is, for instance, in a stable antiphase relation. In this case, the hind girdle is slower than the front girdle thereby leading to a monotonic decrease of ϕ_{rh} and $\phi_{\ell h}$. The drift is modulated. There are two phase values at which the transient slows down. These are the preferred phase relationships. As may be deduced from Fig. 9.23, if the values are comparatively shallow or are about to vanish, noise

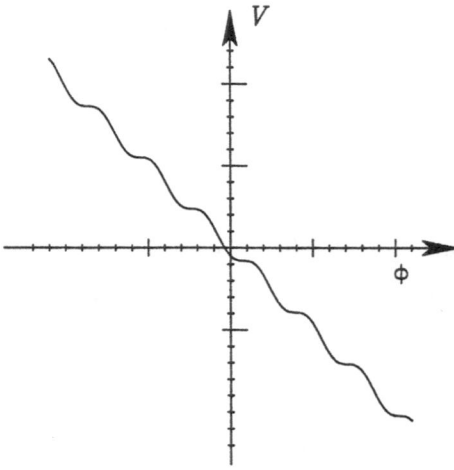

Fig. 9.23. Visualization of frequency locking and loss of frequency locking for too large detuning. To interprete this curve, consider a ball rolling on it. Its position represents the phase angle ϕ. If the overall gradient is small, the ball is trapped in a minimum at some $\phi = \phi_0$ (phase-locking). For a sufficiently large gradient, the local minima vanish and the ball rolls down the curve. In the case of overdamping, its average speed remains constant, but it varies locally

plays an important role in the phase movement. As may directly be seen from the curves of Fig. 9.23, relative coordination appears when the stability of a phase-locked state, i.e. one in a valley of Fig. 9.23, is lost.

9.8 Summary and Outlook

In this chapter, which basically followed the synergetic theory of *Schöner* et al., we have shown how the movement patterns of quadrupeds can be treated by means of a few basic concepts:

Based on synergetics, we introduce the phases of the individual limbs as order parameters and subject them to general order parameter dynamics. To simplify the corresponding equations, we invoke symmetry arguments. At the same time, the resulting patterns can be classified according to their symmetries. We found that patterns can be grouped into classes in which one element is transformed into another by group operations. In order to break the front/hind symmetry, we considered a smaller group and saw that new effects occur, for instance, the coexistence of patterns disappears. It is not difficult to extend all these considerations to cover the cases that are shown in Fig. 9.16, namely walk, rotary and transverse gallops. To this end, we must introduce a direct diagonal coupling and, as we may show, a right/left coupling and a diagonal coupling will be sufficient. Since the procedure is similar to that we have already given, we leave this consideration to the interested reader as a somewhat extended exercise. We might ask what causes the transition between animal gaits. One answer could be the animal's *intention* to move faster. An interesting experiment by *Hoyt* and *Taylor* (1981) may shed more light on the underlying mechanisms. When horses are forced to run in a treadmill at a given speed, their oxygen consumption reaches a relative minimum if their gaits are chosen appropriately (Fig. 9.24).

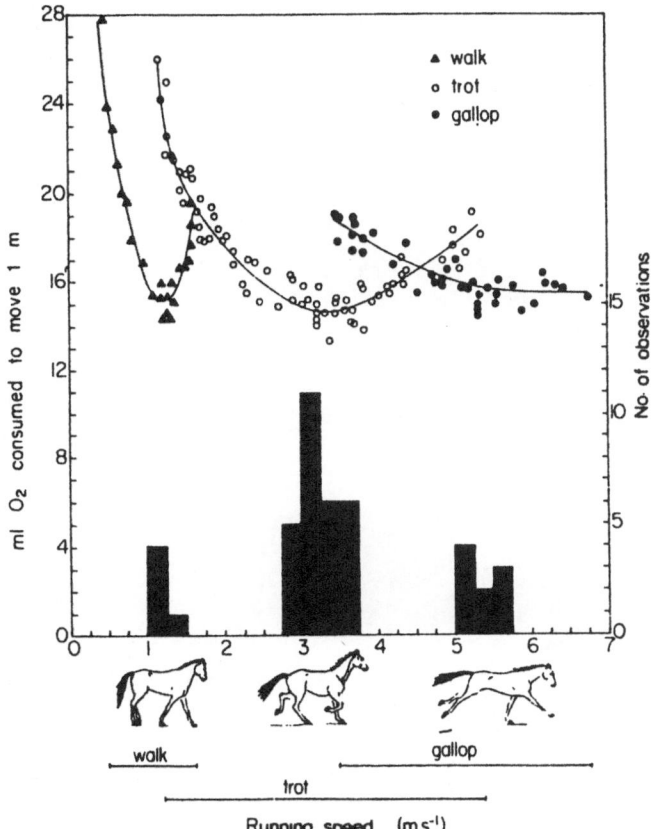

Fig. 9.24. Oxygen consumption of horses moving with different gaits in a treadmill. After *Hoyt* and *Taylor* (1981)

Looking at the finger movement paradigm, we realized what has to be done as a next step, namely to derive the equations for the phases from coupled oscillator equations for the individual limbs in analogy to (7.16), (7.17) of Sect. 7.2. This was done by *Lorenz* (1987), who considered transitions between gaits, but we shall not consider this case any further because of the length of the corresponding calculations. We believe that further experimental studies and a comparison with theoretical results will be of great interest and may shed new light on self-organization in biological systems.

10. Basic Concepts of Synergetics II: Formation of Spatio-temporal Patterns

One of the most striking features of self-organizing systems is their ability to form spatio-temporal patterns. Since we shall conceive of the brain as a self-organizing system that produces spatio-temporal patterns of activity, we wish to deal with the mechanisms of their formation from a general point of view. We have already encountered some simple examples, for instance, the famous convection instability in fluid dynamics (Fig. 4.4). Another example is provided by hexagonal or stripe patterns found in certain chemical reactions. These phenomena and the theoretical concepts underlying their explanation have led to models for the formation of patterns on furs, or on the skin of fish (cf. Fig. 4.7). In this chapter we wish to show how the methods of synergetics allow us to derive such patterns and, in a way, to classify them. Since colored patterns can be described in the same way as black and white patterns, we shall elucidate our basic approach by means of black and white patterns, or more precisely speaking, by grey tone patterns, where the tone may change locally and in the course of time. We note that the grey tone may symbolize, for instance, some local activity, for instance in electroencephalograms. The degree of the grey tone will be described by the variable q. Our reference state q_0 will be homogeneous and time independent and may denote a certain grey level. Then we introduce q as the deviation of the actual grey level from the average one, q_0. q will depend on space and time, at least in general. Therefore we shall write $q = q(\boldsymbol{x}, t)$. The order parameter concept will turn out to be extremely useful for studying the individual patterns and for classifying them.

Let us start with the simplest example possible, namely one order parameter. According to synergetics, the dominant part of q can be split into two factors: one describing the change of the pattern in time, and the other describing the change in space. In general, our hypothesis may be written in the form

$$q(\boldsymbol{x}, t) = \xi_u(t)v_u(\boldsymbol{x}) + \text{higher order}, \tag{10.1}$$

where the second part, called *higher order*, represents a small contribution, at least close to the instability point. Thus to get an insight into the spatio-temporal pattern, we may study the behavior of the first part alone. It turns out that ξ_u plays the role of the order parameter. When the system is above its critical point, a small fluctuation may cause ξ_u to come into existence. The internal dynamics then leads to an increase of ξ_u until ξ_u acquires a steady state. $v_u(x)$ can be determined from the linear stability analysis. The linear stability analysis provides us with the whole set of spatial functions $v(x)$, but here we have picked only that $v(x) = v_u(x)$, whose eigenvalue λ is positive. (For the moment we assume that the eigenvalue is real.) A typical spatial behavior of v_u is shown in Fig. 10.1 (right-hand side); the temporal behavior of ξ_u is shown in the same figure on the left-hand side.

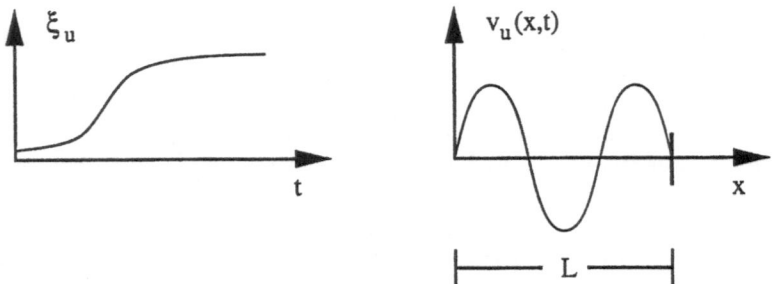

Fig. 10.1. *Left*: The growth of the order parameter ξ_u in the course of time to a stationary constant value. *Right*: Example of the spatial dependence of $v(x)$ as a function of the spatial coordinate x

Let us now study, in one dimension,

$$q(x,t) = \xi_u(t)v_u(x), \tag{10.2}$$

which we shall call the mode skeleton. We shall assume that the system had been below its critical point up to an initial time $t = 0$. Then at that time the control parameter is set to a value somewhat above its critical value, for instance, the fluid is heated accordingly or a certain concentration of chemicals is increased suddenly. At this time, only small fluctuations are present. Then, however, according to Fig. 10.1, $\xi(t)$ increases. Thus, some later time we observe a pattern as shown in the middle part of Fig. 10.2. Finally, at sufficiently large times, a steady state is reached and the pattern has adopted the form shown in the lower part of Fig. 10.2. This example displays a number of interesting features. Ordered states may already be observed in the transient regime and then, of course, in the stable final state. In cases where the potential landscape underlying the order parameter dynamics is symmetric, we may meet the phenomenon of symmetry breaking. Then there are two kinds of evolution of $\xi(t)$. In one case ξ remains positive and grows, in the other case ξ always remains negative and represents a mirror image of the positive branch. Similarly, one may draw mirror images to Fig. 10.2. At those positions where there was an increase of amplitude, there is now a decrease, and vice versa.

Now we move on to consider two order parameters. In this case, the variable q can be decomposed according to

a)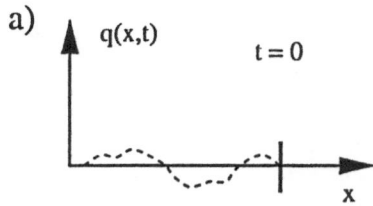

b)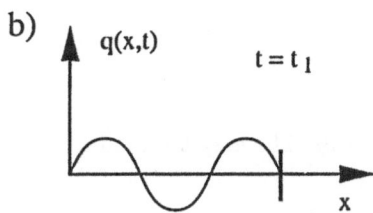

c)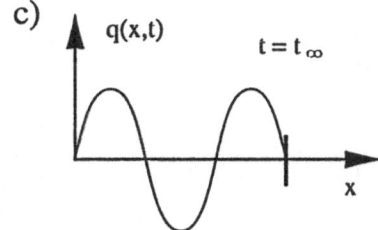

10.2a–c. The growth of the pattern described by (10.2) in the course of time. (a) At $t = 0$ only an initial fluctuation is present (b) At an intermediate time $t = t_1$ a transient pattern has been formed. (c) At sufficiently large time the steady state has been reached and a time independent pattern has evolved

$$q(x, t) = \xi_1(t)v_1(x) + \xi_2(t)v_2(x) + \text{higher order.} \qquad (10.3)$$

In it the leading part is given by the first two terms on the right-hand side, which will again be called the *mode skeleton*. Again the spatial functions v_1, v_2 can be determined from a linear stability analysis. For instance, they may be products of sine-waves in the form $\sin(k_1 x)\sin(k_2 y)$. Our example will show beautifully how the interplay between order parameters ξ_1 and ξ_2 on the one hand, and the spatial modes v_1, v_2 on the other hand leads to a variety of spatio-temporal patterns. They may be analyzed by means of the temporal behavior of the order parameters. To be as concrete and simple as possible, we shall assume very simple spatial patterns v_1 and v_2, as shown in Fig. 10.3.

Let us now distinguish between the well-known kinds of behavior of two order parameters (Sect. 5.2)

a) Stable fixed point (Fig. 10.4, a,b). Let us first consider the case where a fixed point lies on the ξ_1 axis (Fig. 10.4b). Then in the mode skeleton we may put $\xi_1 \neq 0, \xi_2 = 0$. In this case, a competition between modes has occurred. Mode 1 has won the competition and determines the evolving pattern in the form of Fig. 10.3, left-hand side. Quite often, several fixed points may occur, for instance, one lying on the ξ_1 axis, the other one lying on the ξ_2 axis (Fig. 10.5). Depending on the initially given state, the system will follow the

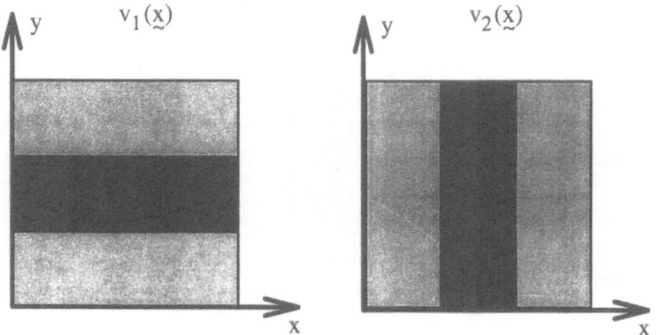

Fig. 10.3. Examples of two basic modes, v_1 and v_2. The grey level indicates the size of the functions

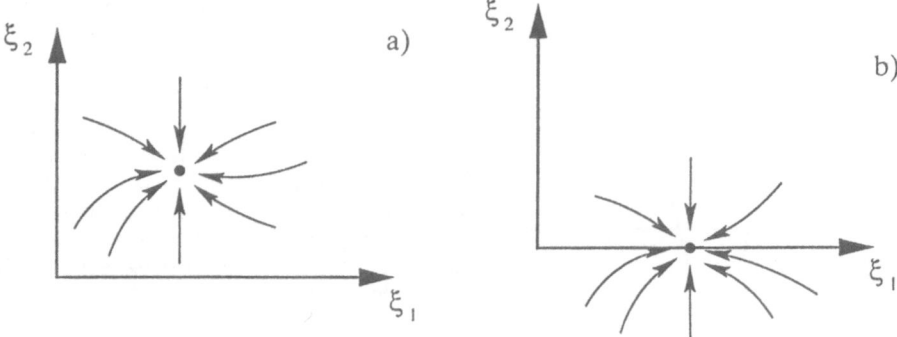

Fig. 10.4. (a) A fixed point in the ξ_1, ξ_2 plane at $\xi_1 \neq 0, \xi_2 \neq 0$. (b) A fixed point in the same plane lying on the ξ_1 axis

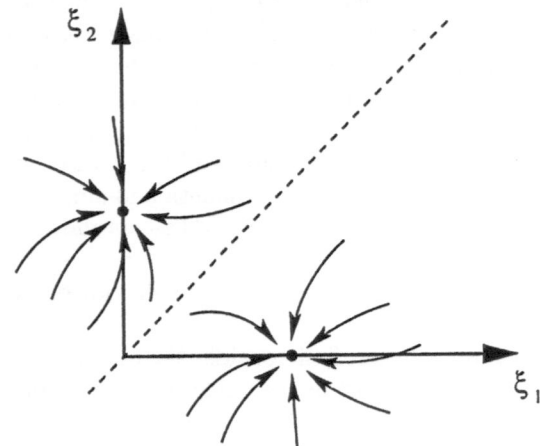

Fig. 10.5. Two fixed points lying on ξ_1 and ξ_2

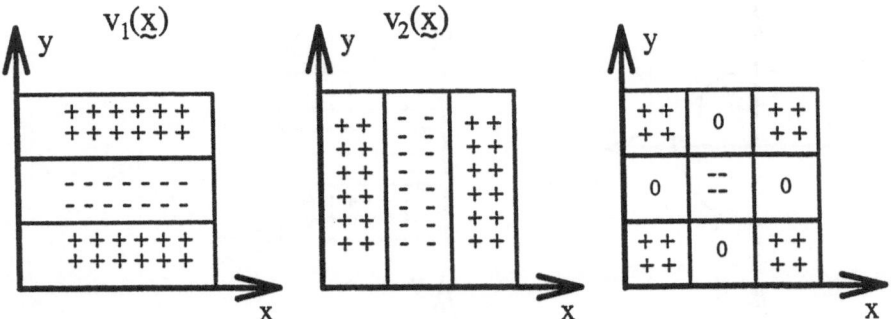

Fig. 10.6. Patterns corresponding to (10.3) and Fig. 10.3 in the case of a fixed point dynamics (see text)

corresponding trajectory and will end up at one or the other fixed point. In such a case, either pattern 1 or pattern 2 will be realized. Finally, the fixed point may lie so that $\xi_1 \neq 0, \xi_2 \neq 0$ (Fig. 10.4a). In this case, a superposition of both patterns 1 and 2 will appear in the steady state, even if originally only one pattern had been present (Fig. 10.6).

In the cases described so far, the patterns that finally evolve are time independent. Things change when we consider

b) A limit cycle (Fig. 10.7). The left column of Fig. 10.7 reminds the reader of a limit cycle in the ξ_1, ξ_2 plane, where the reference point moves around a circle (in the general case this may be a more irregular curve). Let us use the basic spatial patterns v_1, v_2 shown in Fig. 10.3 and let us start at time t_1, where $\xi_2 = 0$. Then we are dealing with the left pattern of Fig. 10.3. At a time t_2 a superposition of both patterns occurs. At a still later time t_3, ξ_1 has become zero and the total pattern is that of pattern 2. In this way, the reader may easily continue to realize Fig. 10.7 and to construct by himself or herself further individual patterns.

Finally let us consider several order parameters. In such a case, the variable q can be written in the form

$$q(\boldsymbol{x}, t) = \sum_u \xi_u(t) v_u(\boldsymbol{x}) + \text{higher order}, \qquad (10.4)$$

i.e., by means of a superposition of spatial modes and amplitudes that are time dependent. Again the patterns may be discussed in terms of the behavior of order parameters. Fixed points will give rise to a static superposition of the elementary patterns in the case of coexistence. In a number of cases competition may also occur and then one of the elementary patterns determines the whole pattern. A nice example of the effect of coexistence is the occurrence of hexagons in fluid dynamics. Actually one may understand the hexagonal patterns as a superposition of three elementary stripe patterns that are oriented at 60° to one another. Here, the same system may show both the stripe and the hexagonal pattern depending on the value of a control

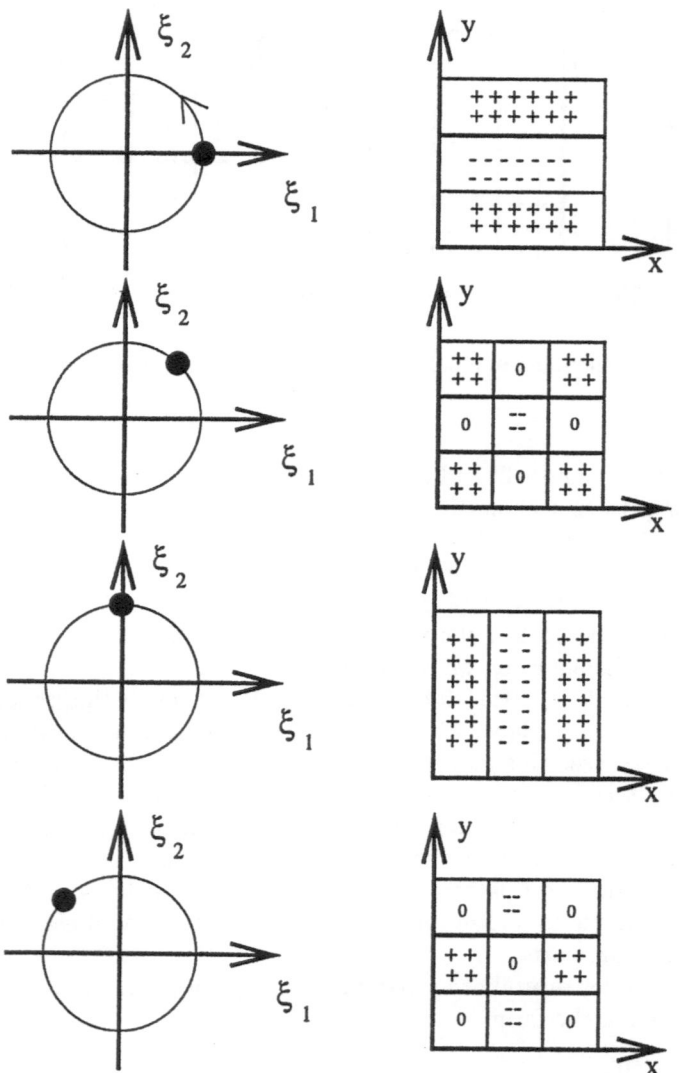

Fig. 10.7. Spatio-temporal patterns in the case of a limit cycle dynamics referring to two variables

parameter. For instance, a fluid heated from below may show, for a heating slightly beyond the critical temperature difference, a hexagonal pattern, which then goes over into a roll pattern for stronger heating. This can easily be understood by a study of the order parameter dynamics. In the first case, the dynamics allows the coexistence of three order parameters; in the other case a competition sets in which is won by one order parameter. Finally and most interestingly, the order parameters may also undergo chaotic motion.

We shall discuss this later in Chap. 13. In this case, we shall find patterns that change locally and in time in a seemingly random fashion. However, one must bear in mind that the total pattern can nevertheless be written as the superposition (10.4) by means of very few spatial modes. Thus, in spite of the seemingly complex changes in time and space, the complexity stems only from the time dependence and not from the spatial dependence.

11. Analysis of Spatio-temporal Patterns *

In the foregoing section we saw how spatio-temporal patterns can be built up from elementary spatial patterns, i.e. from the v's. In the analysis of spatio-temporal patterns in nature, including those occurring in brain activities, we are quite often confronted with the reverse problem. As we have seen in Sect. 2.5, experiments may provide us with spatio-temporal patterns, such as those of the EEG and MEG, and then the question arises of whether these patterns are built up of simpler patterns from which we may gain insight into the underlying time-dependent dynamics. In this section we present two methods that allow us to perform such a decomposition. We start with a decomposition that has been used in EEG and MEG analysis and is also widely used in pattern recognition.[1]

11.1 Karhunen–Loève Expansion, Singular Value Decomposition, Principal Component Analysis – Three Names for the Same Method

All these three expressions describe the same method, as we shall explain in the following. For practical purposes, it will be convenient to decompose a spatial pattern into individual cells, or, as they are sometimes called, pixels. A single cell may have several grey values, such as those depicted in Fig. 11.1. We may attribute a number q to each grey value, which runs from 0 to, for instance, 1, where 0 encodes white, 1 is black and all the values in between

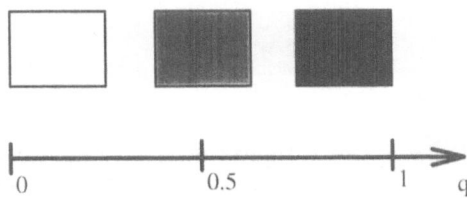

Fig. 11.1. The encoding of grey values by the size of a variable q

---

[1] This chapter is of a purely mathematical nature and can be largely skipped by readers not interested in the mathematical details.

are grey values. So the patterns of Fig. 11.1 can be characterized by quoting a number between 0 and 1. The basic features of the approach which we shall present here can be explained when we consider a pattern composed of two pixels denoted by 1 and 2 (Fig. 11.2). Again each pixel may have grey values ranging from 0 to 1 in complete analogy to Fig. 11.1. We denote the grey value of pixel 1 by q_1, that of pixel 2 by q_2. Any point shown in Fig. 11.3 then corresponds to a specific pattern. In that figure we have shown only the extreme cases in which both pixels are white or black, or one white and the other black. Grey values, except white and black, lie within the square. When we are dealing with a *spatio-temporal* pattern, the grey values of the two pixels will change in time in a coherent, or random, or mixed fashion. Writing the grey values q_1, q_2 in the form of a vector,

$$q = (q_1, q_2),\tag{11.1}$$

we may plot this vector in Fig. 11.4a for a time t. At another time, t_2, we will have another vector. So in the course of time, which we consider at discrete values, a number of reference points will appear in the square (Figs. 11.4b,c). We now wish to capture the distribution of these points in a simple manner. In a first step, we shift the origin of the coordinate system to the center of gravity of these points. To this end, we introduce the center of gravity by means of

$$\bar{q} = \frac{1}{L} \sum_{\ell=1}^{L} q(t_\ell)\tag{11.2}$$

when we are dealing with discrete time points t_ℓ. To facilitate our representation, we will replace (11.2) by a time average according to

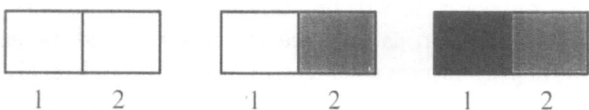

Fig. 11.2. Examples of a pattern with two pixels 1,2 and some grey values

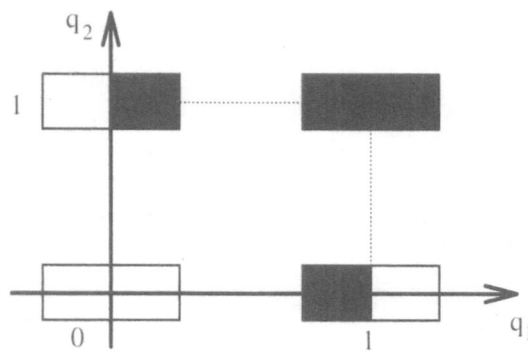

Fig. 11.3. The grey values of a two-dimensional pattern such as that of Fig. 11.2 can be represented as points in the q_1, q_2 plane. If in the two pixels the grey values are zero, the point describing that pattern lies at the origin

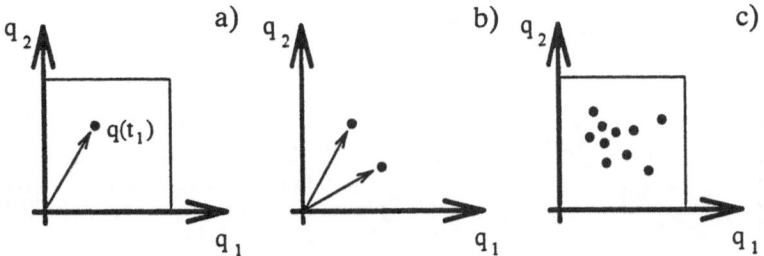

Fig. 11.4. (a) The state vector at time t_1 in the q_1, q_2 plane. (b) The state vectors for two different times. (c) The points of a state vector for several times

$$\frac{1}{L}\sum_j \rightarrow \frac{1}{T}\int_0^T \dots dt,$$
(11.3)

where the summation is replaced by an integration over time. By means of this new definition of the time average, we replace (11.2) by

$$\bar{q}_j = \frac{1}{T}\int_0^T q_j(t)dt, \quad j = 1, \dots, N.$$
(11.4)

In order to shift the coordinate system to the origin, we introduce a new variable h according to

$$h(t) = (q(t) - \bar{q}(t))\tilde{N}.$$
(11.5)

$\tilde{N}$ is a normalization constant so that the vector h acquires unit length. This is achieved by the requirement

$$\frac{1}{T}\int_0^T h(t)^2 dt = 1.$$
(11.6)

It will be our goal to approximate the time-dependent vector h in a manner as simple and efficient as possible. If all points that are reached by $h(t)$ in the course of time lie on a straight line, we can represent $h(t)$ in the form

$$h(t) = \xi(t)v^{(1)},$$
(11.7)

where $v^{(1)}$ points along that line. In many practical cases, the points of $h(t)$ will not lie on a line, but are scattered around it (Fig. 11.5). Thus (11.7) will still be a good approximation if we choose $v^{(1)}$ so that it points in the main direction of the cloud of points represented by $h(t)$.

Let us cast this idea in a more precise mathematical form. If we choose $h(t)$ for a single time, we would like to make the angle between $v^{(1)}$ and $h(t)$ as small as possible or equivalently, their scalar product $hv^{(1)}$ as large as possible. Since the product could be made arbitrarily large if we were to let the length of $v^{(1)}$ grow, we normalize the length of $v^{(1)}$ to unity

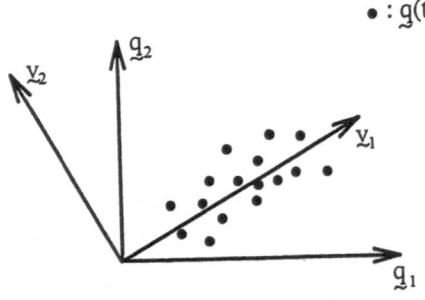

$\bullet : \underline{g}(t_j)$

Fig. 11.5. Example of the scattering of points of state vectors in the q_1, q_2 plane. The vector v_1 is chosen to lie in the direction of the greatest variance of the distribution of the end points of the state vectors

$$v^{(1)2} = 1.$$

Since the scalar product may become negative, we postulate that $(h(t)v^{(1)})^2$ be maximal. For a single time, this problem is, of course, immediately solved, namely $v^{(1)}$ must be chosen parallel to h. But what happens if the vectors $h(t)$ are scattered for different times? Then we require that $(h(t)v^{(1)})^2$ becomes maximal when we average over time:

$$\frac{1}{T} \int_0^T \left(h(t)v^{(1)} \right)^2 dt = \text{max!} \tag{11.8}$$

Because of this construction, $v^{(1)}$ points in the direction of the greatest variance of h. Once we have determined $v^{(1)}$, we may approximate $h(t)$ by (11.7). Because the points scatter, (11.7) is not exact. To find an improvement, we wish to approximate the remainder defined by

$$\Delta = h(t) - \xi_1 v^{(1)}, \tag{11.9}$$

by a second vector, $v^{(2)}$, that points in the direction of the greatest variance of Δ, which is orthogonal on $v^{(1)}$, and where $v^{(2)}$ is again normalized. We thus require

$$\frac{1}{T} \int_0^T \left(\left(h(t) - \xi_1 v^{(1)} \right) v^{(2)} \right)^2 dt = \text{max!} \tag{11.10}$$

But because of the orthogonality of $v^{(2)}$ on $v^{(1)}$, (11.10) reduces to

$$\frac{1}{T} \int_0^T \left(h(t)v^{(k)} \right)^2 dt = \text{max!} \tag{11.11}$$

where now $k = 2$ instead of $k = 1$ as in (11.8). Quite obviously, if the vector h belongs to an N-dimensional space, $N > 2$, we must continue this procedure. Each time we require (11.11), i.e. that $v^{(k)}$ points in the direction of the greatest variance orthogonal to the preceding vectors. Each time we also require

$$v^{(k)^2} = 1. \tag{11.12}$$

In order to solve the variational problem (11.11) with the constraint (11.12), we use Lagrangian multipliers λ_k and require that the variation

$$\delta \left\{ \frac{1}{T} \int_0^T \left(\boldsymbol{h}(t) \boldsymbol{v}^{(k)} \right)^2 dt - \lambda_k \left(\boldsymbol{v}^{(k)} \right)^2 \right\} = 0 \tag{11.13}$$

vanishes. The variation can be performed by differentiating the curly bracket with respect to the individual components of $\boldsymbol{v}^{(k)}$, which can be considered as independent, because we use the method of Lagrangian multipliers,

$$\delta\{...\} = \frac{\partial}{\partial v_j^{(k)}} \{...\}. \tag{11.14}$$

The differentiation can be done immediately and leads to

$$2 \frac{1}{T} \int_0^T h_j(t) \left(\boldsymbol{h}(t) \boldsymbol{v}^{(k)} \right) dt - 2\lambda_1 v_j^{(k)} = 0. \tag{11.15}$$

Dividing (11.15) by 2 and introducing the abbreviation

$$R_{jk} = \frac{1}{T} \int_0^T h_j(t) h_k(t) dt, \tag{11.16}$$

we may write (11.15) in the form

$$\sum_{\ell=1}^N R_{j\ell} v_\ell^{(k)} = \lambda_k v_j^{(k)}. \tag{11.17}$$

We introduce the matrix R by means of

$$R = (R_{jk}), \tag{11.18}$$

which allows us to write (11.17) in the form of a vector equation

$$R \boldsymbol{v}^{(k)} = \lambda_k \boldsymbol{v}^{(k)}. \tag{11.19}$$

Since R is a real symmetric matrix, it possesses real eigenvalues that are positive

$$\lambda_k \geq 0. \tag{11.20}$$

The eigenvectors of (11.19) are orthogonal and can be normalized so that

$$\left(\boldsymbol{v}^{(k)} \boldsymbol{v}^{(k')} \right) = \delta_{kk'} \quad = \begin{cases} 1 \text{ for } k = k' \\ 0 \text{ for } k \neq k' \end{cases}. \tag{11.21}$$

Some important relationships can be derived when we form the trace of R. This is defined, as usual, as the sum of the diagonal elements according to

$$\mathrm{Tr}R \equiv \sum_j R_{jj}. \tag{11.22}$$

Inserting the definition (11.16), we may cast (11.22) into the form

$$\mathrm{Tr}\boldsymbol{R} = \frac{1}{T}\sum_{j=1}^{N}\int_0^T h_j(t)^2 dt = \frac{1}{T}\int_0^T (\boldsymbol{h}(t))^2 dt = 1, \tag{11.23}$$

where we used the normalization of $\boldsymbol{h}$. On the other hand, it is known from algebra that the trace is equal to the sum of the eigenvalues λ_k

$$\mathrm{Tr}R = \sum_{k=1}^{N}\lambda_k. \tag{11.24}$$

A comparison between (11.23) and (11.24) reveals that

$$\sum_{k=1}^{N}\lambda_k = 1. \tag{11.25}$$

To elucidate the significance of (11.25), we proceed in several steps. We first note that by the construction of $\boldsymbol{h}$ outlined above, it may be written as

$$\boldsymbol{h}(t) = \sum_{k=1}^{N}\xi_k(t)\boldsymbol{v}^{(k)}. \tag{11.26}$$

The amplitudes $\xi_k(t)$ can be obtained from the vectors $\boldsymbol{h}$ and $\boldsymbol{v}^{(k)}$ by means of

$$\xi_k(t) = \boldsymbol{v}^{(k)}\boldsymbol{h}(t) \equiv \sum_{j=1}^{N}v_j^{(k)}h_j(t). \tag{11.27}$$

We anticipate the main result for the correlation between $\xi_{k'}$ and ξ_k, namely

$$\frac{1}{T}\int_0^T \xi_{k'}(t)\xi_k(t)dt = \lambda_k\delta_{kk'}, \tag{11.28}$$

with the Kronecker symbol $\delta_{kk'}$ defined in (11.21). For $k \neq k'$ the correlation function vanishes. This means that the amplitudes $\xi_{k'}, \xi_k$ are statistically independent. For $k = k'$, we obtain the result that the time average of ξ_k^2 is equal to the eigenvalue λ_k. Since this average tells us how big the proportion of $\boldsymbol{v}^k$ in $\boldsymbol{h}(t)$ is, the size of the eigenvalues λ_k indicates this amount.

In the following few steps we shall derive the relationship (11.28), the inpatient reader can proceed to formula (11.32), however. In order to prove (11.28), we insert (11.27) into (11.28), which leads us to

$$(11.28) = \frac{1}{T}\int_0^T \sum_{j=1}^{N}v_j^{(k')}h_j(t)\sum_{\ell=1}^{N}v_\ell^{(k)}h_\ell(t)dt. \tag{11.29}$$

We rearrange the sums and integrations and thus write

$$(11.29) = \sum_{j=1}^{N} \sum_{\ell=1}^{N} v_j^{(k')} v_\ell^{(k)} \underbrace{\frac{1}{T} \int_0^T h_j(t) h_\ell(t) dt}_{R_{j\ell}}. \tag{11.30}$$

Using the abbreviation $R_{j\ell}$ and the equation for the eigenvectors and eigen-values (11.17), we obtain

$$(11.29) = \sum_{j=1}^{N} v_j^{(k')} \lambda_k v_j^{(k)} = \lambda_k \delta_{kk'}, \tag{11.31}$$

where the right-hand side was obtained by making use of the orthogonality relation (11.21).

After this little exercise, we continue to consider the significance of (11.28). To this end, we study the mean error that is made when we include only a smaller number of terms, K, than the maximum number, N, required. In this way, we make a mean error that is defined by the square of the difference between h and its approximate decomposition. This mean error is given by

$$E_K = \frac{1}{T} \int_0^T \left(h(t) - \sum_{k=1}^{K} \xi_k(t) v^{(k)} \right)^2 dt. \tag{11.32}$$

From considerations similar to those that led us from (11.28) to (11.31), we readily obtain

$$E_K = 1 - \sum_{k=1}^{K} \lambda_k. \tag{11.33}$$

When we now order the eigenvalues λ_k according to their size,

$$\lambda_1 \geq \lambda_2 \geq \lambda_3..., \tag{11.34}$$

we see that we can minimize the mean error in the most rapid way by in-cluding in (11.33) or correspondingly in (11.32) the largest eigenvalues first. A determination of the eigenvalues thus gives us an insight into the quality of our approximation. It turns out that, in quite a number of practical cases it is sufficient to include the first few terms.

It is tempting, of course, to compare the decomposition (11.26) with the one we use in synergetics (cf. Chap. 10), namely as a decomposition into modes $v^{(k)}$ of the linearization with amplitudes ξ_k that correspond to order parameters. There is both good and bad news for this idea: On the one hand, in a number of cases, patterns evolve in boundaries that possess symmetries, for instance, square symmetry, circular symmetry, or spherical symmetry. Then group theory may be applied to determine the symmetry properties of the linearized equations that determine the patterns of the basic modes, $v^{(k)}$.

The same symmetry principles may be invoked for the modes of the Loève-Karhunen expansion so that we expect an agreement between the two sets of modes. This is actually found in some practical cases that we shall discuss below. On the other hand, we cannot guarantee, at least not in general, that the amplitudes ξ_k that appear in the Loève–Karhunen expansion are identical to the order parameters or amplitudes of enslaved modes. In other words, the Loève–Karhunen expansion is concerned with the frequency of the occurrence of basic patterns, but not with their underlying dynamics. Thus it is highly desirable to develop a method that directly allows the determination of order parameters and enslaved mode amplitudes from measured spatio-temporal patterns. This will be done in the next section.

11.2 A Geometric Approach Based on Order Parameters. The Haken–Friedrich–Uhl Method

In this section we will to show how to analyse the dynamics underlying observed spatio-temporal patterns by means of concepts of synergetics (cf. Chap. 4).

11.2.1 One Real Order Parameter

In order to bring out the essentials of our approach, we consider the case of a single order parameter, ξ_u. The starting point of our analysis is the slaving principle in the form (4.9), which we rewrite and generalize slightly. First of all, we write ξ_u instead of ξ as we did on earlier occasions in this book. Second we shall be satisfied with terms linear and quadratic in ξ_u which is usually a good approximation. Third we note that, in general, q need not vanish when ξ_u vanishes. And finally we combine $q(x_j, t)$ and $v_\ell(x_j), \ell = u, 2, 3, \ldots$ that occur in (4.9) to form the vectors $q(t), v_\ell$, respectively. By means of these steps, we cast (4.9) into the form

$$q(t) = q_0 + \xi_u(t)v_u + \xi_u^2(t)v_2, \tag{11.35}$$

where q_0 is *time independent*. Because v_2 is not normalized and we need for the following normalized vectors v, we put $v_2 = k_s v_s$, where v_s is normalized. Using the abbreviation $\xi_s = k_s \xi_u^2$, we may cast our general hypothesis concerning the state vector q in the form

$$q(t) = q_0 + \xi_u(t)v_u + \xi_s(t)v_s. \tag{11.36}$$

(The index "s" occurring in (11.36) and (11.37) originates from a specific derivation of the slaving principle and refers to "stable" modes, when the index "u" denotes "unstable". In the present context, these notions are of no concern, however.) Neglecting the time-independent vector q_0, we may say that the dynamics of q (11.37) takes place in the two-dimensional space spanned by v_u and v_s. Later in this section [equations (11.75ff.)] we shall consider the case in which we are dealing with a high-dimensional space but still only one order parameter. In this case, the last term in (11.36) must be replaced by

$$\sum_s \xi_s(t)v_s, \tag{11.37}$$

where the sum runs over all stable modes. We wish to show that from the measured data on $q(t)$ we may derive the dynamics of ξ_u, ξ_s and the patterns v_u, v_s. We assume that we are dealing with a nonequilibrium phase transition, where the order parameter obeys an equation of the form [cf. Sect. 5.1, equation (5.1) or (5.2)]

$$\dot{\xi}_u = f(\xi_u). \tag{11.38}$$

In addition to ξ_u, ξ_s, v_u and v_s, the function f and the constant k_s are unknown quantities that we wish to determine. If we deal only with the time-independent state (11.36), we can use only one pattern and not much can be done about our problem. Things change fundamentally, however, when we are dealing with transients, where, for instance, ξ_u changes in the course of time from $t = 0 \rightarrow t \approx \infty$ in a way indicated in Fig. 5.1. Focussing our discussion on the two-dimensional example, we find that q undergoes an evolution in the course of time as indicated in Fig. 11.6. Since q_0 is time independent, we shall introduce the quantity $w(t)$ by means of $w(t) = q - q_0$, where we assume that q_0 coincides with the initial value of $q = q(0)$ which means that the order parameter ξ_u vanishes at $t = 0$. Then $w(t)$ develops in the course of time as shown in Fig. 11.7. This coordinate system differs from that of Fig. 11.6 only by a shift of the origin. e_1, e_2 are unit vectors and $X(t), Y(t)$

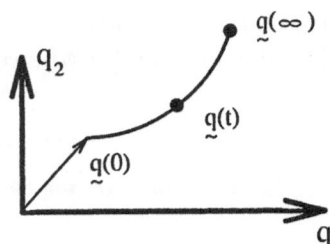

Fig. 11.6. Example of the time evolution of $q(t)$ in the q_1, q_2 plane

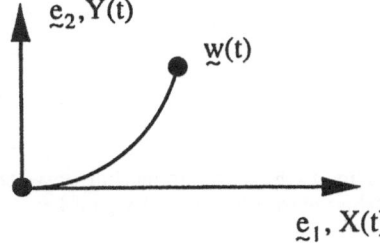

Fig. 11.7. The same as Fig. 11.6, but with a shifted origin

are the vector components of w. We may use two representations of w. The first is

$$w(t) = \xi_u(t)v_u + \xi_s(t)v_s \tag{11.39}$$

with the relationship

$$\xi_s = k_s\xi_u^2. \tag{11.40}$$

Note that v_u, v_s are, in general, not orthogonal on each other, but we shall assume that they are normalized

$$|\, v_u\,|^2 = 1, |\, v_s\,|^2 = 1. \tag{11.41}$$

The other representation of w is given by

$$w(t) = X(t)e_1 + Y(t)e_2. \tag{11.42}$$

While in (11.42) the quantities X and Y are experimentally accessible, the quantities occurring on the right-hand side of (11.39) are unknown. We, therefore, seek a connection between (11.42) and (11.39) (cf. Fig. 11.8) and use a linear transformation given by

$$e_1 = N_{1u}v_u + N_{1s}v_s \tag{11.43}$$

and

$$e_2 = N_{2u}v_u + N_{2s}v_s. \tag{11.44}$$

Inserting (11.43) and (11.44) into (11.42), we obtain after some minor rearrangement

$$w(t) = \underbrace{(X(t)N_{1u} + Y(t)N_{2u})}_{\xi_u} v_u + \underbrace{(X(t)N_{1s} + Y(t)N_{2s})}_{\xi_s} v_s, \tag{11.45}$$

which allows us by comparison with (11.39) to identify the amplitudes ξ_u and ξ_s. Using the definitions of ξ_u and ξ_s given by (11.45), and using (11.40), we obtain the relationship

$$XN_{1s} + YN_{2s} = k_s\left(XN_{1u} + YN_{2u}\right)^2. \tag{11.46}$$

We introduce the differentials dX, dY which lie on the trajectory. Taking the differential of (11.46), we obtain

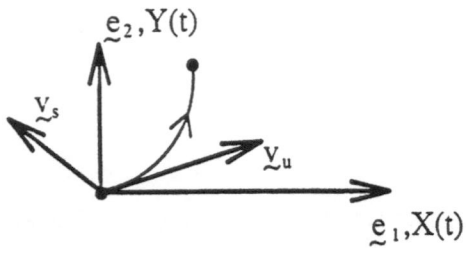

Fig. 11.8. The connection between the vectors e_1, e_2 and v_u and v_s

$$dX N_{1s} + dY N_{2s} = 2k_s \left(X N_{1u} + Y N_{2u} \right) \left(dX N_{1u} + dY N_{2u} \right). \qquad (11.47)$$

At the origin, i.e. at

$$X, Y \to 0, \qquad (11.48)$$

(11.47) reduces to

$$dX N_{1s} + dY N_{2s} = 0. \qquad (11.49)$$

The following discussion still takes place in the e_1, e_2-coordinate system.

We want to convince ourselves that v_u points in the direction of the vector (dX, dY). To this end, we multiply (11.43) by $-N_{2s}$ and (11.44) by N_{1s} and then add the resulting equations. We readily obtain

$$- N_{2s} e_1 + N_{1s} e_2 = \left(-N_{2s} N_{1u} + N_{1s} N_{2u} \right) v_u, \qquad (11.50)$$

where the vector v_s has dropped out, and $v_u \propto (-N_{2s}, N_{1s})$. Thus $v_u \perp (N_{1s}, N_{2s})$ which according to (11.49) is orthogonal on dX, dY. Therefore, in the plane,

$$v_u \| (dX, dY). \qquad (11.51)$$

Thus we have determined v_u which in the following will be assumed to be normalized. In order to determine v_s, we start from the decomposition (11.39), where the relationship (11.40) is assumed to hold. We decompose v_s according to

$$v_s = a \hat{v}_s + b v_u, \qquad (11.52)$$

where we assume

$$\hat{v}_s \perp v_u. \qquad (11.53)$$

Using (11.52) and (11.60) in (11.39), we obtain

$$w(t) = v_u \left(\xi_u + b k_s \xi_u^2 \right) + k_s a \xi_u^2 \hat{v}_s. \qquad (11.54)$$

Making use of the orthogonality and normalization conditions, we obtain

$$(v_u w(t)) = \xi_u + b k_s \xi_u^2 \qquad (11.55)$$

and

$$(\hat{v}_s w(t)) = k_s a \xi_u^2. \qquad (11.56)$$

We also introduce the abbreviations

$$b k_s = c_2, \, a k_s = c_1. \qquad (11.57)$$

In (11.55) and (11.56) we can eliminate the term containing ξ_u^2 and obtain, instead of (11.55),

$$\xi_u = (v_u w) - c (\hat{v}_s w), \qquad (11.58)$$

where we used the definition

$$c = c_2/c_1. \tag{11.59}$$

Inserting (11.58) into (11.56), we finally obtain

$$(\hat{v}_s w) = a k_s \left[(v_u w) - c (\hat{v}_s w) \right]^2, \tag{11.60}$$

a relationship that holds for all times t. Using an obvious abbreviation, we may cast (11.60) into the form

$$\dot{s}(t) = a k_s \left[u(t) - cs(t) \right]^2. \tag{11.61}$$

This relationship allows us to determine k_s and c. To this end we consider (11.61) for two different times (t_1, t_2), i.e.,

$$s_1 = a k_s \left(u_1 - c s_1 \right)^2 \tag{11.62}$$

and

$$s_2 = a k_s \left(u_2 - c s_2 \right)^2. \tag{11.63}$$

Dividing (11.62) by (11.63), we obtain

$$\frac{s_1}{s_2} = \frac{(u_1 - c s_1)^2}{(u_2 - c s_2)^2}. \tag{11.64}$$

Taking the square root of both sides and rearranging (11.64), we obtain

$$(u_1 - c s_1) = \pm \sqrt{s_1/s_2} \, (u_2 - c s_2). \tag{11.65}$$

Quite obviously, c can be immediately determined from (11.65) so that c_2/c_1, or equivalently b/a are determined. Because we require that v_s occurring in (11.52) is normalized, we have the relationship $a^2 + b^2 = 1$ so that a and b are determined. We may resolve (11.62) for k_s, which yields

$$k_s = \frac{s_1}{(u_1 - c s_1)^2} \tag{11.66}$$

allowing us to determine k_s. We note that it is also possible to determine the variables k_s and c of (11.61) by a least squares fit procedure, where

$$W = \int_0^T \left\{ s - k_s [(v_u w) - c (\hat{v}_s w)]^2 \right\}^2 dt$$

is minimized. Such a method is appropriate when the data are noisy.

Let us pause for a moment to consolidate our results. So far we have determined $v_u, \hat{v}_s$, and a, b, k_s. By means of (11.52) and knowing a and b, we determined v_s. Thus we reached our goal of determining the spatial modes and the relationship between ξ_s and ξ_u according to (11.40). Our last task is to determine the equation of motion for ξ_u. To this end we introduce the adjoint vector $\bar{v}_u$ such that

$$(\bar{v}_u v_u) = 1 \tag{11.67}$$

and

$$(\bar{v}_u v_s) = 0. \tag{11.68}$$

Multiplying (11.53) by $\overline{v}_u$, we obtain

$$(\overline{v}_u w(t)) = \xi_u + bk_s\xi_u^2. \tag{11.69}$$

Since the left-hand side of (11.69) is known, so is ξ_u. In order to verify the equations of motion, we differentiate both sides of (11.69) with respect to time and require that

$$\frac{d}{dt}(...) \equiv \dot{\xi}_u = f(\xi_u). \tag{11.70}$$

An example is provided by $f(\xi_u) = \lambda\xi_u - \beta\xi_u^3$, where λ and β are still unknown coefficients. For their determination we note that, quite obviously,

$$\frac{d\dot{\xi}_u}{d\xi_u} = \lambda \quad \text{for} \quad \xi_u = 0. \tag{11.71}$$

Then, having a potential function of the form Fig. 11.9 in mind, which leads to a phase portrait as shown in Fig. 11.10, we obtain a condition for $\dot{\xi}_u = 0$ in the form

$$\lambda\xi_u - \beta\xi_u^3 = 0, \tag{11.72}$$

which we may immediately resolve to obtain

$$\beta = \lambda/\xi_u^2, \tag{11.73}$$

where ξ_u is the stationary solution of (11.70). Thus we are able to deduce all unknown quantities and our program is finished. We note that there are other possibilities for equations of motion: For instance, depending on the type of bifurcations, we may have equations of the form

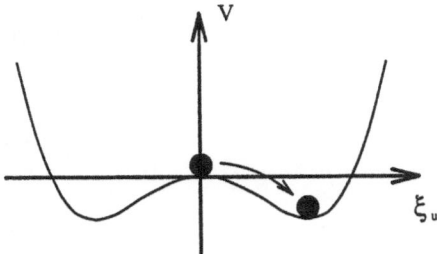

Fig. 11.9. A potential landscape

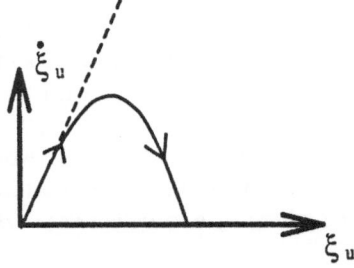

Fig. 11.10. Movement of ξ_u in the phase plane. A trajectory produced by the potential landscape of Fig. 11.9

$$\dot{\xi}_u = a + b\xi_u + \gamma\xi_u^2 - \beta\xi_u^3 \tag{11.74}$$

or their corresponding normal forms. The coefficients can be determined by means of a plot in the $(\xi_u, \dot{\xi}_u)$-plane.

We finally discuss the general case of a high-dimensional space of q, where we still have only one order parameter but many enslaved modes, v_s; cf. (11.37). Our starting point is the decomposition of w in the form

$$w(t) = \xi_u(t)v_u + \sum_s \xi_s(t)v_s, \tag{11.75}$$

where we again assume the relationship

$$\xi_s = k_s\xi_u^2, \quad s = 1, 2, \dots \tag{11.76}$$

between enslaved mode amplitudes ξ_s and the order parameter ξ_u. Inserting (11.76) into (11.75), we immediately find

$$w(t) = \xi_u v_u + \xi_u^2 \underbrace{\sum_s k_s v_s}_{V_s}. \tag{11.77}$$

Quite obviously, the space in which $w(t)$ moves is spanned by the vectors v_u and V_s, i.e. it is again two dimensional. Thus we may immediately apply the method developed in the foregoing. This two-dimensional vector space can be spanned by w taken at a time t_1 and at another time t_2, where t_1 and t_2 can be chosen conveniently. Formally we may cast the representation of the vector space into the form

$$w(t) = X(t)e_1 + Y(t)e_2, \tag{11.78}$$

where the unit vectors are given by

$$e_1 = \frac{w(t_1)}{|\,w(t_1)\,|} \tag{11.79}$$

and

$$e_2 = N_2\left[w(t_2) - e_1\left(e_1 w(t_2)\right)\right] \tag{11.80}$$

with the normalization condition

$$N_2^{-2} = |\,w(t_2) - e_1\left(e_1 w(t_2)\right)\,|. \tag{11.81}$$

Thus the high-dimensional problem is totally reduced to the formerly treated two-dimensional problem. It should be noted that from the data we can determine v_u but not the individual v_s, only the total sum V_s. This is not of great concern because in pattern formation we are mainly interested in the so-called mode skeleton, which is represented by the first term on the right-hand side of (11.77).

Another method that is of great practical use and allows us to find a two-dimensional basis for w consists in applying the Karhunen–Loève method in a first step. Then we may deal with w in a space where e_1 and e_2 correspond to the Karhunen–Loève modes u_1 and u_2.

11.2.2 Oscillations Connected
with One Complex Order Parameter

In this section we wish to analyze spatio-temporal patterns that are governed by a single order parameter that undergoes oscillations. For the sake of mathematical clarity, we shall use a complex order parameter that we write in the form

$$\xi_u = ae^{i\Omega t} + \text{h.o.} \tag{11.82}$$

Ω is its basic frequency. We shall allow the complex amplitude a to have a time dependence, but one that is slow compared to that of $e^{i\Omega t}$. We may also allow higher order harmonics as indicated by h.o. We subtract from the observed spatio-temporal pattern $q(t)$ its temporal mean value, $\overline{q}$. It then remains to analyze the function $w(t) = q(t) - \overline{q}$. Because of the slaving principle in its lowest approximation, we may write w in the form

$$w(t) = \xi_u v_u + \xi_u^* v_u^* + k_s \xi_u^2 v_s + k_s^* \xi_u^{*2} v_s^*. \tag{11.83}$$

We first determine the vector of the unstable mode, v_u. This is easily achieved by means of

$$\frac{1}{2T} \int_{-T}^{T} e^{-i\Omega t} w(t) dt = a v_u, \tag{11.84}$$

because the terms that are connected with the enslaved modes start with higher frequencies according to

$$\xi_u^2 = a^2 e^{2i\Omega t} + \text{h.o.} \tag{11.85}$$

Provided the term denoted by "h.o." in (11.82) is sufficiently small, the only dependence of $w(t)$ on $e^{\pm 2i\Omega t}$ in (11.83) can stem from the third and fourth terms on the right of (11.83). This allows us to determine, for instance, v_s by means of

$$c v_s = \frac{1}{2T} \int_{-T}^{T} e^{-2i\Omega t} w(t) dt \tag{11.86}$$

to within a multiplicative constant c. This procedure forms the basis of a method which we shall follow up in Sect. 12.6 to determine the order parameter equation for ξ_u. If the "h.o" terms in (12.10) cannot be neglected, a more elaborate procedure is required.

12. Movements on a Pedalo

12.1 The Task

In the preceding chapters on movement coordination we have been concerned with situations in which, because of the experimental conditions, we were dealing with only very few degrees of freedom. In this chapter we wish to analyse movements in which, at least in principle, quite a number of degrees of freedom may participate. In particular we shall study how the individual degrees of freedom become coordinated during learning. The experiment we have in mind uses the pedalo shown in Fig. 12.1. It consists of two steps fixed

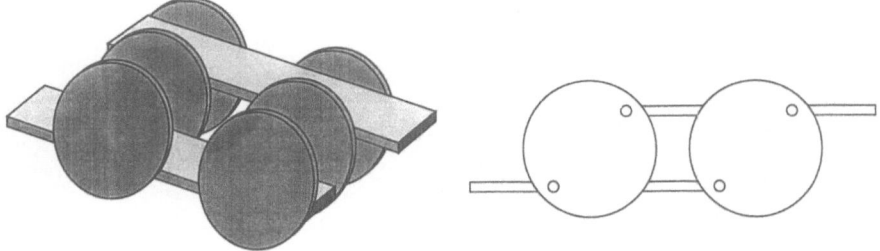

Fig. 12.1. Schematic drawing of a pedalo. The pedalo consists of two steps that are excentrically mounted on wheels. The resulting movement of the left and right leg is shifted by 180°. The right-hand side shows the side view

eccentrically to four wheels. The task to be learned involves moving forwards by lifting and pushing one's feet. This task is by no means trivial, but can be accomplished after a comparatively short learning period in which we can study the subject's progress in movement control. A task of general interest will be to evaluate the progress in learning, i.e. to establish criteria for the performance of the required movement. To this end, we first introduce an appropriate description of the movement patterns. Then we shall proceed along two different lines. The first will be based on an analysis of the movement patterns in terms of the Karhunen–Loève expansion that we explained in Sect. 11.1. As we shall see, the evolving patterns are dominated by few Karhunen–Loève modes. This will be a strong indication that the movement patterns are governed by a few order parameters only.

The other line of research will be an analysis of the movement patterns in terms of order parameters. There we shall find the quite surprising result that the whole movement pattern is governed by a single order parameter, at least after learning. This means that the numerous degrees of freedom, in particular the arm movements, are governed, or in other words, enslaved by a single variable.

12.2 Description of the Movement Pattern

In the experiment light emitting diodes (LEDs) are attached to the joints of the person as shown in Fig. 12.2. Figure 12.3 shows the links between the joints and their enumeration. The pedalo rider drives the pedalo in a direction perpendicular to the plane of the observer. The coordinates of this plane are given by X, Y. The motion of the LEDs is registered by a camera. The sequence of images can be digitized and one obtains a temporal movement

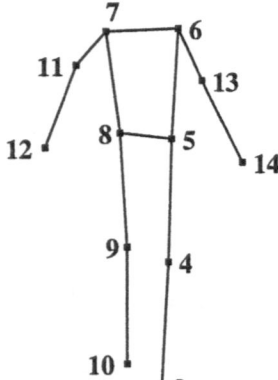

Fig. 12.2. Positioning and numbering of the LEDs

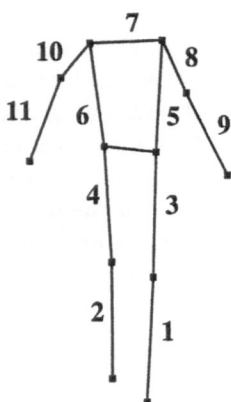

Fig. 12.3. Numbering of the connecting links of the pedalo rider

of the points $(x'(t), y'(t))$. Because the pedalo rider moves in the direction of the camera, his image grows in the course of time. To compensate for this effect, two LEDs are fixed to the pedalo. (The corresponding points 1 and 2 are not shown in Fig. 12.2.) By means of the distance between x'_1 and x'_2 the other points of the overall movement may be rescaled ($x' \rightarrow x$). Because the space in which these experiments are performed is limited, only ten periods can be registered during a run. The experiments were performed by *Körndle* (1992) at the University of Regensburg.

12.3 Quantification of the Pedalo Movement

In order to process the measured data, we have to cast the movement into an appropriate form. To this end we calculate the angle ϕ_i between each link ℓ_i and the plane of the observer, and the angle θ_i between ℓ_i and the vertical axis in the plane of the observer. We denote the differences of the coordinates of two directly connected joints with coordinates (x_j, y_j) and (x_k, y_k) by $\Delta x, \Delta y$

$$\Delta x = x_j - x_k$$
$$\Delta y = y_j - y_k. \tag{12.1}$$

Then using simple geometry, we may immediately determine ϕ_i, θ_i from

$$\theta_i = \arcsin\left(\frac{\Delta x_i}{\sqrt{(\Delta x_i)^2 + (\Delta y_i)^2}}\right), \tag{12.2}$$

$$\phi_i = \arccos\left(\frac{\sqrt{(\Delta x_i)^2 + (\Delta y_i)^2}}{\ell_i}\right). \tag{12.3}$$

The index i labels all pairs of connected joints. Because Δz was not determined, θ is not uniquely determined, but the uncertainty is only in sign. To resolve this ambiguity we shall assume θ to be positive. The temporal change of some angles is shown in Fig. 12.4. In that figure we also show the smoothed reference movement. This movement was calculated by means of the movement of the two steps of the pedalo, i.e., using the points 3 and 10 of Fig. 12.2.

$$y_{\text{ref}} = y_{10} - y_3. \tag{12.4}$$

y_{ref} is then normalized to the interval 0,1 and smoothed. By means of this reference movement, we divide the movement into its periods. Using the 22 angles defined above, we may form a 22-dimensional vector $q(t)$

$$q(t) = \begin{pmatrix} \phi_1(t) \\ \theta_1(t) \\ \vdots \\ \phi_{11}(t) \\ \theta_{11}(t) \end{pmatrix}. \tag{12.5}$$

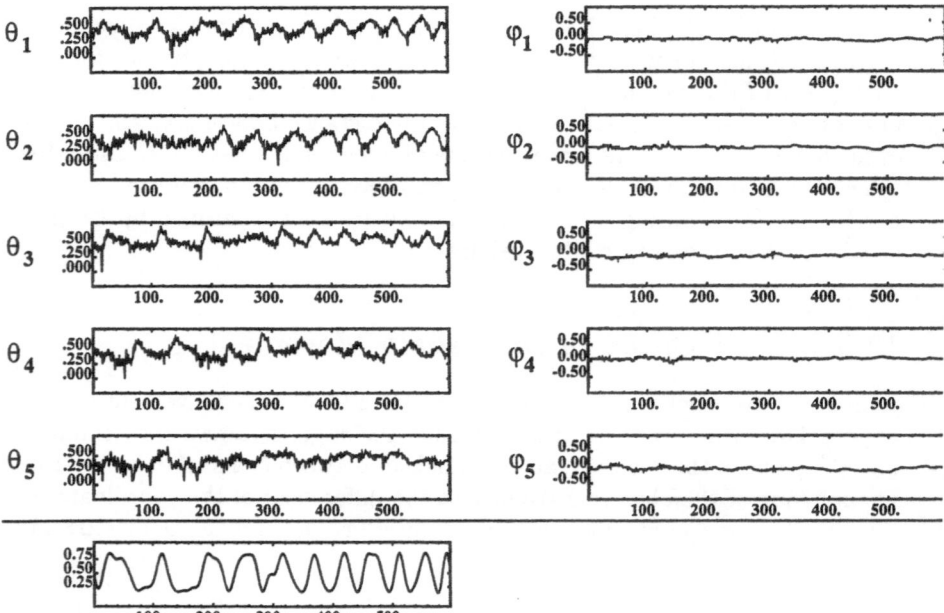

Fig. 12.4. Temporal evolution of the first five angles. In addition, the smoothed reference movement is shown on the lower left

In the following we shall analyse the movement in terms of this pattern vector.

12.4 Analysis of the Movement Using the Karhunen–Loève Expansion

The full experiment consists of five test movements, denoted by HERM 1, HERM 2, HANN 4, RAIN 3, MARK 3. The data gained from HERM 1 and HERM 2 are of particular interest because they stem from the same subject before and after learning, respectively. The other data stem from subjects with differing skills. In this section we study the movement using the Karhunen–Loève expansion. A typical result that refers to HERM 1 is shown in Fig. 12.5.

The first row shows the reference movement which can be derived directly from the pedalo. The second row represents the eigenvalues of the Karhunen–Loève expansion. The last three rows show the Karhunen–Loève modes for the first three eigenvalues. The angles are amplified to bring out the characteristic features. The reference curve provides us with a first impression of the smoothness of the movement. The maximum eigenvalue, λ_1, gives us a hint at how well the movement can be represented by a single mode. It

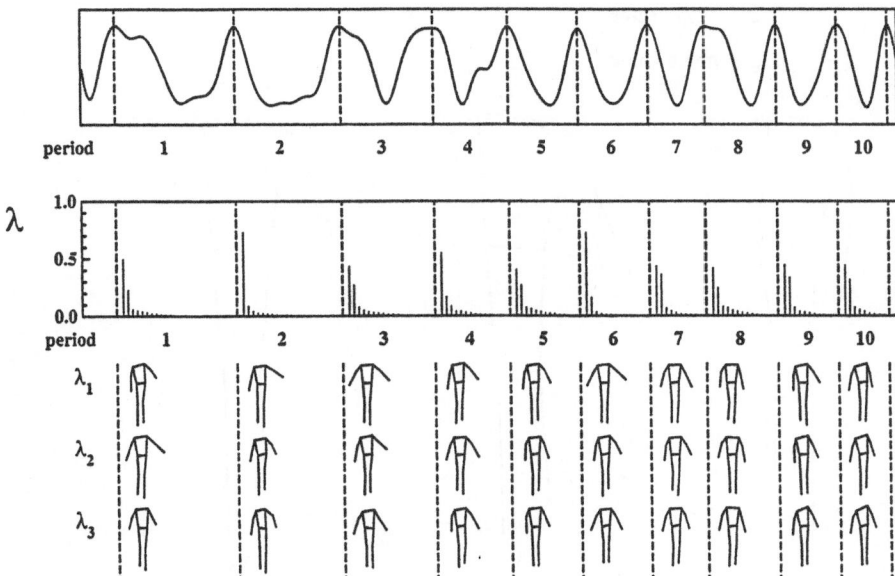

Fig. 12.5. Results for subject HERM 1. The upper row shows the reference movement. The middle row shows the eigenvalues of the Karhunen–Loève decomposition. The last three rows show the resulting eigenvectors for each period corresponding to the first three dominant eigenvalues λ_1, λ_2 and λ_3

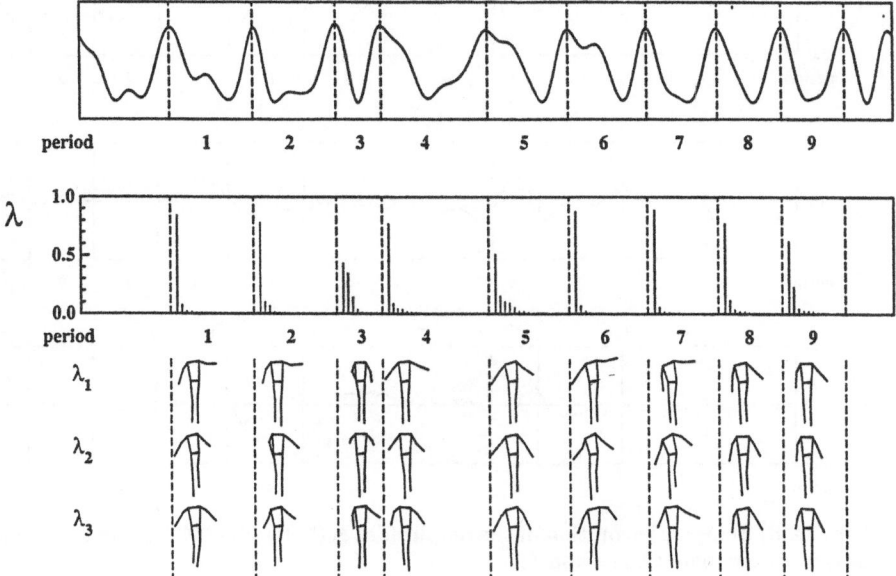

Fig. 12.6. Same as Fig. 12.5, but for a second test ride HERM 2

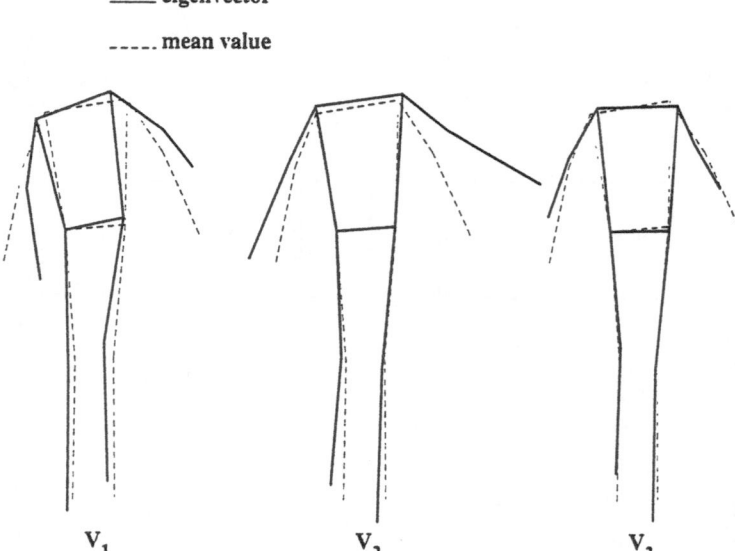

Fig. 12.7. An enlarged representation of the three eigenvectors of the set of data HERM 1 and the first period

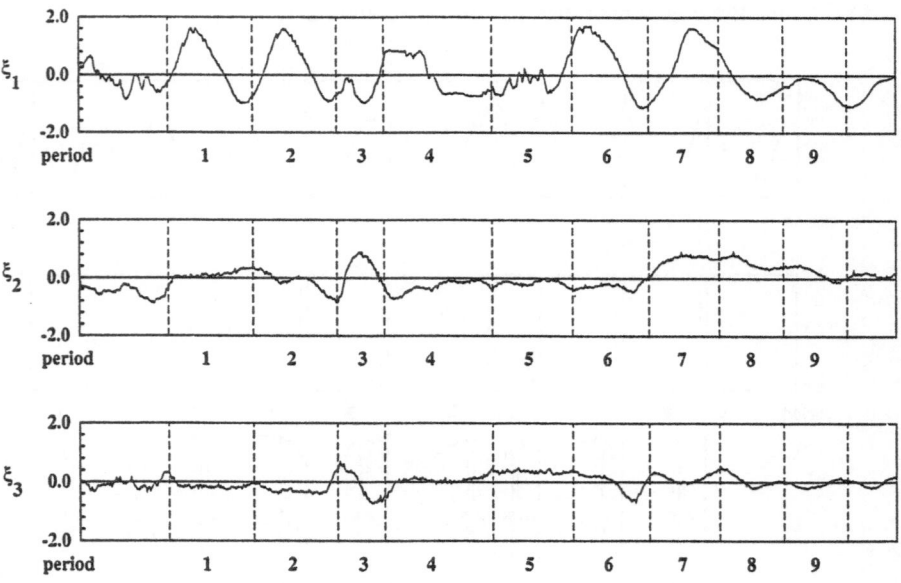

Fig. 12.8. Temporal evolution of the amplitudes $\xi_k(t)$ for HERM 1. Note the smallness of ξ_3 compared to ξ_1 and ξ_2

is interesting to note that after the learning period this value is markedly increased and the other eigenvalues are small or even negligible (Fig. 12.6). The results for the other subjects are basically similar as far as the number of dominant modes is concerned, though two subjects showed pronounced arm movements. Figure 12.7 represents the shape of the movement more precisely.

By means of the Karhunen–Loève modes v_k, the amplitudes $\xi_k(t)$ that occur in the expansion

$$\hat{q}(t) \equiv q(t) - \overline{q} = \sum_k \xi_k(t) v_k \tag{12.6}$$

can be calculated for each period

$$\xi_k(t) = v_k \hat{q}(t). \tag{12.7}$$

Figure 12.8 shows the temporal evolution of $\xi_k(t)$ for the first three modes, corresponding to the data of Fig. 12.5. The results of this figure (and of figures that are not shown here and correspond to the other subjects) substantiate our main conclusion that the movement can be well represented by the first, or first and second mode.

12.5 A Detailed Analysis of the Movements of Arms and Legs

When we study the movement of the pedalo rider more closely, we quickly recognize that the movement of the legs is mainly prescribed by the movement of the steps of the pedalo. Thus there is only a small variability of the movement of the legs. Thus differences in the movement of arms and body should be particularly pronounced. To study these individual movements, we split the vector of (12.5) into the following two vectors. The connections are numbered as in (12.5).

$$q_L(t) = \begin{pmatrix} \phi_1(t) \\ \theta_1(t) \\ \vdots \\ \phi_4(t) \\ \theta_4(t) \end{pmatrix} \qquad q_A(t) = \begin{pmatrix} \phi_5(t) \\ \theta_5(t) \\ \vdots \\ \phi_{11}(t) \\ \theta_{11}(t) \end{pmatrix}. \tag{12.8}$$

We now perform a Karhunen–Loève analysis for the time series q_L of the movement of the legs and q_A of the movements of arms and body separately, i.e. we form

$$\hat{q}_A(t) = q_A(t) - \overline{q}_A$$
$$\hat{q}_L(t) = q_L(t) - \overline{q}_L. \tag{12.9}$$

We then can calculate the corresponding eigenvalues λ_k^A and λ_k^L. The results are shown in Fig. 12.9, jointly with the individual patterns corresponding to

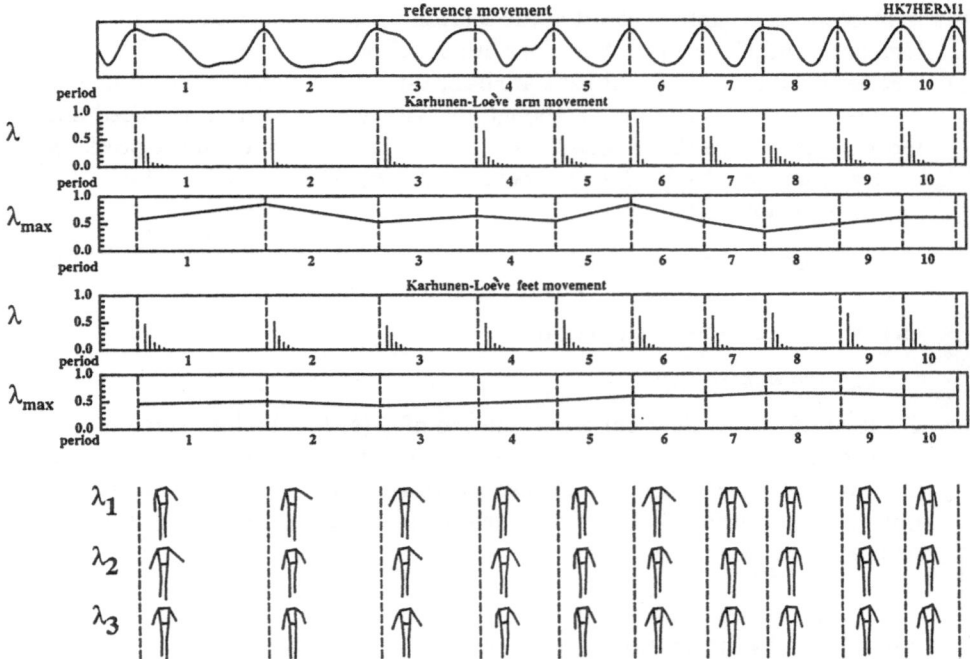

Fig. 12.9. A separate analysis of the pedalo movement for legs and arms. First row: Reference movement. Second row: Eigenvalue spectrum of the arm movement for each period. Third row: Value of the biggest eigenvalue. The last two rows show the eigenvalues of the movement of the legs. Note the striking constancy of the eigenvalues of the leg movement. (Data from HERM 1 before learning.)

the eigenvalues $\lambda_1, \lambda_2, \lambda_3$. While in the case of the movements of the arms, the maximum eigenvalue shows appreciable variations over the individual periods, the maximum eigenvalue of the movements of the legs remains rather constant. Figure 12.10 shows the results after learning. Again we find that the maximum eigenvalue of the movement of the legs remains practically constant, whereas the eigenvalue of the arm movement increases appreciably.

12.6 Haken–Friedrich–Uhl Order Parameter Analysis

The results of the preceding sections support the idea that the movement can be described by one or a few order parameters, at least after the learning period. To this end we make the hypothesis that the motion is governed by a single complex order parameter that we write in the form

$$\xi_u = Ae^{i\omega t} + \text{h.o.,} \tag{12.10}$$

where ω is the fundamental frequency $= 2\pi/T$ and T the duration of one period. We assume that A may be time dependent, but that it is slow compared

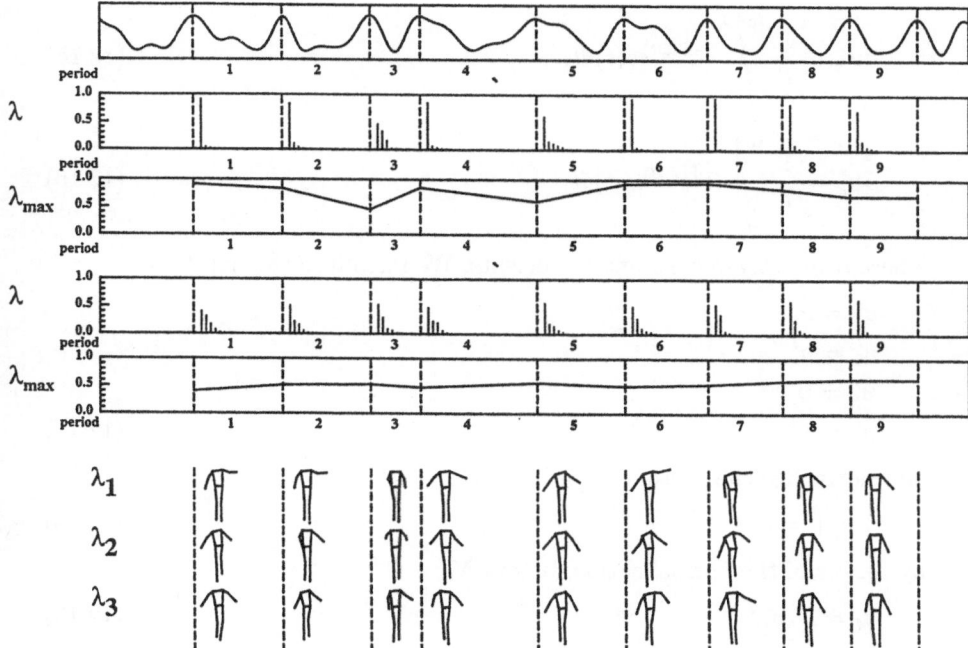

Fig. 12.10. Same as Fig. 12.9, but with data from HERM 2, i.e. after learning

to exp($i\omega t$). The higher harmonics are indicated by h.o. and are assumed to be negligible. In the following we proceed in accordance with Sect. 11.2 and, in a first step, form the difference between the pattern vector q and its time average

$$w(t) = q(t) - \overline{q}. \tag{12.11}$$

We write the pattern vector w in the form

$$w(t) = \xi_u v_u + \xi_u^* v_u^* + k_s \xi_u^2 v_s + k_s^* \xi_u^{*2} v_s^*. \tag{12.12}$$

Note that v_u is the (complex) mode belonging to the order parameter ξ_u and v_s is a superposition of enslaved modes. Because of the assumed time dependence (12.10), we can easily calculate the modes that occur in (12.12) by means of

$$\alpha v_u = \frac{1}{T} \int_{t_0}^{t_0+T} e^{-i\omega t} w(t) dt, \tag{12.13}$$

$$\alpha v_u^* = \frac{1}{T} \int_{t_0}^{t_0+T} e^{+i\omega t} w(t) dt, \tag{12.14}$$

$$\beta \boldsymbol{v}_s = \frac{1}{T} \int\limits_{t_0}^{t_0+T} e^{-i2\omega t} \boldsymbol{w}(t)dt, \qquad (12.15)$$

$$\beta \boldsymbol{v}_s^* = \frac{1}{T} \int\limits_{t_0}^{t_0+T} e^{+i2\omega t} \boldsymbol{w}(t)dt, \qquad (12.16)$$

where α and β are constant coefficients. We introduce the notation

$$\boldsymbol{v}_1' = \boldsymbol{v}_u$$
$$\boldsymbol{v}_2' = \boldsymbol{v}_u^*$$
$$\boldsymbol{v}_3' = \boldsymbol{v}_s$$
$$\boldsymbol{v}_4' = \boldsymbol{v}_s^* \qquad (12.17)$$

and normalize the vectors according to

$$\mid \boldsymbol{v}_i \mid^2 = 1 \qquad (12.18)$$

by means of the normalization factors N_i,

$$\boldsymbol{v}_i = N_i \boldsymbol{v}_i'. \qquad (12.19)$$

Because these vectors are not orthogonal, at least in general, we need the adjoint vectors

$$(\boldsymbol{v}_i^+ \boldsymbol{v}_j) = \delta_{ij} \qquad (12.20)$$

and decompose them in the form

$$\boldsymbol{v}_i^+ = \sum_j a_{ij} \boldsymbol{v}_j. \qquad (12.21)$$

Then the amplitudes of the individual modes are determined by

$$\xi_i(t) = (\boldsymbol{v}_i^+ \boldsymbol{w}(t)). \qquad (12.22)$$

Figure 12.11 shows the amplitudes of the set of data of HERM 1. Since we expect the order parameters to obey a differential equation of first order, we also plot the first derivative of ξ_1 versus time (Fig. 12.12). We now follow up our hypothesis that the motion is governed by a single order parameter. To do this, we establish a *phase portrait* in which we plot $\dot{\xi}_1$ versus ξ_1. Because ξ_1 is a complex quantity, we exhibit this phase portrait for the real part and the imaginary part separately. This may be considered as a generalization of the phase portraits (or phase planes) discussed in Chap. 5. Because the motion is (more or less) periodic, we expect closed phase portraits as indeed is the case for the phase portraits shown in Fig. 12.13. While the phase portraits of Fig. 12.13 correspond to the patterns shown in Fig. 12.5, those of Fig. 12.14 correspond to the data of Fig. 12.6. Provided the motion is governed by a single order parameter, we may seek the equation for that order parameter. As a first attempt, we make the hypothesis

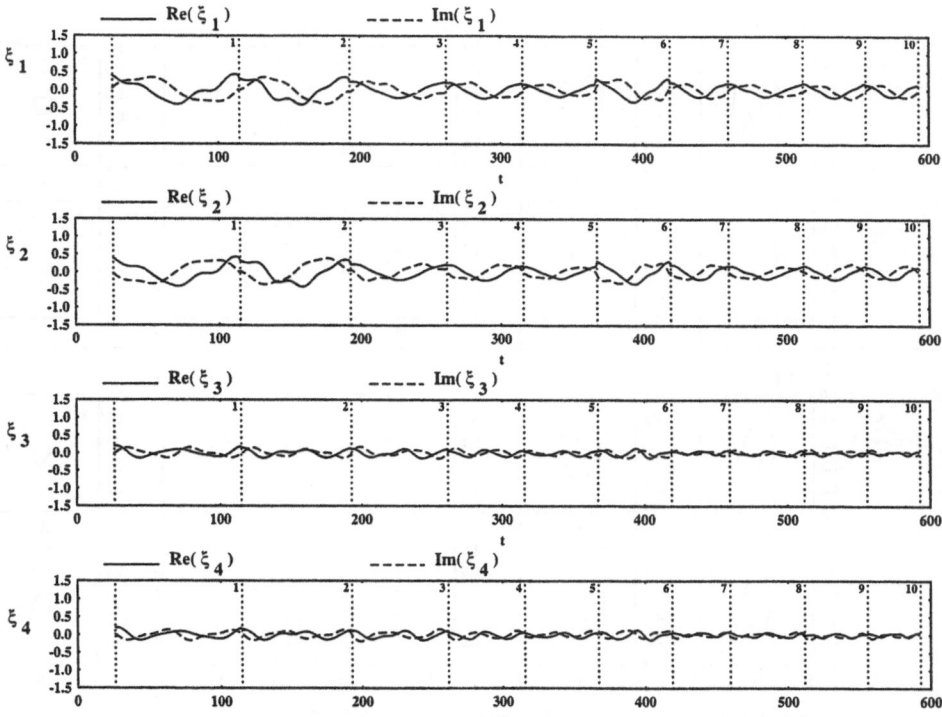

Fig. 12.11. Time dependence of the amplitudes ξ_k, $k = 1, 2, 3, 4$, for HERM 1

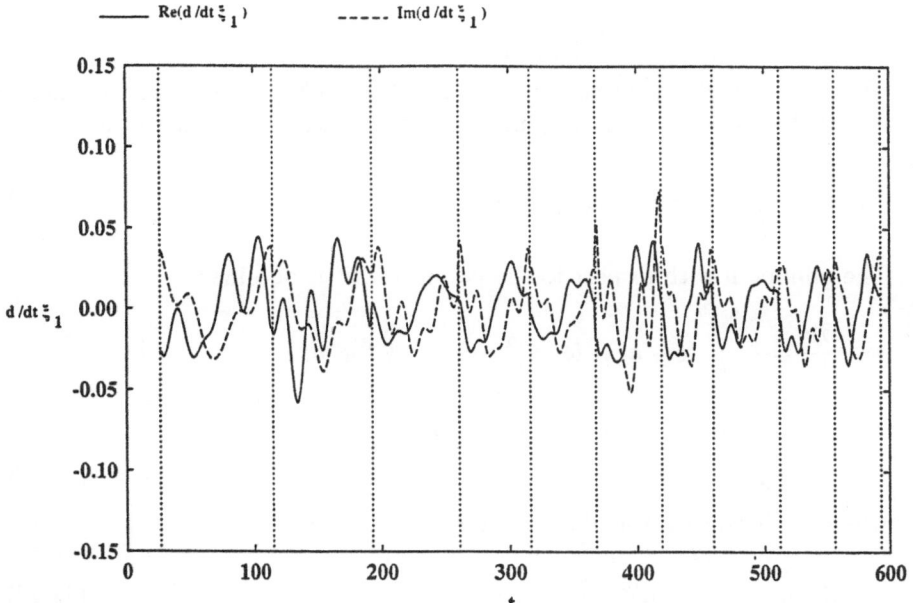

Fig. 12.12. Time evolution of the temporal derivative of the first mode

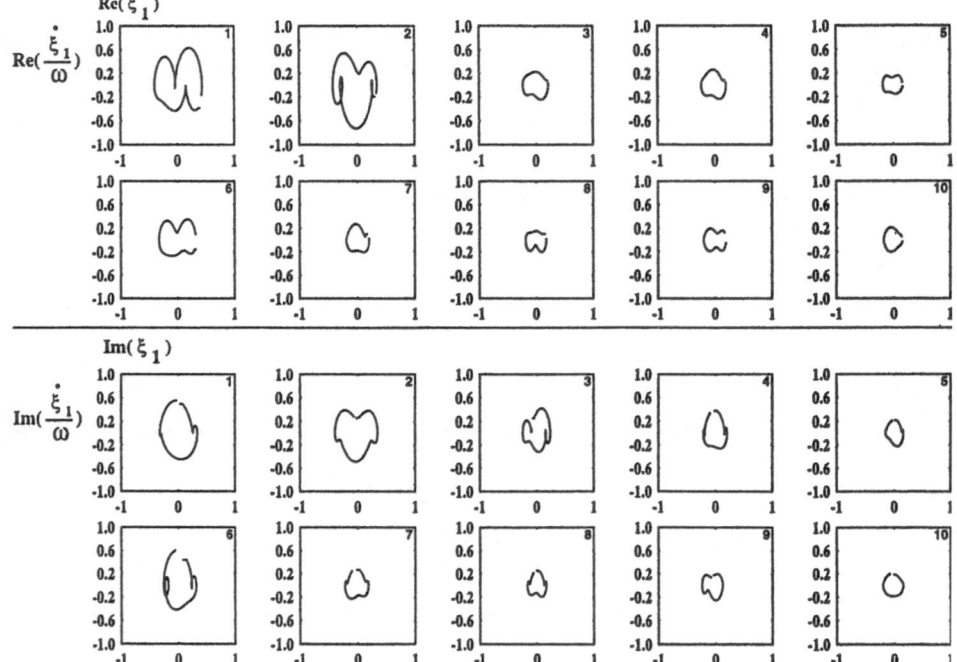

Fig. 12.13. Phase portraits for the first mode v_1. We plot $\dot\xi/\omega$ against ξ. The upper two rows show the real part, the lower two rows the imaginary part. (Data from HERM 1)

$$\dot\xi = a\xi + b\xi\,|\xi|^2 + c\xi^3 + d\xi^{*3}, \tag{12.23}$$

where the complex parameters a, b, c and d are unknown and must be determined. We determine them by minimizing the mean square deviation

$$V = \int_{t_0}^{t_0+T} \left|(\dot\xi - a\xi - b\xi\,|\xi|^2 - c\xi^3 - d\xi^{*3})\right|^2 dt. \tag{12.24}$$

The minimum with respect to the parameter a^* is obtained by

$$\frac{\partial V}{\partial a^*} = -\int_{t_0}^{t_0+T} (\dot\xi - a\xi - b\xi\,|\xi|^2 - c\xi^3 - d\xi^{*3})\xi^*\,dt$$

$$= -\langle\dot\xi\xi^*\rangle + a\langle|\xi|^2\rangle + b\langle|\xi|^4\rangle + c\langle|\xi|^2\,\xi^2\rangle$$

$$+ d\langle\xi^{*4}\rangle \overset{!}{=} 0, \tag{12.25}$$

where we have used the abbreviation

$$\langle(...)\rangle = \int_{t_0}^{t_0+T} (...)dt. \tag{12.26}$$

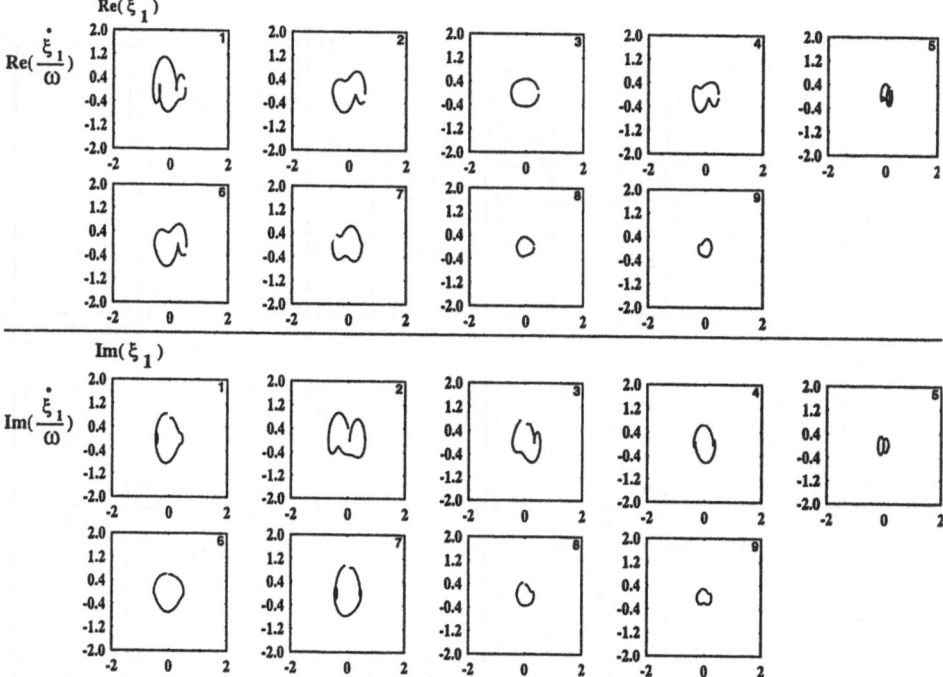

Fig. 12.14. Same as Fig. 12.13, but for HERM 2, i.e. after learning

The other equations can be found in analogy to (12.25) by using

$$\frac{\partial V}{\partial b^*} = \frac{\partial V}{\partial c^*} = \frac{\partial V}{\partial d^*} = 0. \tag{12.27}$$

This leads us to a set of linear equations

$$Ax = y, \tag{12.28}$$

where we have introduced the following abbreviations

$$x = \begin{pmatrix} a \\ b \\ c \\ d \end{pmatrix}, \tag{12.29}$$

$$y = \begin{pmatrix} \langle \dot{\xi}\xi^* \rangle \\ \langle \dot{\xi}\,|\xi|^2\,\xi^* \rangle \\ \langle \dot{\xi}\xi^{*3} \rangle \\ \langle \dot{\xi}\xi^3 \rangle \end{pmatrix}, \tag{12.30}$$

$$A = \begin{pmatrix} \langle |\xi|^2 \rangle & \langle |\xi|^4 \rangle & \langle |\xi|^2\,\xi^2 \rangle \langle \xi^{*4} \rangle \\ \langle |\xi|^4 \rangle & \langle |\xi|^6 \rangle & \langle |\xi|^4\,\xi^2 \rangle \langle |\xi|^2\,\xi^{*4} \rangle \\ \langle |\xi|^2\,\xi^{*2} \rangle \langle |\xi|^4\,\xi^{*2} \rangle \langle |\xi|^6 \rangle & \langle \xi^{*6} \rangle \\ \langle \xi^4 \rangle & \langle |\xi|^2\,\xi^{*4} \rangle \langle \xi^6 \rangle & \langle |\xi|^6 \rangle \end{pmatrix}. \tag{12.31}$$

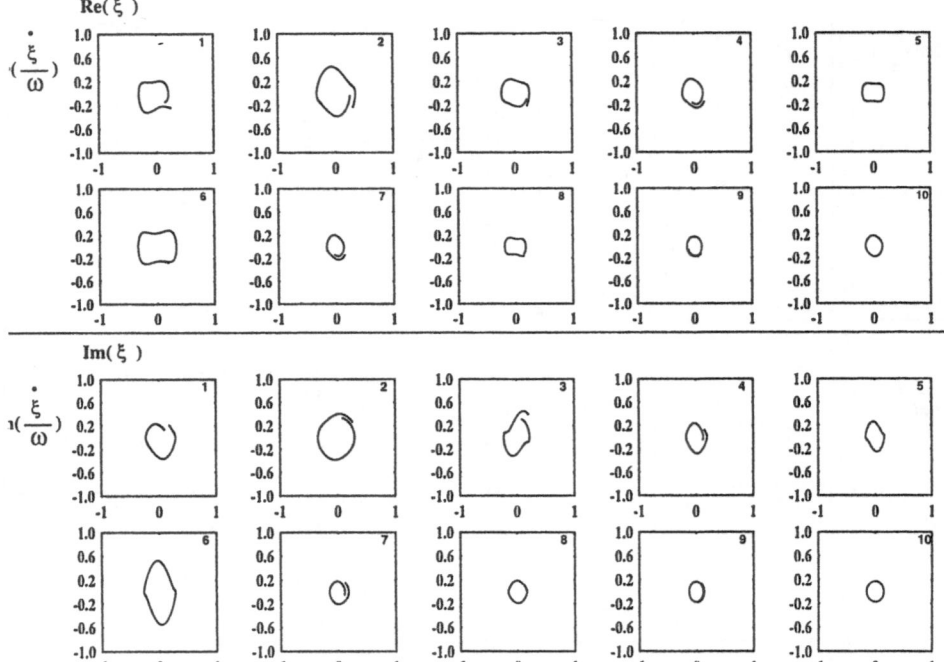

Fig. 12.15. Phase portraits computed by means of an integration of the order parameter equation (12.23) by means of the parameters that were calculated according to (12.28) with (12.29)–(12.31)

This set of equations can be easily solved for each period. By means of the resulting parameters a,b,..., we may reconstruct the time series by integrating (12.23) over a time t, the initial values being given by $\xi(t_0)$ of the measured data. The results are shown in Figs. 12.15 and 12.16 and should be compared with the experimental results shown in Figs. 12.13 and 12.14, respectively. At first sight, the agreement between the theoretical and experimental data is not too bad, but there are a number of features that deserve improvement. As a closer inspection of the experimental data of Figs. 12.13 and 12.14 reveals, especially in the learning phase we are not dealing with a smooth circle or a smooth curve, but evidently a curve that is composed of individual segments. To this end we decompose each period into four equal parts and try to reconstruct the time series by means of displaced nonlinear oscillators. The order parameter equations then read

$$\dot{\xi} = a(\xi - \xi_0) + b \mid (\xi - \xi_0) \mid^2 (\xi - \xi_0), \tag{12.32}$$

where a, b and ξ_0 are complex parameters that must be adjusted for each quarter of a period. To this end we form the potential

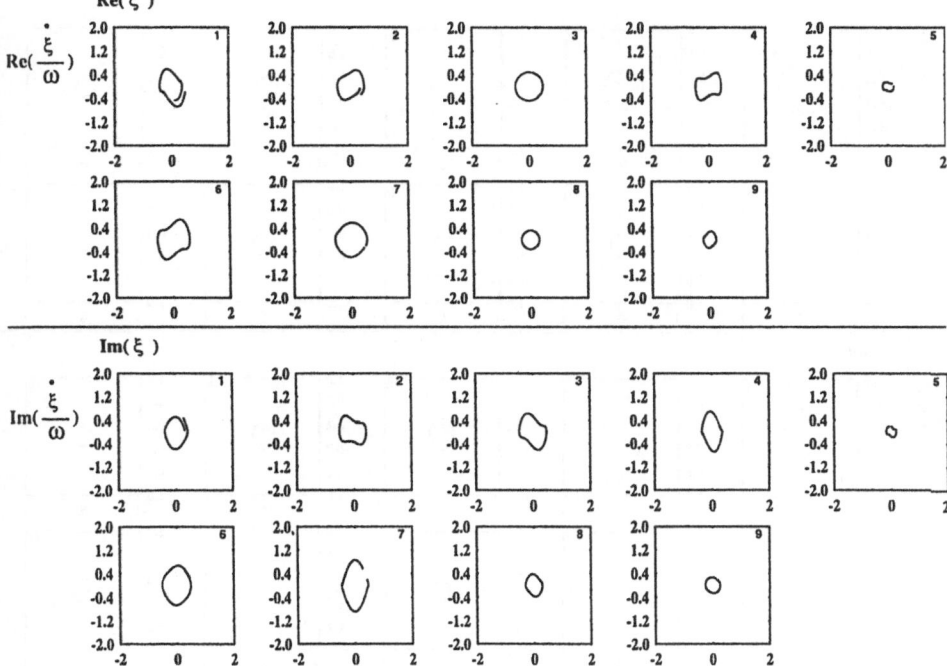

Fig. 12.16. Same as Fig. 12.15, but for HERM 2

$$V = \int_{t_0}^{t_0+T/4} \Big[\Big(\dot{\xi}^* - a^*(\xi^* - \xi_0^*) - b^* \mid (\xi - \xi_0) \mid^2 (\xi^* - \xi_0^*) \Big)$$

$$\times \Big(\dot{\xi} - a(\xi - \xi_0) - b \mid (\xi - \xi_0) \mid^2 (\xi - \xi_0) \Big) \Big] dt \qquad (12.33)$$

and minimize (12.33) by means of a gradient dynamics

$$\dot{a} = -\frac{\partial V}{\partial a^*} = \langle \dot{\xi}(\xi^* - \xi_0^*) \rangle - a \langle \mid \xi - \xi_0 \mid^2 \rangle - b \langle \mid \xi - \xi_0 \mid^4 \rangle, \qquad (12.34)$$

and similarly,

$$\dot{b} = -\frac{\partial V}{\partial b^*} \qquad (12.35)$$

$$\dot{\xi}_0 = -\frac{\partial V}{\partial \xi_0^*}, \qquad (12.36)$$

where we have omitted the somewhat lengthy expressions for $\partial V/\partial b^*$ and $\partial V/\partial \xi_0^*$. The improved data are shown in Fig. 12.17 and may be compared with the experimental data (Fig. 12.13). Concerning both the size and the shape of the theoretical phase portraits, the agreement with the experimental data is now quite good, especially in later phases of learning. The deviations

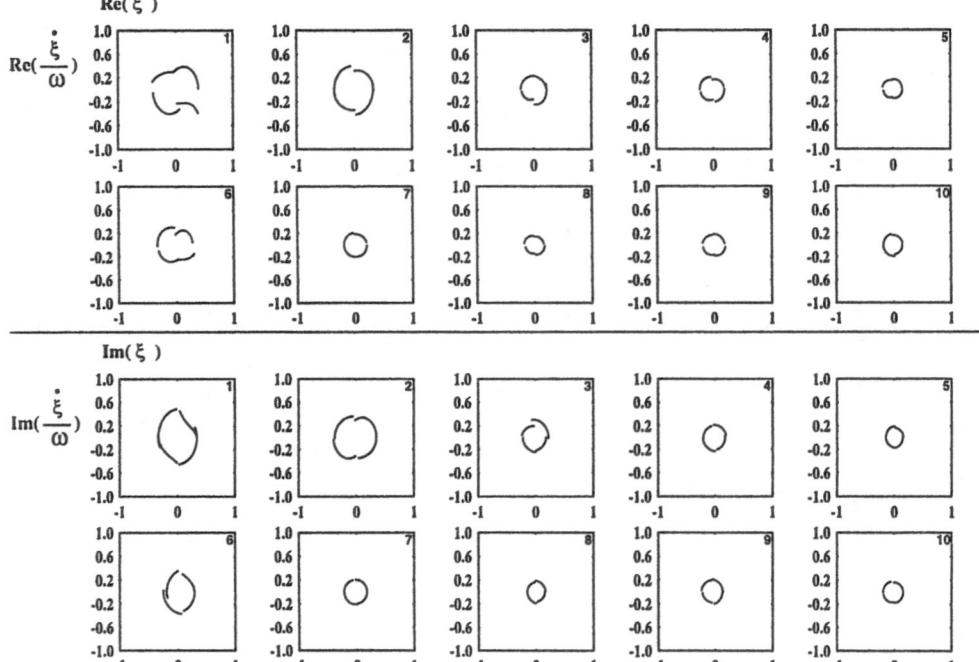

Fig. 12.17. Phase portraits resulting from an integration of the order parameter equation (12.32) over periods T/4. (Data from HERM 1)

are attributed to the possibility that the movement is governed by more than one order parameter. On the other hand, the good agreement in a number of cases is a strong indication that here the movement is in fact governed by a single order parameter. We believe that this result is indeed remarkable. It shows that a complex coordination task is accomplished by humans in that the coordination is governed by a single order parameter, but for each quarter of a period the parameters of the corresponding order parameter equation are set anew by the subject. While Fig. 12.17 represents the results for a subject whose performance was fairly good, it is perhaps still more elucidating to present the results for a subject whose performance was pretty bad and had only scattered episodes of somewhat better performance. Figure 12.18 shows the phase portrait of the first mode. Quite clearly, besides good performances, such as in episodes 4 and 5, there were irregular performances, for instance during the periods 3, 7 and 8. While in these cases the good periods could be reproduced to an excellent degree, the periods of bad performance could not be covered at all by the single order parameter hypothesis. This supports our conclusion that in bad performances, during the learning period, a number of order parameters are involved. As we shall see later in our general discussion, this new setting of control parameters seems to be a general feature of the

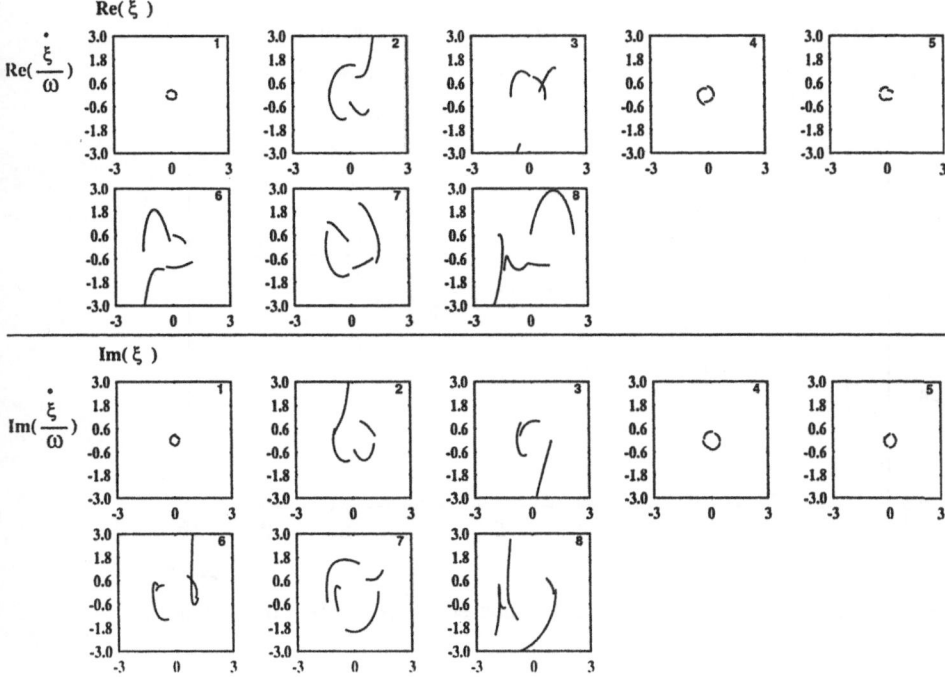

Fig. 12.18. Example of irregular phase portraits. (Data from HANN 4)

self-organization of movement coordination. Quite obviously, both the experimental set-up and the theoretical analysis can be further refined, but I think these experiments and their theoretical modeling are of central importance to the general issue of our book.

It is interesting to note that (12.23) and in particular (12.33) are closely related to the Van der Pol oscillator equation in the rotating wave approximation. This becomes particularly evident when we introduce the new variable $\eta = \xi + \xi_0$ into (12.32). Then the equation for η is identical with the Van der Pol equation in the rotating wave approximation. This is remarkable because in the finger movement studies, the finger movement could also be well approximated by that equation (though an additional term in the form of the Rayleigh oscillator was necessary). All in all we recognize that this type of equation seems to play a fundamental role, not only in finger movements but also in far more complicated coordination tasks.

The results of this chapter lead us to the insight that even a rather complicated coordination task is solved by humans in that the whole movement is governed by a single order parameter. The coefficients of the corresponding order parameter equations are, in a way, fixed by the individuals. We might suspect that skilled drivers have found a unique recipe for how to fix these parameters. Surprisingly, this is not the case. Figure 12.19 shows the time

Re(a)——   Im(a)---   Re(b)······   Im(b)---   Re(ξ)-----   Im(ξ)---

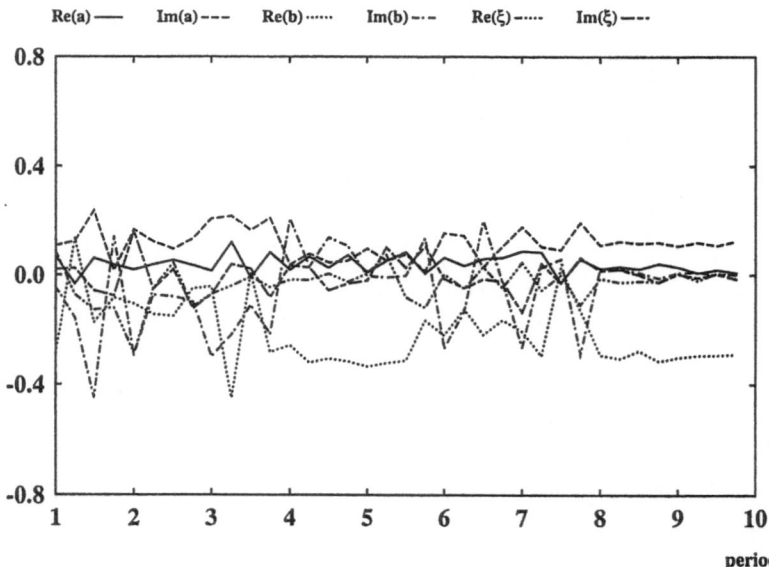

Fig. 12.19. Temporal evolution of parameters a, b, ξ_0. (Data from HERM 2)

Re(a)——   Im(a)---   Re(b)······   Im(b)---   Re(ξ)-----   Im(ξ)---

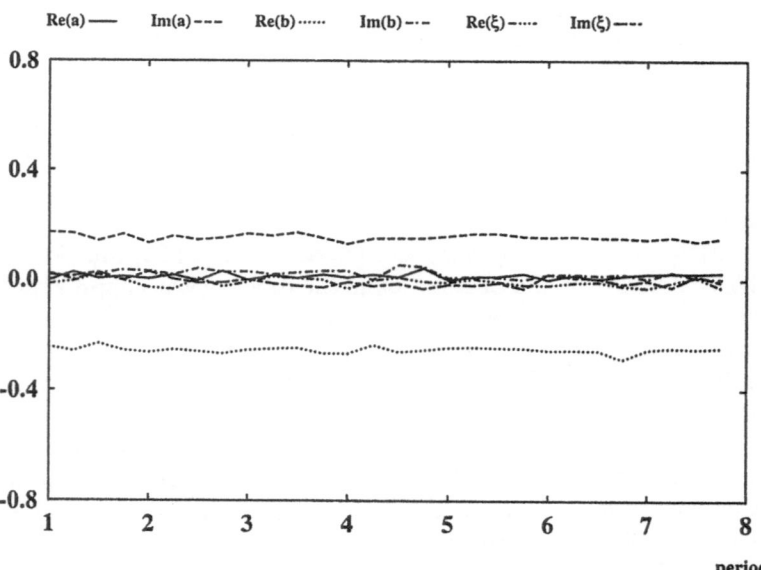

Fig. 12.20. Same as Fig. 12.19, but for data from MARK 3

evolution of the coefficients of HERM 2, i.e., after learning, while Fig. 12.20 represents the corresponding results for a skilled driver (MARK 3). While HERM 2 shows strong fluctuations when fixing the parameters, MARK 3 is absolutely steady. At present we may draw two different conclusions: Either good performance does not require a unique parameter set, or MARK 3 is a still better driver than HERM 2. This latter interpretation is supported by the phase portraits (not shown here) of MARK 3: They are uninterrupted circles!

12.7 Concluding Remarks on Part II

In Chap. 12 a specific learning task was analysed, leading us to the conclusion that after learning the movement is governed by a single complex order parameter. The same may be expected of any other periodic movement, such as walking, running, riding a bicycle, and so on. This strong coordination, or enslavement by a single order parameter, may also include breathing and, occasionally, the heart beat.

At the beginning of Part II the paradigmatic nature of movement coordination with respect to the general problem of behavior was stressed. Though an enormous amount of research work lies ahead for behavioral science in the light of synergetics, I think it is worthwhile – to say the least – to study behavioral patterns with respect to their control and order parameters. The existence and perhaps the nature of corresponding order parameters, even of complex behavioral patterns, is also suggested by our daily language, for example words such as "joy" or "anger", etc. These words indicate special states of humans with highly correlated external and internal symptoms such as facial expressions, blood pressure, and so on. The central question will be to cast these concepts into a scientific form that includes procedures for quantifying them.

Part III **EEG and MEG**

13. Chaos, Chaos, Chaos

The repetition of the word *chaos* in the title is to indicate that this word is used with quite different meanings. The famous artist *Escher* once drew a picture with the title *Order and Chaos*. In the middle of that picture we may see a beautiful regular crystal that is surrounded by all kinds of trash, such as broken bottles, opened cans, and so on. The interpretation of the words *order* and *chaos* is obvious. According to this interpretation, we are dealing with states that do not change in the course of time. In science, especially physics and mathematics, the word chaos has a quite different meaning; it is connected with changes. But here again we have to distinguish between two different kinds of chaos: the microscopic and the macroscopic or deterministic chaos. We came across an example of microscopic chaos when dealing with the emission of light by lamps. The light field consists of many individual waves that are emitted quite irregularly by the atoms. Another example of microscopic chaos is provided by the motion of the molecules of a gas (Fig. 13.1). They collide again and again and change their directions and velocities of motion at every instant in an entirely irregular fashion. Thus microscopic chaos stems from the irregular motion or behavior of very many individual parts. By contrast, macroscopic chaos originates from the irregular behavior of very few degrees of freedom or variables. For instance, the motion of a pendulum can be described by two variables, namely its displacement and velocity. When we couple the pendulum to a rotating motor by means of a spring, it is as if we are adding a third variable. If the pendulum is driven strongly enough, it may show an entirely irregular motion that cannot be described as motion on a limit cycle or torus (cf. Sect. 5.3). In fact,

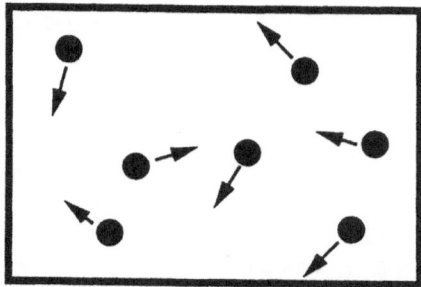

Fig. 13.1. Illustration of microscopic chaos. Gas atoms move in random directions with random velocities

quite a new kind of motion occurs that is called *deterministic chaos*. The word deterministic is added, because the laws governing the motion of the pendulum are those of deterministic Newtonian mechanics.

Historically, deterministic chaos was first studied by the famous French mathematicien *Poincaré* around the turn of the 20th century. But the properties of chaos were so complicated that none of his successors continued to work on this problem. Only in 1963 by means of computer solutions of simplified hydrodynamic equations did the meteorologist *Lorenz* rediscover chaotic motion. The equations he treated are astonishingly simple

$$\dot{x} = \sigma(y - z), \tag{13.1}$$

$$\dot{y} = x(r - z) - y, \tag{13.2}$$

$$\dot{z} = xy - bz, \tag{13.3}$$

where x, y, z are time-dependent variables and σ, r and b are constants. When he plotted his solutions in phase space, he found trajectories of the form shown in Fig. 13.2. The point (x, y, z) that characterizes the solution of the equations (13.1)–(13.3) circles around in a certain area of space for some time, then suddenly jumps to another region, where it circles around, jumps back, and so on. These jumps back and forth occur quite irregularly and depend sensitively on the position of the representative point. Here we speak of a sensitivity of the system to initial conditions. This sensitivity is quite counter-intuitive to our understanding of mechanical motions. For instance, when we drop a stone to earth and repeat this experiment with a slightly different initial position, the stone will hit the ground only a short distance from

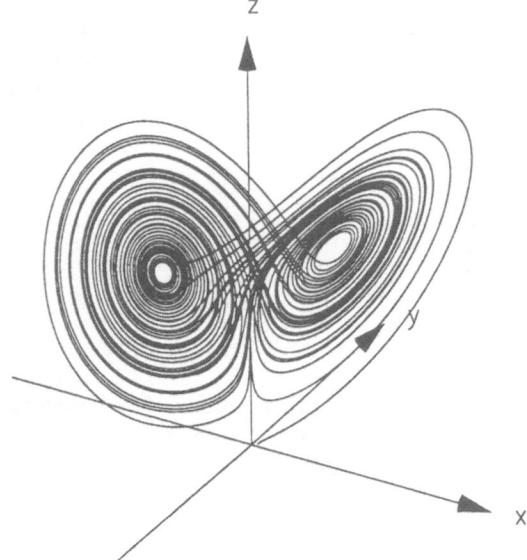

Fig. 13.2. The Lorenz attractor

where it hit the ground before. Loosely speaking, in this case a small change in the initial conditions will cause only a small deviation of the trajectory. On the other hand, there are, indeed, mechanical processes in which a sensitivity to initial conditions appears naturally. Imagine a small steel ball that we let fall on a vertical razor blade (Fig. 13.3). Quite evidently, the trajectory of the steel ball will go to the left even if the center of gravity of the steel ball falls only a tiny bit to the left of the tip of the razor blade and in the other case it will go to the right. This sensitivity to initial conditions had been used for centuries in gambling machines such as the Galton device (Fig. 13.4).

An important feature of chaotic motion is the occurrence of nonlinear terms in the equations of motion. Indeed, in (13.1)–(13.3) we observe two nonlinearities, namely $-xz$ in (13.2) and xy in (13.3). A still simpler system was found by *Rössler* (1977), and is described by the equations

$$\dot{x} = -y - z, \tag{13.4}$$

$$\dot{y} = x + ay, \tag{13.5}$$

$$\dot{z} = b + z(x - c). \tag{13.6}$$

It contains only one nonlinearity, namely xz in (13.6).

There are a few features which can be used to define the properties of a chaotic attractor. When we let the system start at some point in phase space, it will be attracted to a certain region, namely the attractor region from which it will thence never escape. When we consider the motion of two reference points on trajectories of a chaotic attractor that are initially close

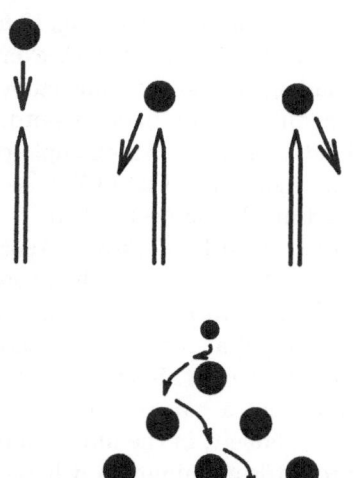

Fig. 13.3. A steel ball falling on a vertical razor blade illustrates the sensitive dependence of the trajectories on initial conditions

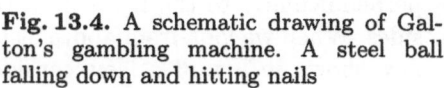

Fig. 13.4. A schematic drawing of Galton's gambling machine. A steel ball falling down and hitting nails

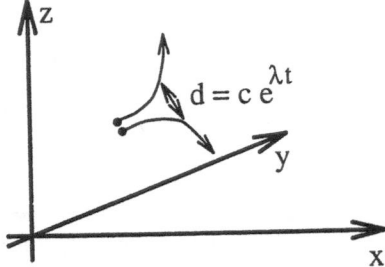

Fig. 13.5. Illustration of the exponential growth of two initially neighboring trajectories

together, the reference points will diverge quickly, their separation d growing exponentially with time t (Fig. 13.5)

$$d \propto e^{\lambda t}. \tag{13.7}$$

The constant λ appearing in the exponential function is called the *Lyapunov* exponent. A positive Lyapunov exponent indicates that we are dealing with a chaotic attractor, because it reflects the sensitivity towards initial conditions. When we let time flow, the trajectories will fill the attractor, but surprisingly enough not entirely, only to a limited degree. In order to get a quantitative measure of the degree to which the trajectories fill the attractor space, we introduce the notion of a fractal dimension. We shall come back to this concept in Appendix A.

Another important property of chaotic trajectories is their self-similarity. To explain this, let us consider a famous example already discovered by *Koch* in 1904. Let us consider the triangle of Fig. 13.6 and let us split each side into three equal parts. Then on the middle part we erect a new triangle and continue this procedure indefinitely. These steps are indicated in Fig. 13.6. When we consider the resulting Koch curve under a microscope with ever-increasing degree of resolution, we shall again and again find the same curve. The shape of the curves is independent of the magnification or, in other words, of the scaling. Such scale invariance is called *self-similarity*. An example of self-similarity with respect to chaotic attractors is shown in Fig. 13.7. When we magnify this figure as indicated, we again obtain the same structure.

An important class of chaotic motion was discovered by *Shilnikov* (1965, 1970). To explain this kind of chaos, let us consider fixed points in the phase plane. Fixed points are those points from which trajectories start or where they terminate, as we have seen in Chap. 5. There is yet another kind of fixed point at which some trajectories terminate and other start. Such a point is called a *saddle* point, because when we imagine that a ball is rolling on a saddle, it can do so in two preferential directions. Namely, in the direction of the horse back, the ball will roll on the saddle towards its minimum, whereas perpendicularly to the horse back the ball will run away to one of the two sides. When we look from above at the possible trajectories, we shall see the flow shown in Fig. 13.8. Now consider a generalization to three dimensions. Then we may have a saddle point with two unstable directions in the plane

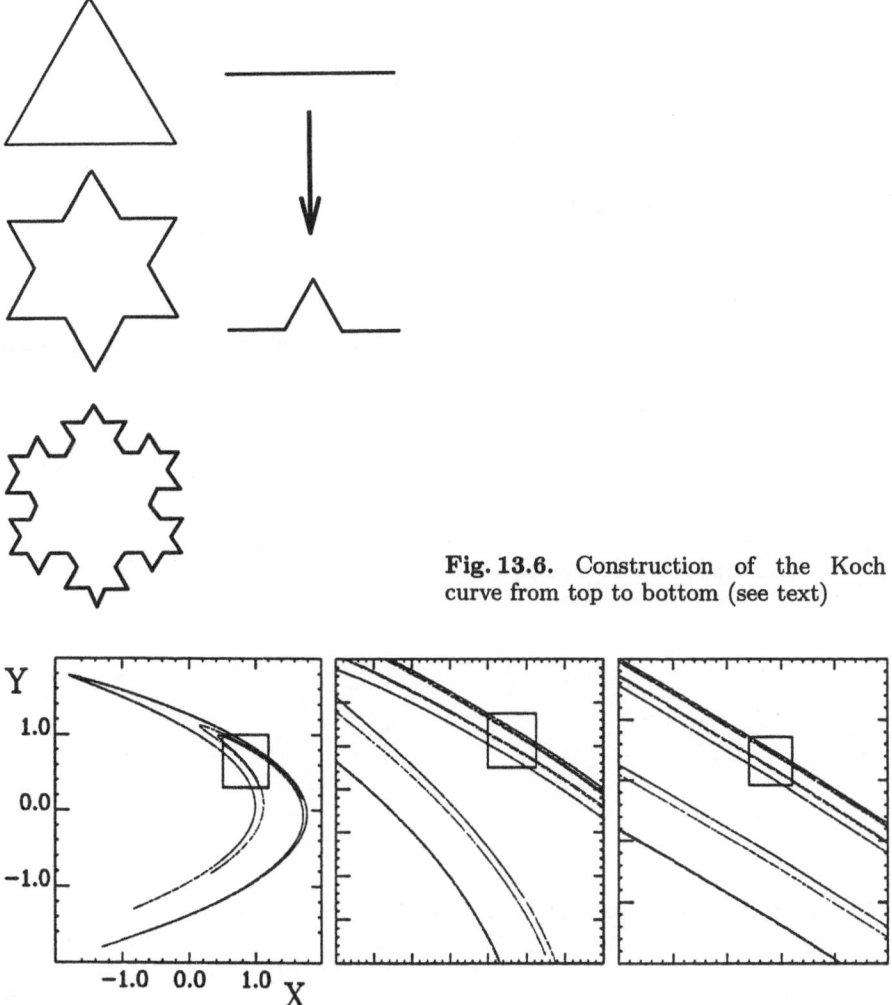

Fig. 13.6. Construction of the Koch curve from top to bottom (see text)

Fig. 13.7. Illustration of the phenomenon of self-similarity by means of the so-called Hénon–Heiles attractor. This figure should be viewed from left to right. The first square on the left shows the original trajectory, the litte square in it is then magnified to give the next square in the sequence, and so on. As may be seen, the individual squares can no longer be distinguished from each other despite their different magnifications

and perpendicular to it a stable direction along the z-axis (Fig. 13.9, right-hand side). The unstable trajectories may spiral outwards, as we know from the motion close to a focus. Thus close to this saddle point, the motion of the representative point can be described by a time dependence of the coordinates in the form

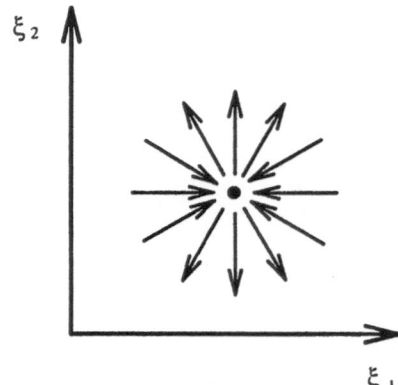

Fig. 13.8. The trajectories at a saddle point in two dimensions

$$\Gamma_0^+ \qquad\qquad\qquad \Gamma_0^-$$

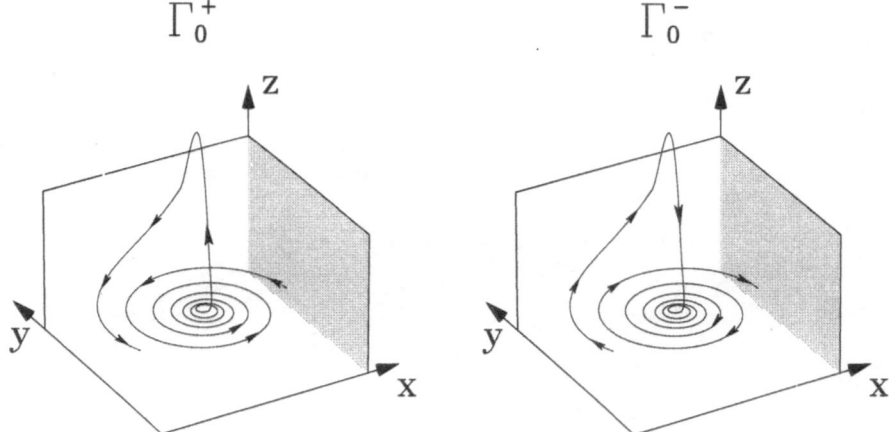

Fig. 13.9. The trajectories underlying Shilnikov chaos (see text)

$$x(t) = a\cos(\omega t)e^{\lambda_r t}, \tag{13.8}$$

$$y(t) = a\sin(\omega t)e^{\lambda_r t}, \tag{13.9}$$

$$z(t) = be^{\lambda_z t}, \tag{13.10}$$

where $\lambda_r > 0$ and $\lambda_z < 0$. What happens to the point that spirals outwards? *Shilnikov* considered the case in which its trajectory returns to the saddle point (Fig. 13.9). An orbit that runs away and then returns to the same saddle point is called a *homoclinic orbit*. *Shilnikov* was able to show that under the condition

$$\mid \lambda_z \mid > \mid \lambda_r \mid \tag{13.11}$$

there is an infinite set of unstable limit cycles around the trajectory just considered. The representative point can jump in an irregular fashion from one unstable limit cycle to another, and in this way produces chaos. When we reverse time, then we find precisely the same picture, but the arrows of the motion have to be reversed. Thus instead of outwards spiraling, we find inwards spiraling in the xy-plane, and instead of an incoming trajectory along the z-axis, we find an outgoing trajectory (Fig. 13.9, left-hand side). Thus we have two kinds of Shilnikov chaos.

But what have all these results and concepts to do with synergetics, and more specifically, with brain functions? In synergetics we are dealing with systems composed of very many individual parts so that we would expect microscopic chaos rather than deterministic chaos. But here essential findings of synergetics come in: As we have seen, through the cooperation of many individual parts, macroscopic states governed by order parameters can be generated. The number of order parameters may be very small, e.g. three or a few more. Thus the complex dynamics of the whole system is described by a few order parameters that in this way may obey exactly the equations of deterministic chaos. Synergetics explains why complex systems can show deterministic chaos. Does deterministic chaos occur in the brain? The answer is definitely yes, and we will come back to these exciting questions in the next chapter.

14. Analysis of Electroencephalograms

14.1 Goals of the Analysis

In this chapter we wish to show how the concepts and methods of synergetics make possible a new approach to the analysis of electroencephalograms (EEGs). We came across EEGs in the introductory chapter on the exploration of the brain. The microscopic origin of the electric fields that are measured in EEGs is not entirely clarified. Thus in the spirit of synergetics, we shall consider these fields as macroscopic phenomena that we wish to study by means of the concepts of this discipline.

First of all we should point out that, traditionally, a main impetus of EEG research has been to find loci of specific brain activities, such as vision, or certain kinds of thoughts. Such research is also important for finding centers of epileptic activity, because in a number of cases these centers can be removed by surgery and the disease can be cured. In accordance with this local description of activities, concepts have been developed in order to locate such centers. To visualize such an approach, let us consider several radio stations that emit radio waves which are received across a certain region. Then we may ask whether it is possible to reconstruct the location of the radio transmitters from the spatio-temporal distribution of the radio waves received. Similarly, we may try to devise methods enabling us to locate centers emitting electric fields by means of the measured spatio-temporal field distributions across the scalp. We do not want to exaggerate our criticism of this approach, but there are two difficulties at least: Quite often the determination of the location of these centers is by no means unique, and secondly it may happen that the centers seem to lie outside the brain which, of course, is a meaningless result. Therefore, other approaches will be needed and we have to discuss when the above-mentioned traditional method applies and when not.

To be more specific, let us consider epileptic seizures, because their signals are well defined. It is known that these seizures can usually be divided into two classes. In the case of partial seizures, the epileptic activity is localized at one or several epileptogenic foci. This is a case where the above-mentioned analysis is, or should be applicable. On the other hand, generalized seizures are characterized by global seizure activity involving both cerebral hemispheres. In particular, the so-called *petit–mal epilepsy* belongs to the class of generalized seizures. It is usually related to an absence which lasts a few seconds. It shows up in EEG signals as a pronounced spike wave behavior. In the following we wish to show how concepts of synergetics allow us to deal with these generalized seizures, which are, quite evidently, nonlocalized. Our basic idea will be as follows: As we know from numerous examples in synergetics, the spatio-temporal patterns that evolve in systems close to instability points may be conceived as superpositions of a few spatial patterns (or *modes*) with time-dependent amplitudes. We came across a few examples in Chaps. 4 and 10. If we take seriously the idea that our brain operates close to instability points, we can expect that EEGs can be treated as superpositions of a few fundamental spatial modes with time-dependent amplitudes.

14.2 Identification of Order Parameters and Spatial Modes

Here we show that in the analysis of epileptic seizures or of some other brain-waves, we may, indeed, determine the spatial modes and the dynamics of their time-dependent amplitudes. Quite obviously, this opens a new kind of EEG analysis, and, as we shall see later, also of MEG analysis. The experimental data on *petit-mal epileptic* seizures were provided by *Lehmann*. The maps were derived at locations based on the standard international 10 – 20 system. In the analysis by *Friedrich* and *Uhl* (1992, 1995), which we shall present in the following, two data sets referring to persons A and B were studied. Figure 14.1 shows the time-series generated by the potential difference between the individual electrodes on the scalp and a reference electrode. The data exhibit a pronounced spike-wave behavior, which in clinical applications is the symptom of *petit-mal epilepsy*. Usually there are three spike-wave cycles per second. Figure 14.2 demonstrates the evolution of the spatial pattern during one cycle. These images were reconstructed from the time-series from the invididual electrodes, as explained in Sect. 2.4. The patterns appear to be spatially coherent. In case of the data set A, the patterns consist of two regions of opposite polarity. During the spike segment, small oscillations of the patterns around a mean orientation are observed. During the time segment of the relaxation wave, the pattern undergoes a reversal of polarity. Subsequently, a second polarity change drives the system back to a state roughly similar to the initial one, and then the next cycle is initiated. In case of the data set B, the evolution of the patterns is quite different although the time-series of Fig. 14.1 exhibits similar spike-wave characteristics.

To start our analysis, we have to define an appropriate state vector $q(t)$. Since the data set consists of the time-dependent potentials $V_j(t)$ measured at the individual electrodes j this suggests identifying q with $V(t) = (V_1, ..., V_n)$. We then have to make an assumption about the number of order parameters ξ_j which allow a representation of $V(t)$ in the form

DATA SET A

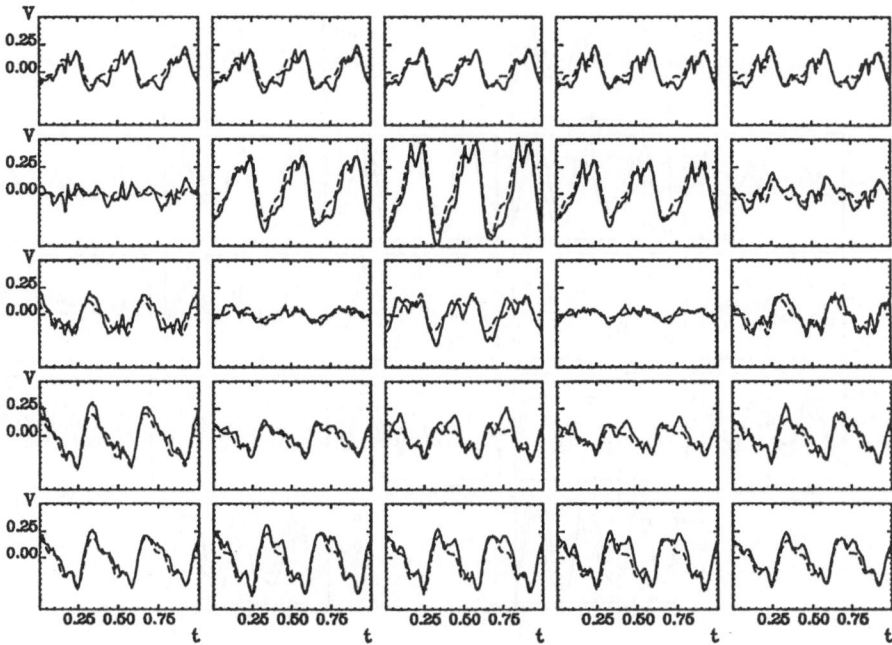

Fig. 14.1. Spike-wave behavior of the electric potential during *petit-mal epilepsy* at the various electrodes: data set A, data set B. *Solid line:* Experimental time-series. *Dashed line:* Time-series obtained by the reconstruction [Figs. 14.1–14.6 after *Friedrich, Uhl*, (1995)] (Data set B see next page)

$$V(t) = \sum_{j=1}^{m} \xi_j(t) v_j(r) + \text{higher order} \tag{14.1}$$

in the sense of Chap. 10. To this end we may proceed in two ways:

1) We may determine the fractal dimension D (Appendix A) by a time-series analysis, which yields a value of D between two and three.
2) We perform a Karhunen–Loève analysis of $V(t)$.

As it turns out, the first three modes of the Karhunen–Loève decomposition cover more than 97 percent of the total potential $V(t)$.

Both results indicate that three order parameters ξ_1, ξ_2, ξ_3 should be sufficient to represent $V(t)$, i.e.,

$$V(t) = \xi_1(t) v_1 + \xi_2(t) v_2 + \xi_3(t) v_3. \tag{14.2}$$

Our task will it be to determine the spatial modes v_j and the dynamics of the order parameters. For our analysis it will be useful to introduce the adjoint modes $v_j^+, j = 1, 2, 3$ which fulfill the orthogonality condition

DATA SET B

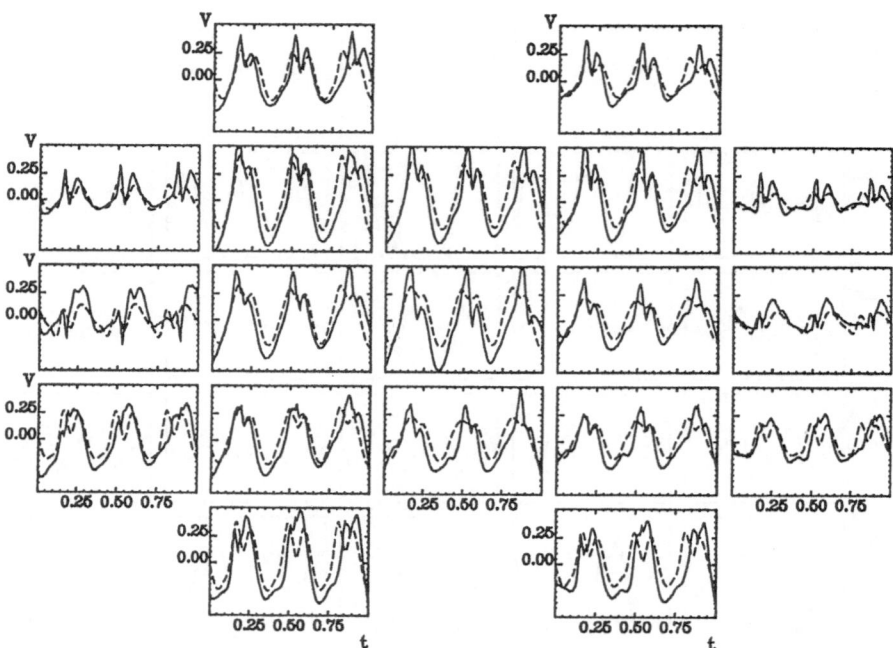

Fig. 14.1. (continued)

$$v_j^+ v_k = \delta_{jk}. \tag{14.3}$$

Multiplying (14.2) from the left by $v_j^+, j = 1, 2, 3$, and using (14.3), we immediately obtain

$$\xi_j(t) = v_j^+ V(t), \quad j = 1, 2, 3. \tag{14.4}$$

In the spirit of Sect. 11.2, we must have some idea of what the dynamical equations for the order parameters look like. To this end, in a first approximation, we identify v_j in (14.2) with the first three leading Karhunen–Loève modes and denote them by $v_j^{(0)}$. Because by construction they are orthogonal, $v_j^{(0)} v_k^{(0)} = \delta_{jk}$, the relations (14.4) can be specialized to yield

$$\xi_j^{(0)}(t) = v_j^{(0)} V(t). \tag{14.5}$$

When we plot $\xi_j^{(0)}(t)$ versus time, we find by inspection [this is the (only) place where intuition comes in!] that to a good approximation the amplitudes of the first and third Karhunen–Loève mode are closely related by a relationship of the form

$$\dot{\xi}_1^{(0)}(t) \approx \text{const.} \times \xi_3^{(0)}. \tag{14.6}$$

DATA SET A

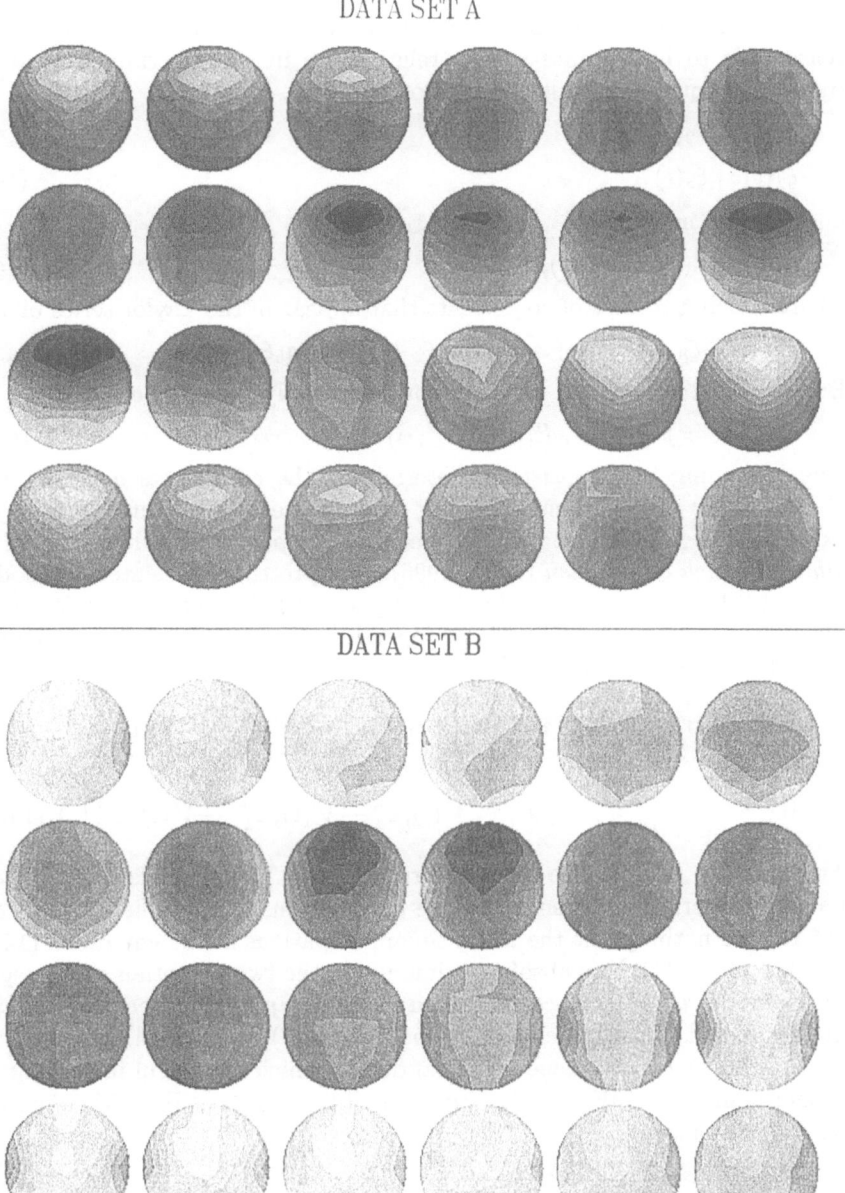

DATA SET B

Fig. 14.2. Temporal evolution of the electric field patterns during one spike-wave cycle; data set A, data set B. The patterns are obtained by interpolation of the multigrid EEG-potential $V(t)$

This relationship seems to be crucial for the pattern dynamics. It is thus reasonable to incorporate such a relationship in the determination of the dynamical equations. In generalization of (14.6), we assume

$$\dot{\xi}_1(t) = \xi_2(t), \qquad (14.7)$$

$$\dot{\xi}_2(t) = \xi_3(t) \qquad (14.8)$$

and require, in addition,

$$\dot{\xi}_3(t) = f(\xi_1, \xi_2, \xi_3, \boldsymbol{a}), \qquad (14.9)$$

where $\boldsymbol{a}$ denotes a set of coefficients that appear in the Taylor series of f,

$$f(\xi_1, \xi_2, \xi_3, \boldsymbol{a}) = a_1 + a_2\xi_1 + a_3\xi_2 + a_4\xi_3 + a_5\xi_1^2 + \dots. \qquad (14.10)$$

Because of (14.7)–(14.9), we may represent the dynamics also by

$$d^3\xi_1/dt = f\left(\xi_1, d\xi_1/dt, d^2\xi_1/dt^2, \boldsymbol{a}\right). \qquad (14.11)$$

The remaining task consists in determining the coefficients $\boldsymbol{a}$ of the Taylor expansion (14.10) simultaneously with the vectors $\boldsymbol{v}_i^+$ directly from the experimental time-series. To this end, we adopt a method developed by *Uhl, Friedrich* and *Haken* (1993, 1995). [We presented a related method in Sect. 11.2.1, cf. in particular the mean square fit following (11.66)]. We introduce the functions W_1, W_2, which depend on the adjoint modes $\boldsymbol{v}_i^+$ as well as on the coefficients $\boldsymbol{a}$:

$$W_1 = \langle \left[\boldsymbol{v}_1^+ \frac{d}{dt}\boldsymbol{V}(t) - \boldsymbol{v}_2^+ \boldsymbol{V}(t)\right]^2 \rangle + \langle \left[\boldsymbol{v}_1^+ \frac{d^2}{dt^2}\boldsymbol{V}(t) - \boldsymbol{v}_3^+ \boldsymbol{V}(t)\right]^2 \rangle, \quad (14.12)$$

$$W_2 = \langle \left[\boldsymbol{v}_1^+ \frac{d^3}{dt^3}\boldsymbol{V}(t) - f\left(\boldsymbol{v}_1^+ \boldsymbol{V}(t), \boldsymbol{v}_1^+ \frac{d}{dt}\boldsymbol{V}(t), \boldsymbol{v}_1^+ \frac{d^2}{dt^2}\boldsymbol{V}(t), \boldsymbol{a}\right)\right]^2 \rangle. (14.13)$$

The brackets $\langle .... \rangle$ denote the temporal average. The potentials W_1, W_2 can be easily interpreted when we invoke the relations (14.4). Then (14.12) and (14.13) are nothing but the mean square deviations of the equations (14.7), (14.8) and (14.11). The absolute minimum of the two potentials with respect to the modes $\boldsymbol{v}_i^+$ and the coefficients $\boldsymbol{a}$ yields an approximation to the spatial modes as well as to the temporal dynamics (14.7), (14.8), (14.11).

Now let us discuss some technical details which are useful in seeking the absolute minima of W_1, W_2. By variation of the potential W_1 with respect to $\boldsymbol{v}_2^+, \boldsymbol{v}_3^+$, we obtain

$$C_1 \boldsymbol{v}_2^+ = C_2 \boldsymbol{v}_1^+$$

$$C_1 \boldsymbol{v}_3^+ = C_3 \boldsymbol{v}_1^+. \qquad (14.14)$$

Here, the matrices C_1, C_2, C_3 are the following correlation matrices:

$$(C_1)_{ij} = \langle V_i(t)V_j(t)\rangle$$

$$(C_2)_{ij} = \langle V_i(t)\frac{d}{dt}V_j(t)\rangle$$

$$(C_3)_{ij} = \langle V_i(t)\frac{d^2}{dt^2}V_j(t)\rangle. \qquad (14.15)$$

Provided the vector v_1^+ is known, we can express v_2^+, v_3^+ by v_1^+. The potential W_2 is minimized by numerical means. It depends on the rather high-dimensional vector v_1^+. The dimensionality of this vector can be reduced by performing a Karhunen–Loève decomposition and using as $V(t)$ a representation by the first few dominant Karhunen–Loève modes. As a result, the adjoint vector v_1^+ depends on the number of KL-modes considered. In the present case the first ten KL-modes were used, which yields an accurate representation of the time signal $V(t)$. The nonlinear function f was approximated by polynomials in ξ_1 up to order four, quadratic in ξ_2 and linear in ξ_3. Since the adjoint modes v_j^+ are determined, we may calculate the time dependence of ξ_j by means of the relations (14.4) and the experimentally known time dependence of $V(t)$.

It remains to determine the modes v_j. We achieve this by a least squares fit to (14.2). This yields:

$$v_1 = \langle \xi_1(t) V(t) \rangle / \langle \xi_1(t)^2 \rangle$$
$$v_2 = \langle \xi_2(t) V(t) \rangle / \langle \xi_2(t)^2 \rangle$$
$$v_3 = \langle \xi_3(t) V(t) \rangle / \langle \xi_3(t)^2 \rangle. \tag{14.16}$$

Results are presented in the next section.

14.3 Results

Let us summarize the results. Figure 14.3 shows the adjoint modes v_j^+, (first line $j = 1$, second line $j = 3$, third line $j = 2$), together with their amplitudes calculated from the experimental time series according to (14.4). The figure demonstrates that the relations (14.7)–(14.9) are approximated sufficiently well. Figure 14.4 shows the spatial modes v_i. The corresponding time signals are obtained from a numerical integration of the dynamical system (14.7)–(14.9) using the parameter values a determined by the method described above. Appropriate initial conditions were chosen. With these solutions it is possible to calculate the complete time signal $V(t)$ according to (14.2). Figure 14.1 shows that this reconstruction is in close accordance with the experimental data.

Now let us consider the dynamical behavior generated by the order parameter dynamics. Figure 14.5 represents the trajectories in the phase space spanned by the amplitudes obtained by a numerical integration. Figure 14.6 exhibits the same kind of plot but now for the amplitudes extracted from the experimental data.

When we look more closely at the theoretical and experimental results shown in Figs. 14.5 and 14.6, respectively, we find the following behavior of the trajectories: In case A they seem to spiral away, more or less in a plane, from one point in space and then return from the third dimension. In case B the trajectories look qualitatively the same as in case A, but their direction

DATA SET A

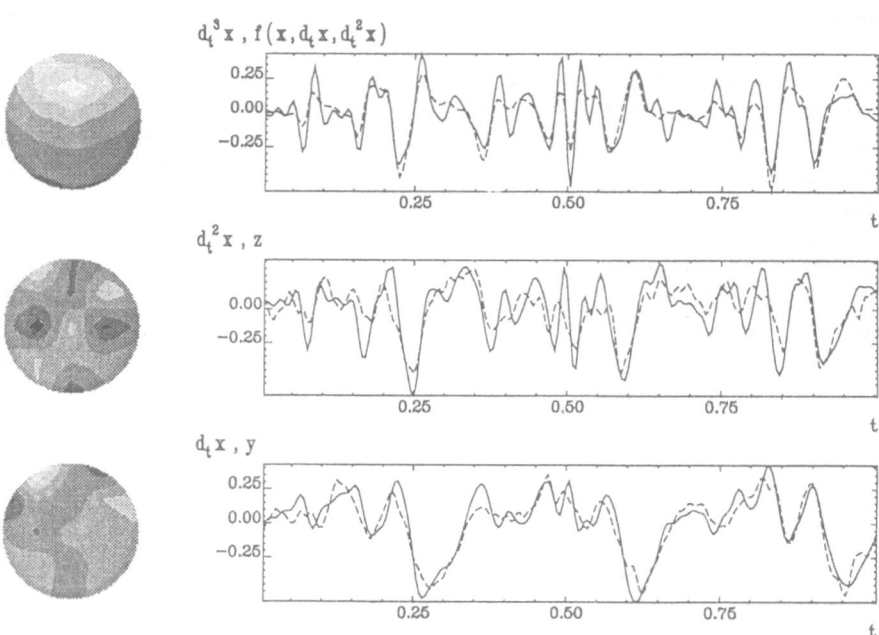

Fig. 14.3. The adjoint modes v_1^+, v_3^+, v_2^+. The time-series show the amplitudes $d_t^3 x(t), d_t^2 x(t), d_t x(t)$ (*solid lines*). The dashed lines show the quantities calculated from the time series which are related with $d_t^j x(t)$ through the order parameter equations; data set A, data set B (see next page)

is reversed. Both cases are, of course, strongly reminiscent of Shilnikov chaos, which we discussed in Chap. 13. Equations (14.7)–(14.9) allow us to study this problem in more detail. Depending on the parameters of cases A and B, we indeed find that in case A there is a saddle point from which the trajectories spiral outwards and return from the stable direction. We may calculate the linear (positive or negative) growth rates λ_r, λ_z (cf. Chap. 13) and find the condition for Shilnikov chaos fulfilled. Though the trajectories in case B look similar to those of case A, the Shilnikov condition is not fulfilled and thus case B requires a further analysis in the future.

The analysis of this chapter has revealed a number of remarkable features of brain activity. There are cases in which brain activity is governed by only few order parameters. This low dimension of the order parameter space has been found previously by *Babloyantz* (1985) in her pioneering work using a time-series analysis. As we shall explain in the Appendix (see also References to the present chapter), this analysis encounters the difficulty that experimental time-series are not stationary for long enough that reliable data, in particular precise numbers for fractal dimensions, can be obtained. The

DATA SET B

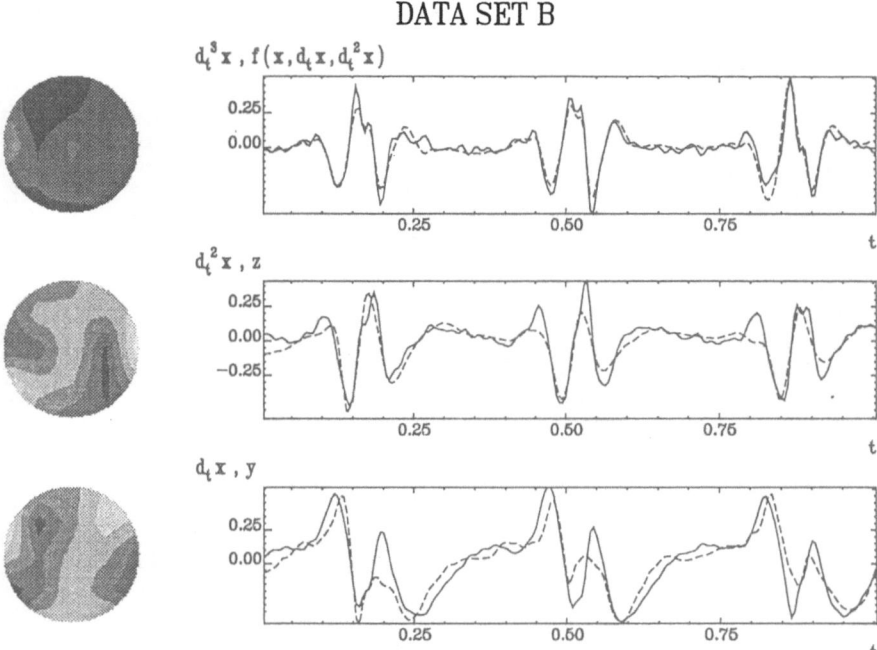

Fig. 14.3. (continued)

present analysis, based on multi-electrode experiments and a new evaluation method, leads us directly to the topology of the trajectories and even allows the determination of the kind of chaos. In the case of *petit–mal epilepsy* theoretically two kinds of epilepsy are found. It remains to be seen whether they exhibit features that show up in other clinically observable phenomena.

The analysis by *Friedrich, Fuchs* and *Haken* (1987, 1991, 1992) of multi-electrode experiments on α-waves also revealed low-dimensional dynamics. The number of order parameters was found to be five. Here again the underlying dynamics is chaotic, but the type of chaotic attractor has not yet been determined. Low-dimensional dynamics implies that huge amounts of neurons fire coherently. It will be an exciting task for the future to study the kind of brain stimulation by which this chaotic dynamics can be broken.

In conclusion, a word on the method of analysis may be in order. While the Karhunen–Loève analysis may serve as a starting point, this analysis is not sufficient to explore the dynamics more deeply. The above method based on synergetics seems to be more adequate.

DATA SET A

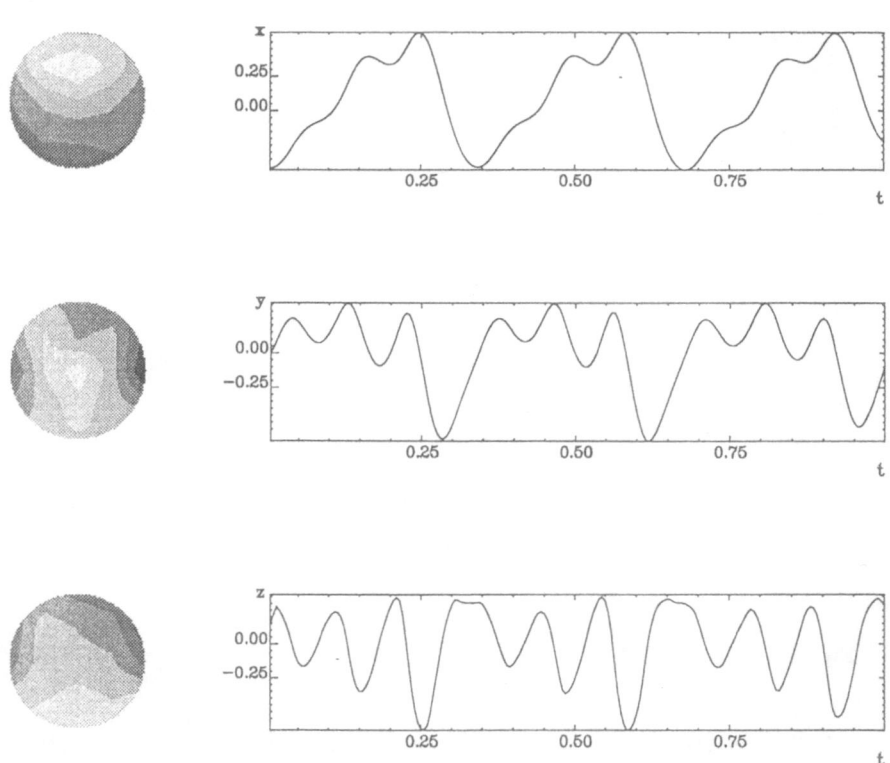

Fig. 14.4. The modes v_1, v_2, v_3. The time signals are the amplitudes of the modes obtained by a numerical integration of the order parameter equations; data set A, data set B (see next page)

DATA SET B

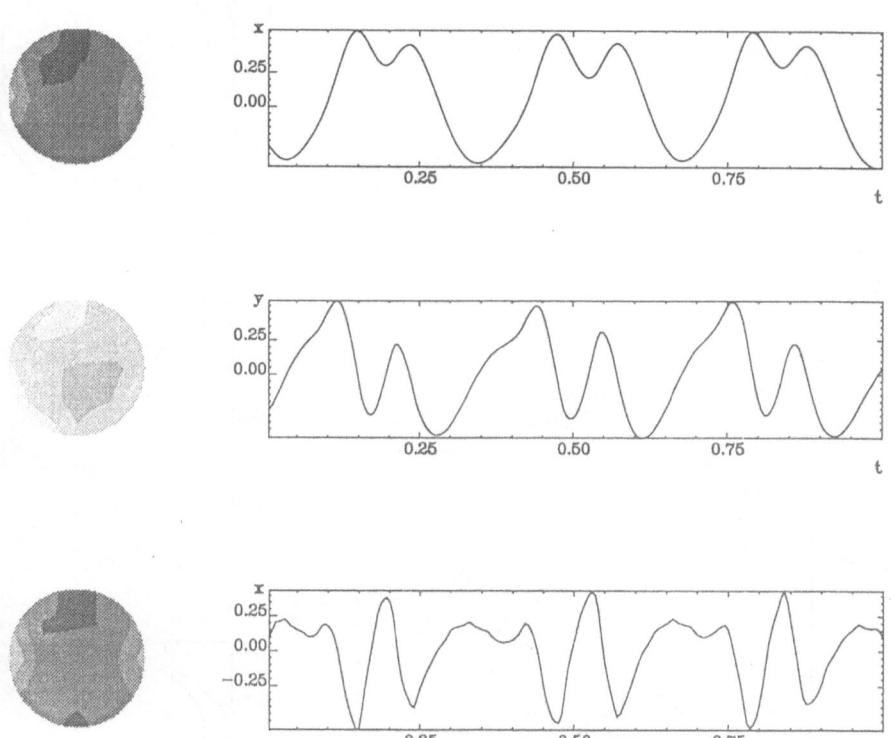

Fig. 14.4. (continued)

DATA SET A

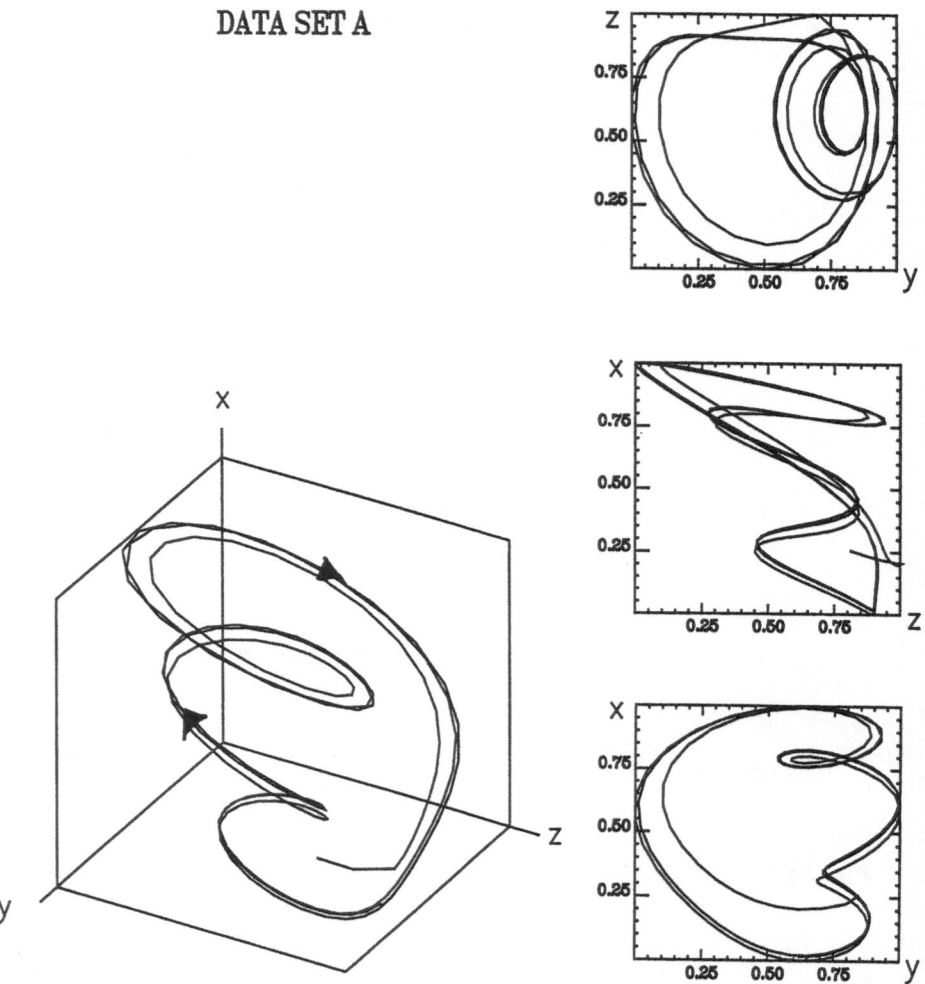

Fig. 14.5. Reconstruction of the trajectories obtained from a numerical integration of the order parameter equations in the phase space spanned by the coordinates $x, y = d_t x(t), y = d_t^2 x(t)$; data set A, data set B (see next page)

DATA SET B

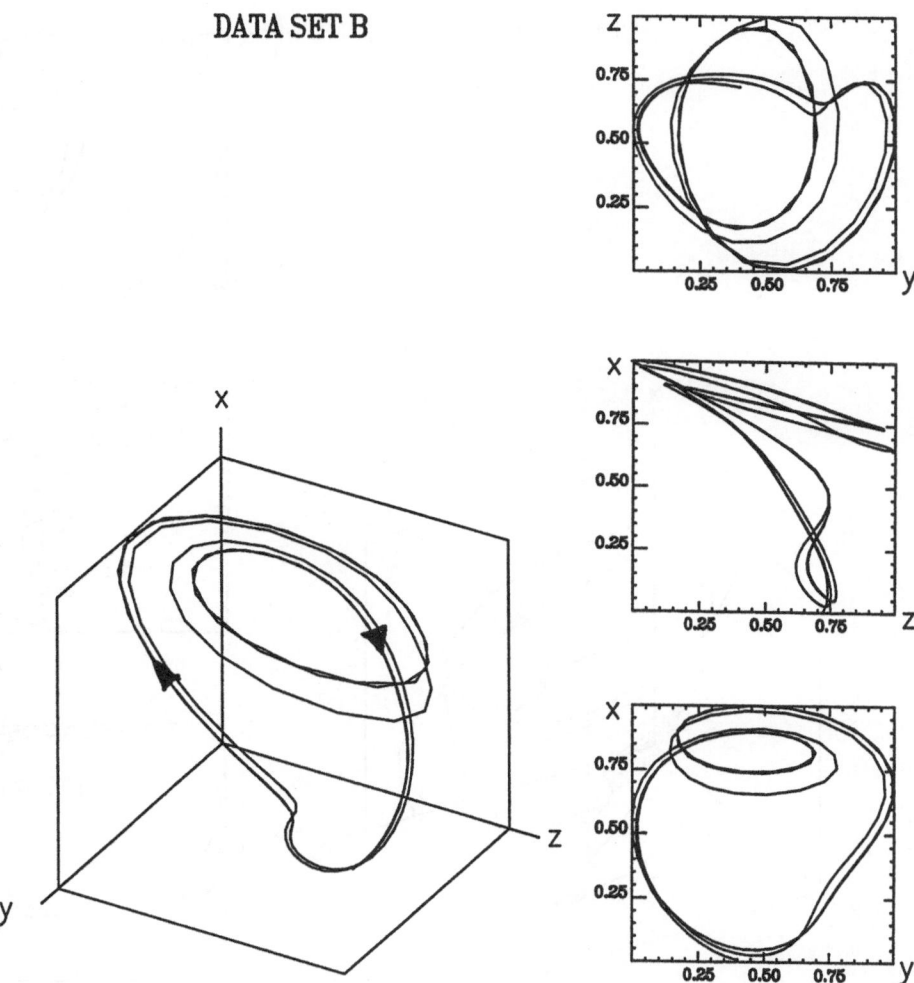

Fig. 14.5. (continued)

DATA SET A

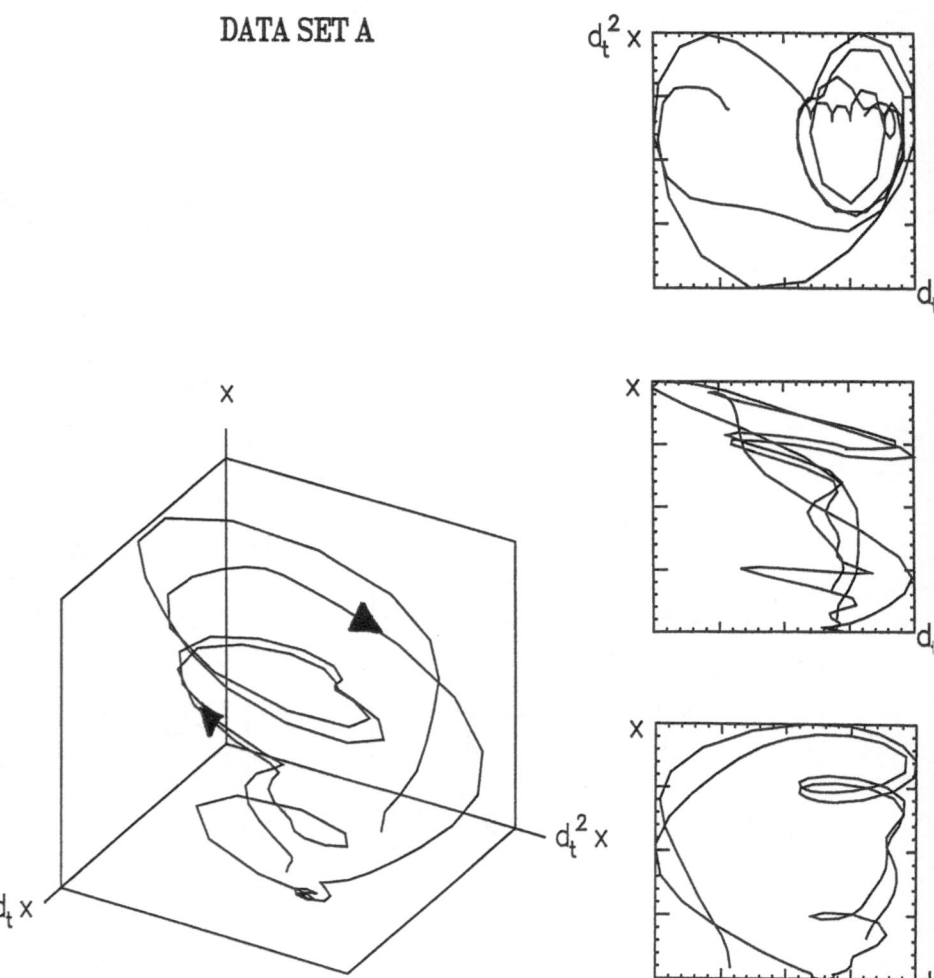

Fig. 14.6. Same as Fig. 14.5 but for the amplitude $x(t)$ determined from experimental data; data set A, data set B (see next page)

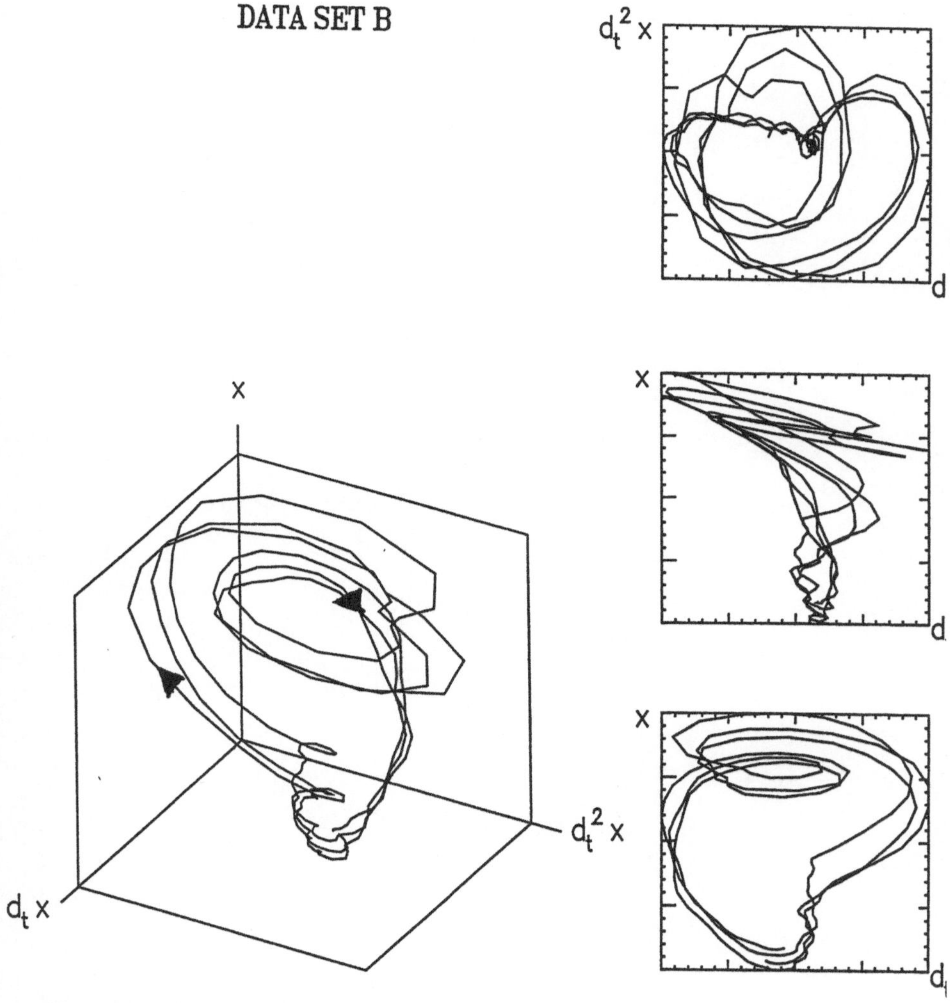

DATA SET B

Fig. 14.6. (continued)

15. Analysis of MEG Patterns

15.1 Experimental Results

The experimental and theoretical study of magnetic fields produced by the brain is currently of great interest and numerous studies are being performed. It is not our goal, however, to give a survey of all these various studies. Rather we want to pick an experiment and its theoretical analysis that is closely related to other experiments reported on in our book and also to the concepts of synergetics. We refer to MEG experiments done by *Kelso* and his co-workers (1991, 1992), in which SQUIDs, i.e. sensors of magnetic fields, were fixed on the left parietal temporal cortex so that the devices covered parts of the motor and auditory cortex.

Before we present the highly interesting results, we mention a few technical details, because they shed light on the spatial and temporal resolution. The SQUID system of Biomagnetic Technologies Inc. consisted of 37 axially symmetric first order gradiometers, each 20 mm in diameter and spaced 22 mm apart. The SQUID array had a diameter of about 13 cm and was centered 2 cm from C3 (international 10-20 system) in the posterior direction (cf. Fig. 2.7). Each squid sensor was read with a sampling frequency of 862 Hz and bandpass filtered between 0.1 and 100 Hz. The raw data from the 29 runs that were taken into account were averaged, filtered (0.3 Hz $< f < 50$ Hz) and downsampled to 8192 data points per channel leading to an effective sampling frequency of 212 Hz. All the analyses by *Fuchs, Kelso, Haken* (1992), and the models to be outlined below are based on the averaged data.

Now let us turn to the experiment by *Kelso* et al. (1991, 1992); see also *Fuchs, Kelso, Haken* (1992). An acoustic signal in the form of equidistent bleeps at a frequency of Ω was given to a subject whose task was to push a button with his finger in a way that was called *syncopic* by *Kelso*. The pushes of the button had to occur in between the acoustic stimuli. The experiment started at $\Omega = 1$ Hz and after ten stimulus cycles, this frequency was raised by 0.25 Hz and another ten cycles were performed. This procedure was continued until $\Omega = 2.25$ Hz was reached. The periods in which the stimulus was kept constant will be called *plateaus*. There were six plateaus, I–VI, altogether. The experiment showed a very interesting outcome; up to 1.75 Hz, i.e. within the plateaus I – III, the subject could react in the syncopic manner. At 1.75 Hz and above, a synchronous movement was observed. In the transition

region the finger movement showed critical fluctuations and critical slowing down. Thus a phase transition similar to the finger movement paradigm
treated in Sect. 6.2 and the following sections took place. But what happens
in the brain as measured by the MEG? The results are shown in Fig. 2.9 for
plateau II, i.e. before the phase transition, and in Fig. 2.10 for plateau V, i.e.
after the phase transition. Each of the individual boxes shows the measured
and averaged time-series stemming from the individual sensors with their
corresponding positions on the scalp. In the left lower corner the stimulus
(upper part) and the response (lower part) are shown. At first sight, these
results look rather confusing, but we shall see that they contain remarkable
information. In the next section we shall analyse them and in Sects. 15.3 and
15.4 we shall model them. While the brain signals oscillate predominantly
with Ω in the plateaus I – III, in the plateaus IV – VI they oscillate predominantly with 2Ω. When one Fourier analyses the brain signal, the first
component oscillates with Ω, but a phase-jump occurs with π. The phase
difference between the first component of the brain signal and the acoustic
signal remains constant before the transition undergoes a phase-jump at the
transition and then remains again constant afterwards. In addition, critical
fluctuations and critical slowing down of the MEG are observed.

15.2 Temporal and Spatial Analysis

15.2.1 Temporal Analysis

In this section I shall essentially follow an analysis by *Fuchs, Kelso, Haken*
(1992). A useful method for the study of time-series is, of course, the
Fourier analysis. Since the stimulus is periodic with a time that we call
T_c (c = "cycle"), the Fourier decomposition of a signal $g(t)$ reads $g(t) =
\sum_n c_n \exp(i2\pi nt/T_c)$. c_n is the Fourier coefficient connected with the fre-
quency $\nu_n = n/T_c$ so that we put $c_n = \hat{g}(\nu_n)$. According to Fourier's theorem,
$\hat{g}(\nu_n)$ is given by

$$\hat{g}(\nu_n) = \frac{1}{T_c} \int_0^{T_c} g(t) \exp(-i2\pi \nu_n t) dt. \tag{15.1}$$

Because the sampling takes place at N discrete values of time $t \equiv t_\ell =
\ell/\nu_s, \ell = 0, ..., N - 1$, where ν_s is the sampling frequency, we have to replace
the integral by a sum:

$$\hat{g}(\nu_n) = \sum_{\ell=0}^{N-1} g(t_\ell) \exp(-i2\pi \nu_n t_\ell). \tag{15.2}$$

The first interesting question we may ask is whether the transition from the
syncopated to the synchronized finger movement is reflected by the MEG

data. To this end we compare the difference of the phase inherent in the MEG signals and the phase inherent in the acoustic signals. We define the corresponding phase ϕ by means of the Fourier coefficients of the fundamental frequency, ν_1 :

$$g_s(\nu_1) = |g_s(\nu_1)| \exp(i\phi_s), \tag{15.3}$$

$$g_r(\nu_1) = |g_r(\nu_1)| \exp(i\phi_r), \tag{15.4}$$

where the indices s and r refer to (acoustic) signal and (magnetic or finger) response, respectively. The resulting phase differences $\phi_r - \phi_s$ are plotted in Fig. 15.1. To explain this figure, let us start with the lower left corner, which shows the phase difference between the finger response and the acoustic signal (ordinate). The dotted horizontal lines have a separation of π. Along the abscissa the time intervals, each over two cycles (over which was actually Fourier–analysed to obtain less noisier results) are plotted and the plateaus are marked by small vertical lines.

Now let us turn to the other boxes in that figure. Each of them corresponds to a square of Fig. 2.9 and the corresponding results for the other plateaus I, III–VI. In complete analogy to the lower left corner of Fig. 15.1, in each box

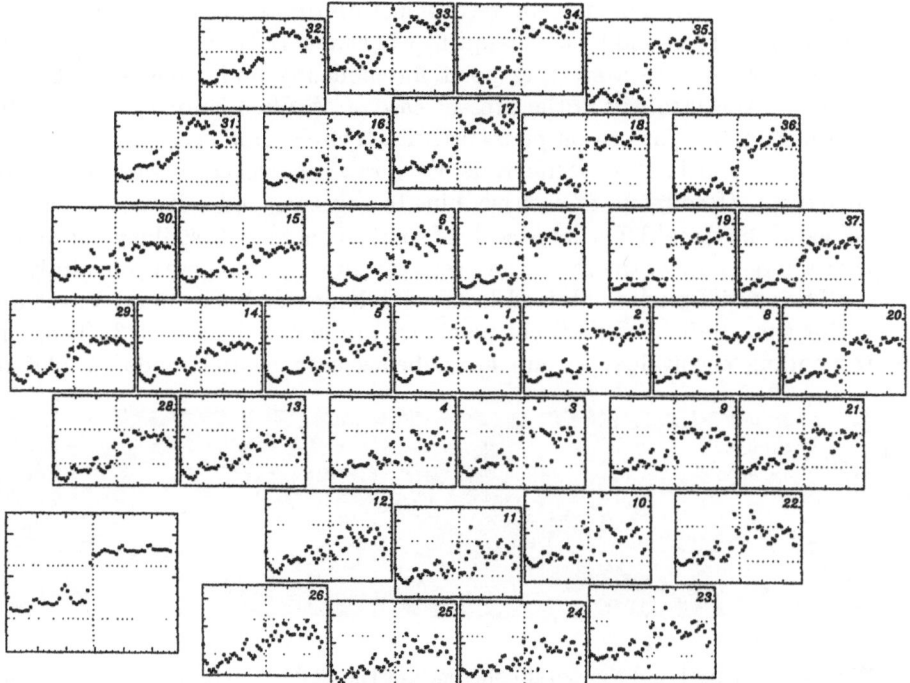

Fig. 15.1. Phase difference between the brain (SQUID area) and manual response signals (*lower left*), relative, in each case to the stimulus. The data are averaged over two cycles. (For details compare text.) (After *Fuchs, Kelso, Haken* (1992))

the phase difference of the magnetic response of that box and the acoustic signal is shown. The abscissa is the same as in the lower left box. Phase jumps are again visible, but depending on the position of the sensors, the jumps are more or less pronounced. This different kind of behavior is probably due to the fact that the positions belong partly to the motor cortex and partly to the sensory cortex, but still awaits further analysis.

Another important kind of information is in the power spectrum, which is given by the Fourier coefficients $\hat{g}(\nu_n)$ by means of $\mid \hat{g}(\nu_n) \mid^2$. Without presenting the corresponding plots, we mention the most salient features:

In plateau II, i.e. before the transition, in almost all channnels (i.e. boxes) the power is concentrated at the frequency ν_1 of the signal. On plateau V after the transition, in many of the channels at the left side of the array the largest component corresponds to the first higher harmonic, i.e. at a frequency twice the fundamental frequency, ν_1. There are some subtleties that we shall not discuss here, such as the occurrence of dominant triple frequencies and so-called subharmonics, i.e. at frequencies $\nu_1/2$.

15.2.2 Spatio-temporal Analysis

As we have seen in Chap. 14, in the case of EEGs, we may gain considerable insight into brain dynamics by reconstructing the spatio-temporal patterns (cf. Figs. 2.6, 14.2) and by a subsequent Karhunen–Loève decomposition. The latter may serve as a starting point for modeling brain dynamics. Let us follow up this strategy in the present case of MEGs. Figure 15.2 shows the spatio-temporal pattern on plateau II. Each circle shows the distribution of the magnetic field at a specific time. Each row is a cycle down-sampled to 14 time steps. (Red refers to positive, blue to negative, and yellow to vanishing values of the signal.) The large scale of spatial coherence is striking and shows that a large part of the motor and sensory cortex participates coherently.

The spatial modes obtained from a Karhunen–Loève expansion over the entire time series (over all six plateaus) are shown in Fig. 15.3. The corresponding eigenvalues λ_k are given below each spatial mode. A plot of

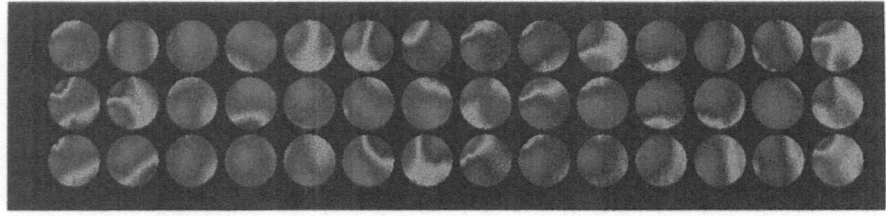

Fig. 15.2. Spatial and temporal coherence of the SQUID data on plateau II. The patterns show the distribution of the magnetic field at a specific time. Each row is a cycle down sampled to 14 time steps. The signal is almost periodic with a positive amplitude shortly after the stimulus and a mainly negative amplitude appearing after the response. (Section of a figure of *Fuchs, Kelso, Haken* (1992))

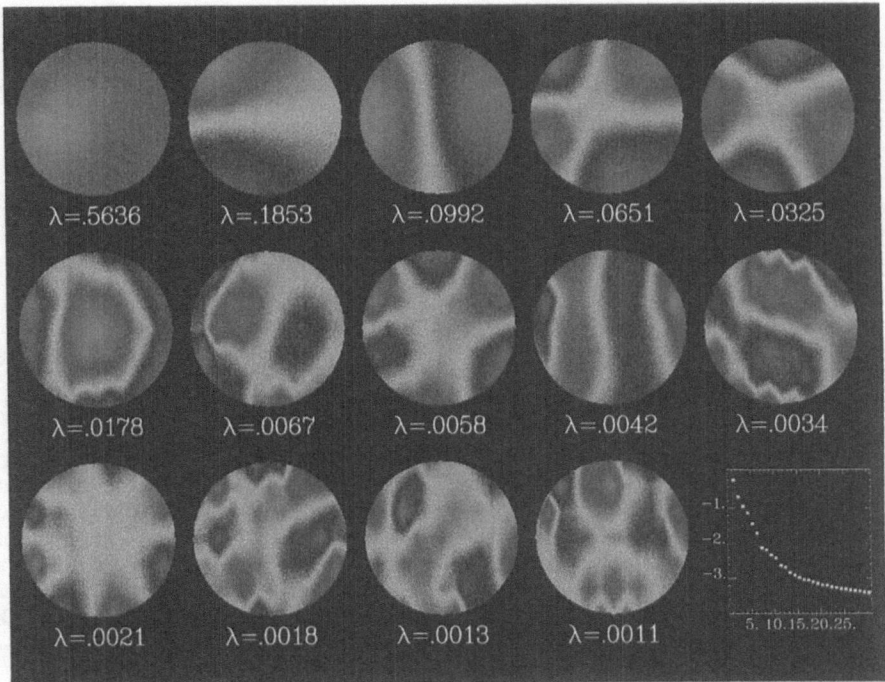

Fig. 15.3. The spatial Karhunen–Loève modes over the entire time series. λ is the eigenvalue of the Karhunen–Loève equations and describes the contribution of the corresponding mode to the entire signal. In the lower right corner a plot of $\log \lambda_k$ versus k, is shown. After *Fuchs, Kelso, Haken* (1992)

$\log \lambda_k, k = 1, 2...$ is shown in the lower right corner. One of the most striking features is the symmetry exhibited by the KL-modes. Analogous behavior had been found previously in the analysis of EEG patterns (*Friedrich, Fuchs, Haken* (1992)). The solutions of the equation that describes standing waves in circular geometry (a drum!) or spherical geometry (oscillations of the surface of the sun) exhibit the same patterns! This is shown in Fig. 15.4. What is the reason for this striking similarity? Does it stem from symmetries of the brain (the brain as sphere?), of the circular arrangement of the SQUIDs, or from intrinsic properties of the brain? These questions are not easy to answer. A clue to a possible answer lies in the size of the eigenvalues λ_k that are a measure of the amount by which a mode $v_k \equiv v_k(x)$, cf. (11.26), is contained in the total spatio-temporal pattern (averaged over time). As we see, λ_k decreases with increasing number of nodal lines of $v_k(x)$, i.e. lines where $v_k(x)$ vanishes (number of yellow lines in Figs. 15.3,4). As is known from the solutions of the wave equations, an increasing number of nodal lines indicates a decrease of wavelengths. Thus the Karhunen–Loève expansion shows that shorter and shorter wavelengths of brain activity are less and less important.

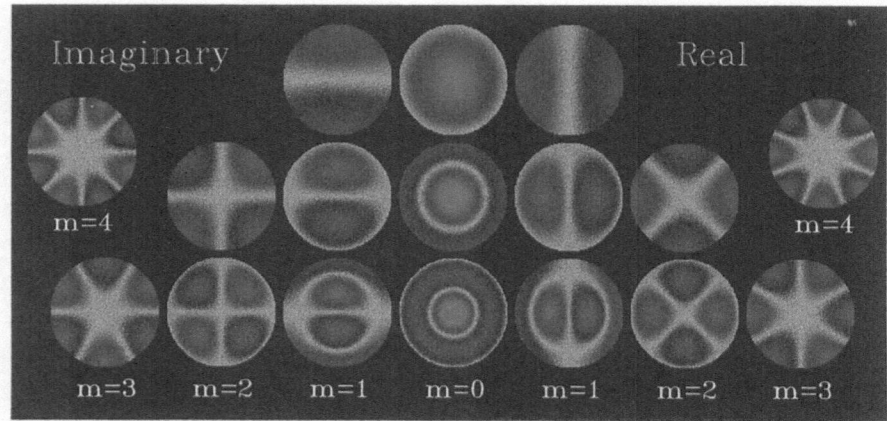

Fig. 15.4. Classification of the spatial modes with respect to spherical symmetry. The upper half shows the real (going from the middle column to the right) and imaginary (*left side*) part of spherical harmonics in a projection of the upper part of the sphere to an equatorial plane. The rows corresponds to $\ell = 1, \ell = 2, \ell = 3$ from top to bottom. Additionally the function for $\ell = 4, m = 4$ is shown. The patterns surrounded by a white circle approximately coincide with those of the Karhunen-Loève modes shown in Fig. 15.3. The 9th mode of Fig. 15.4 is a superposition of the functions $\ell = 2, m = 1$ and $\ell = 3, m = 3$. After *Fuchs, Kelso, Haken* (1992)

If we are asked to decompose a spatial function into a superposition of modes containing smaller and smaller frequencies, where the modes are compatible with the circular (or spherical) symmetry and are orthogonal on each other, we are automatically led to the modes shown in Fig. 15.4. As a more detailed analysis reveals, a classification for spherical geometry is still preferable over circular geometry.

Thus, all three reasons contribute:

1) brain dynamics – longer wavelengths prevail
2) circular geometry of array
3) spherical geometry of brain

A study of the temporal dynamics of the spatial modes was, in the case of EEG analysis (Chap. 14), an important starting point for model building. Therefore, we write the spatio-temporal signal $q(\boldsymbol{x}, t)$ in the form

$$q(\boldsymbol{x}, t) = \sum_k \xi_k(t) v_k(\boldsymbol{x}). \tag{15.5}$$

Knowing q and v_k, we can determine the time-dependent amplitudes $\xi_k(t)$. Because of the complexity of the problem, it is wise to be modest and to study the time dependence of the first two leading modes only. This is shown in Fig. 15.5, lower and upper part. Let us consider this figure in more detail. The upper right corner shows the leading mode in its spatial dependence. The first row on the left-hand side shows the time dependence of $\xi_1(t)$. The

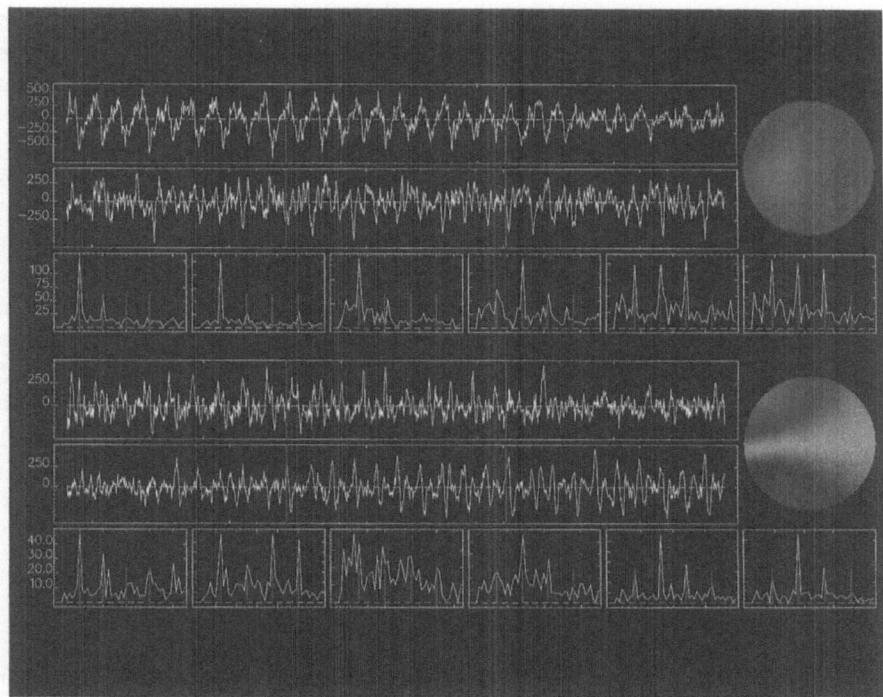

Fig. 15.5. Dynamics of the spatial modes. *Upper right corner*: Spatial distribution
of the first Karhunen–Loève mode according to Fig. 15.3. *First row*: Amplitude of
this most dominant mode on plateaus I–III (pre-transition) versus time. The signal
is nearly periodic. *Second row*: The same but for plateaus IV–VI (post-transition).
The signal is less periodic and the amplitude has dropped significantly. *Third row*:
Power spectra on the different plateaus. Before the transition, there is a strong
component at the frequency of the stimulus. *Lower part, right-hand side*: Second
largest Karhunen–Loève mode. The three lower rows correspond to the three upper
rows, but the time series and spectra show an almost opposite behavior. The spectra
are diffuse and low in power before the transition, whereas on plateaus V and VI
the signal becomes periodic and the spectra show clear peaks. The largest peak
is not at the stimulus frequency, but at its first harmonics. (Redrawn after *Fuchs,
Kelso, Haken* (1992))

red vertical lines indicate the end or beginning of the corresponding plateaus
(I–III). The second row shows the same but for the plateaus IV–VI. The
third row shows the power spectrum corresponding to the individual plateaus.
Before the transition, the signal is nearly periodic at a dominant frequency
equal to the fundamental frequency ν_1. Above the transition, the signal is less
periodic and the amplitude has dropped significantly. The lower part of the
figure shows the behavior of the second biggest mode with the corresponding
plots as shown in the upper part of Fig. 15.5. Before the transition occurs, the
spectra are diffuse and low in power, whereas on plateaus V and VI the signal
becomes periodic and the spectra show clear peaks. However, in contrast to

the most dominant mode, the largest peak is not at the stimulus frequency, but its first harmonic. The time-series belonging to the other leading modes up to $m = 2$ have also been calculated (cf. *Fuchs, Kelso, Haken* (1992)), but we shall not present these results here.

In the following it will be our goal to present models that account for the spatio-temporal behavior of brain activity, at least as represented by the two leading modes. To this end we shall proceed along two entirely different lines. In the first approach that we shall present in Sect. 15.3, we start from the known fundamental spatial modes, v_1 and v_2, as shown in the upper and lower part, right-hand side of Fig. 15.5. We then shall devise equations for the mode amplitudes ξ_1 and ξ_2 that reflect the observed behavior. In that section we shall also present a refined model that takes into account more Karhunen–Loève modes. In Sect. 15.4 we shall develop an entirely different approach where we shall conceive of the brain as a nonlinear medium in which waves can propagate. We then devise a nonlinear wave equation for $q(x, t)$. Thus we have two entirely different methods at hand, and shall compare these at the end of this chapter.

15.3 Modeling the Dynamics

In this section we want to model the spatio-temporal patterns that were experimentally found in the MEG as described in the previous sections. In order to get a first insight into the dynamics, we take into account only the first two Karhunen–Loève modes $v_1(x), v_2(x)$ so that we describe the spatio-temporal pattern $q(x, t)$ in the form

$$q(x, t) = \xi_1(t)v_1(x) + \xi_2(t)v_2(x). \tag{15.6}$$

Our goal will be to establish equations for the mode amplitudes ξ_1 and ξ_2 that capture the most salient features of the time series shown in Fig. 15.5 for the first two modes.

Let us consider the mode amplitude ξ_1 first. Because the time-series shows a pronounced periodic behavior, a good starting point is the equation of a harmonic oscillator of frequency ω_{01}

$$\ddot{\xi}_1 + \omega_{01}^2\xi_1 = 0. \tag{15.7}$$

Since according to the experiments the amplitude is stable, we expect there to be some kind of balance between a damping force and the driving force. As usual we assume that the damping is proportional to the speed with which ξ_1 changes, i.e. proportional to $\dot{\xi}_1$. Furthermore we have seen on various occasions, for instance in the case of finger movements, that the dynamics is governed by a Van der Pol oscillator, in which the damping force is proportional to the square of the amplitude. This leads us to introduce the damping force in the form

$$A_1\xi_1^2\dot{\xi}_1. \tag{15.8}$$

Ultimately the driving force stems from the acoustic signal, which we represent in our model by its leading Fourier term

$$F \cos(\Omega t + \phi). \tag{15.9}$$

Decomposing the cosine function into the exponential function in the usual way, we may write instead of (15.9)

$$\frac{1}{2} \left(F e^{i\Omega t + i\phi} + \text{complex conjugate} \right). \tag{15.10}$$

When we insert (15.10) on the right-hand side of (15.7), we find that the amplitude ξ_1 oscillates with a phase which coincides with that of the driving field

$$\xi_1 \sim e^{i\phi}. \tag{15.11}$$

This would preclude the pronounced transition of phases by about π, which is found experimentally. The subject is, indeed, able to choose his or her manual response arbitrarily in-phase or out of phase with the driving signal, at least for small enough frequencies. This leads us to the idea of considering a parametric driving, or, in other words, to the equations of a parameteric oscillator. Such a parametric oscillator is, for instance, a swing on which a person moves up and down. It is well known that the swing can be set in motion if the person on it moves his or her weight up and down with double the frequency at which the swing oscillates. Thus we represent the driving force in the form

$$E_1 \sin(2\Omega t)\xi_1, \tag{15.12}$$

where E_1 is a constant. Taking the terms (15.7), (15.8) and (15.12) together, we find an equation of motion in the form

$$\ddot{\xi}_1 + A_1 \xi_1^2 \dot{\xi}_1 + E_1 \sin(2\Omega t)\xi_1 + \omega_{01}^2 \xi_1 = 0. \tag{15.13}$$

This equation of motion does not take into account any coupling between the amplitudes ξ_1 and ξ_2, which is indispensable for the explanation for the experimental results. In order to get an idea of what the coupling terms could look like, we are guided by the finger movement paradigm of Chap. 7. There we have seen that the finger displacements were coupled by a term given by

$$(\dot{x}_1 - \dot{x}_2)[\alpha + \beta(x_1 - x_2)^2]. \tag{15.14}$$

In the following we shall make the replacements

$$x_1 \to \xi_1 \quad x_2 \to \xi_2, \tag{15.15}$$

$$x_1 \to \xi_1 \quad x_2 \to F \cos \Omega t, \text{ and } x_1 \to F \cos \Omega t \quad x_2 \to \xi_2. \tag{15.16}$$

We now proceed as follows: In (15.14) we multiply the brackets out to obtain the individual terms. Then we replace some of these terms according to (15.15) or (15.16). In principle we could replace all terms according to the rules (15.15) and (15.16). By a closer study of the resulting expressions, it turns out, however, that not all terms are needed to eventually represent the experimental results. Therefore, in the list we show only those terms that are actually taken into account.

$$\dot{x}_1, \tag{15.17}$$

$$\dot{x}_1 x_1^2 \rightarrow \dot{\xi}_1 \xi_1^2, \tag{15.18}$$

$$\dot{x}_1 x_2^2 \rightarrow \dot{\xi}_1 \xi_2^2; \rightarrow \xi_2^2 (F \cos \Omega t)^\bullet = -\Omega \xi_2^2 F \sin \Omega t, \tag{15.19}$$

$$\dot{x}_1 x_1 x_2, \tag{15.20}$$

$$\dot{x}_2 \rightarrow \dot{\xi}_2, \tag{15.21}$$

$$\dot{x}_2 x_1^2, \tag{15.22}$$

$$x_1 x_2 \dot{x}_2 \rightarrow \xi_1 F \cos \Omega t (F \cos \Omega t)^\bullet = -\frac{\Omega}{2} \xi_1 F^2 \sin 2\Omega t, \tag{15.23}$$

$$\dot{x}_2 x_2^2. \tag{15.24}$$

Let us discuss these terms in detail: (15.18) leads to a nonlinear damping that has already been taken into account in (15.13). (15.19) introduces a coupling term with ξ_2. So does (15.21). Finally it is interesting to note that (15.23) leads us back to the parametric driving term (15.12) that we introduced in (15.13). Incorporating the new terms (15.19) and (15.21) in (15.13), we obtain

$$\ddot{\xi}_1 + f_1 = 0. \tag{15.25}$$

In a similar way we can derive an equation for ξ_2

$$\ddot{\xi}_2 + f_2 = 0. \tag{15.26}$$

In these equations f_1 and f_2 are given by

$$f_1 = (A_1 \xi_1^2 + B_1 \xi_2^2)\dot{\xi}_1 + \omega_{01}^2 [1 + \epsilon_1 \sin(2\Omega t)]\xi_1 \\ + C_1 \xi_2^2 \sin \Omega t + D_1 \xi_2 \tag{15.27}$$

and

$$f_2 = (A_2 \xi_1^2 + B_2 \xi_1^2)\dot{\xi}_2 + \omega_{02}^2 [1 + \epsilon_2 \sin(2\Omega t)]\xi_2 \\ + C_2 \xi_1^2 \sin \Omega t + D_2 \dot{\xi}_1. \tag{15.28}$$

Here we have introduced ϵ_1 via

$$E_1 = \omega_{01}^2 \epsilon_1. \tag{15.29}$$

The equations (15.25), (15.26) with (15.27), (15.28) can be approximately treated in an analytical fashion and can also be integrated numerically (*Jirsa, Friedrich, Haken* (1994)). In these calculations the following parameter values were used:

$A_1{=}2$, $A_2 = 3$, $B_1 = 1.3$, $B_2 = 1$, $C_1 = -0.05$, $C_2 = 1.5$, $D_1 = 0.1$,

$D_2{=}1$, $\epsilon_1 = 0.02$, $\epsilon_2 = -0.6$, $\omega_{01} = 0.32$, $\omega_{02} = 1$. $\qquad(15.30)$

Let us discuss a few properties of these equations. The parametric excitation possesses the characteristic property of having stable and unstable solutions, depending on the ratio between ω_{0i} and Ω with $i = 1, 2$. The unstable solutions are obtained for $\omega_{0i}/2\Omega = k$, where $k = 1/2, 1, 3/2, ...$ This property of the parametric excitation provides a selection of frequencies depending on the chosen parameter space. The dynamics of the model described by (15.25), (15.26) with (15.27), (15.28) is as follows: In the pretransition region the oscillator ξ_1 is in its unstable region of $k = 1/2$ and oscillates with the stimulus frequency Ω. The oscillator ξ_2 is in its stable region and hence damped. When the stimulus frequency Ω is increased, the oscillator ξ_2 reaches the unstable region $k = 1$ at a critical frequency and starts to increase. Due to this increase of ξ_2, the relative phase between the first Fourier components of ξ_1 and ξ_2 and the stimulus signal perform the sudden transition by π. The amplitude of the oscillator ξ_2 mainly oscillating with twice the stimulus frequency becomes large and the oscillator ξ_1 becomes damped via the cross coupling to ξ_2. We shall not present the explicit time series here, because we want to do still better. Before I present the corresponding results by *Jirsa, Friedrich, Haken* (1995) I should quote the famous mathematicien *Gauss*. It is reported that he once said that a mathematical theorem is like a cathedral; one must remove the scaffolding after the cathedral was completed so that one can see the beauty of the building. I do not quite share his opinion, because especially for students it is important to learn how a theorem has originally been obtained (often in a lousy way!). But in the present case, the detailed reasoning of how the final result (that I shall present below) has been obtained would be too lengthy. I can just indicate a few points. In a first step, the spatial modes v_1 and v_2 that occur in (15.6) were improved, using the Uhl–Friedrich–Haken method, which we applied in Chap. 14. Then with these improved spatial functions, v_1 and v_2, we form the difference between the actually observed spatio-temporal pattern and the approximate pattern by means of

$$\tilde{q}(\boldsymbol{x}, t) = q(\boldsymbol{x}, t) - \xi_1(t)v_1(\boldsymbol{x}) - \xi_2(t)v_2(\boldsymbol{x}). \qquad (15.31)$$

In the next step of the analysis this residual pattern is again decomposed into its Karhunen–Loève modes $u_j(\boldsymbol{x})$ taken over all six plateaus

$$\tilde{q}(\boldsymbol{x}, t) = \sum_{j=1}^{4} \eta_j(t)u_j(\boldsymbol{x}), \qquad (15.32)$$

where the leading four modes are taken into account. A detailed discussion of the frequencies occurring before and after the transition paves the way to the final result that we immediately write down, namely

$$q(\boldsymbol{x}, t) = \xi_1(t)v_1(\boldsymbol{x}) + \xi_2(t)v_2(\boldsymbol{x}) + \tilde{\xi}_1(t)\tilde{v}_1(\boldsymbol{x}) + \xi_s(t)v_s(\boldsymbol{x}) + n(\boldsymbol{x}, t) (15.33)$$

where

$$\tilde{\xi}_1(t) = 0.5\xi_1(t - T/4), \tag{15.34}$$

and

$$\xi_s(t) = -1.5\xi_1^2(t) - 1.5\xi_1(t)\dot{\xi}_2(t) + 0.7\dot{\xi}_1^2(t) - 0.5\xi_2^2(t). \tag{15.35}$$

The time T is the period of the acoustic signal. The numerical coefficients were determined by a least mean squares fit. The result (15.33) tells us that the total spatio-temporal pattern can be written as a superposition of three order parameter terms $v_1, v_2, \tilde{v}_1$, and one enslaved mode v_s with amplitude ξ_s that can be expressed by the order parameters ξ_1, ξ_2. $n(x, t)$ denotes a residual small "noise" term. Because of the relation (15.34), only two order parameters are involved in (15.33). The term

$$\xi_1(t)v_1(x) + 0.5\xi_1(t - T/4)\tilde{v}_1(x) \tag{15.36}$$

can be interpreted as a traveling wave, while the term

$$\xi_2(t)v_2(x) \tag{15.37}$$

represents a standing wave. Because of the relations (15.34) and (15.35), the total dynamics is indeed governed by the order parameters ξ_1 and ξ_2 which

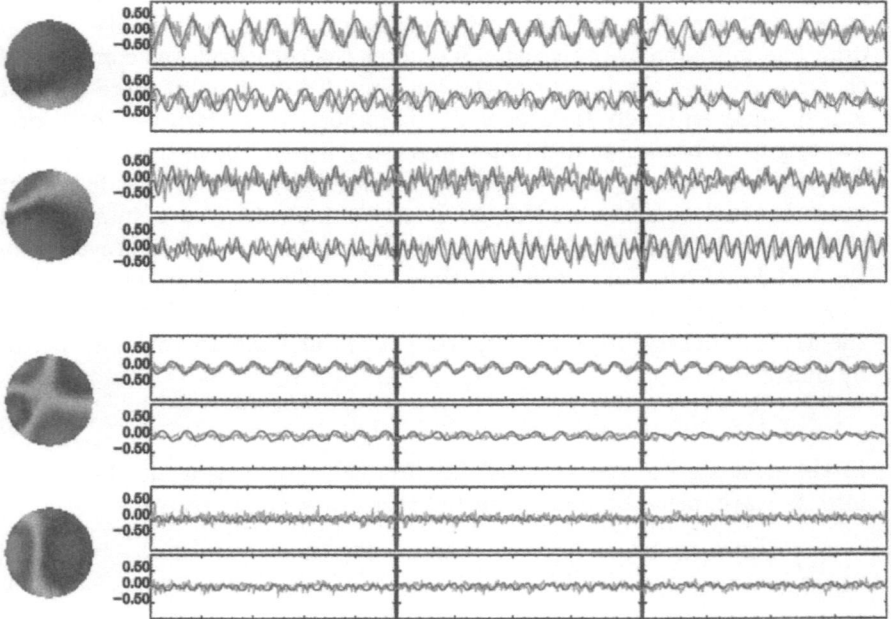

Fig. 15.6. The left row shows the spatial modes $v_1, v_2, \tilde{v}_1, v_s$. The right-hand side shows the corresponding amplitude versus time. The green curves correspond to the experimental results, the red curves show the theoretical results stemming from an integration of the equations, as described in the text. After *Jirsa, Friedrich, Haken*, unpublished

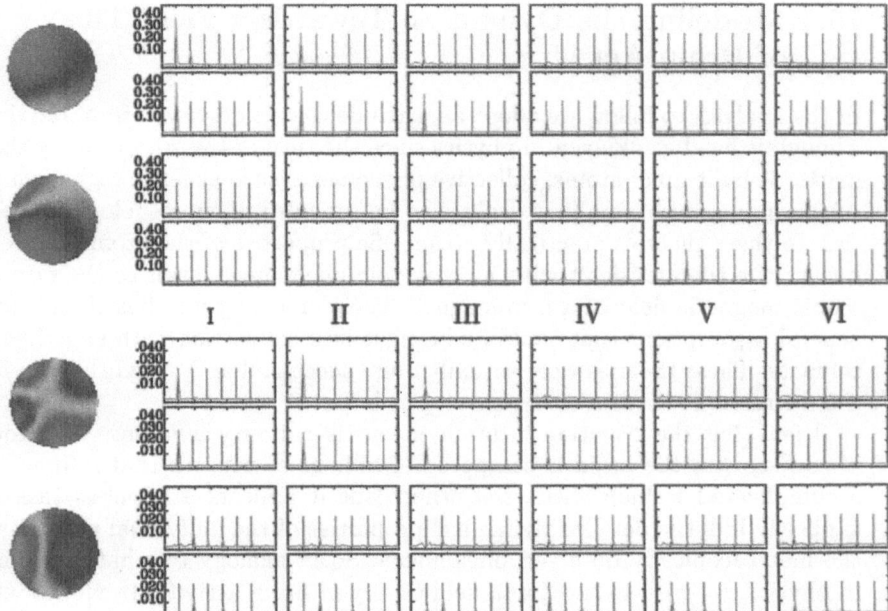

Fig. 15.7. The power spectra corresponding to the time-series of Fig. 15.6. After *Jirsa, Friedrich, Haken*, unpublished

obey the original equations (15.25)–(15.28). What had to be determined from the experimental data was, on the one hand, the coefficients occurring in the order parameter equations, and also the spatial modes. The spatial modes $v_1, v_2, \tilde{v}_1$ and v_s are shown on the left-hand side of Fig. 15.6. The right-hand side shows the temporal evolution of the corresponding amplitudes ξ for the different plateaus. The green curve shows the experimental results when projected onto the corresponding spatial modes. The red curve shows the results obtained by a direct integration of (15.25)–(15.28) and using the relations (15.34) and (15.35). Figure 15.7 shows a corresponding plot for the power spectra. We believe that the agreement is quite remarkable.

In spite of the success of this model, a warning should be added: It turns out that, in a number of cases, these models are not uniquely determined, neither concerning the numerical values of the coefficients nor the specific form of the individual terms. So in such cases there may be whole classes of systems that give rise to the same spatio-temporal patterns. We shall come back to this point in the next section.

15.4 Modeling the Dynamics: Towards a Field Theory of Brain Activity

In the analysis to follow we adopt an attitude that is borrowed from physics. Though it has been known in physics since the turn of the 20th century that matter is built up of atoms, collective phenomena, such as ferromagnetism or superconductivity, were first dealt with by so-called phenomenological theories. In these theories practically no specific properties of the atoms entered; rather, the theories dealt with macroscopic phenomena, such as the macroscopic magnetic field of a ferromagnet. Their modeling was based on a few general assumptions that could be brought into connection with experimental facts. These theories were actually quite successful and paved the way for the microscopic theories to be developed later.

I feel that the situation in brain theory is not very different. We know, of course, that the brain is composed of billions of neurons that interact in a complicated fashion with each other. But it appears a hopeless task to explicitly link the detailed dynamical properties of the individual neurons to the macroscopically observed phenomena. So in analogy to what was done in physics, we wish to develop a field theory of brain activity. In view of the great variety of actions performed by the brain, we have to be specific. So our field theory will be devoted to modeling the action of the brain within the Kelso experiment described above. It is hoped, however, that it will provide a more general basis also applicable to other processes occurring on macroscopic scales.

First we shall assume that brain activity is based on processes in the brain without external sources, such as the acoustic signal, or other external signals. In other words, we first consider the intrinsic dynamics. Since there is evidence from both EEG and MEG data that waves of activity based on electrochemical and magnetic processes propagate in the brain, we shall start from a wave-like equation

$$\ddot{q}(\boldsymbol{x}, t) + Lq(\boldsymbol{x}, t) = 0. \tag{15.38}$$

In a typical wave equation L is a linear operator, which may be, for instance,

$$L = -c^2 \Delta + \alpha_0^2, \tag{15.39}$$

where c and α_0 are some constants and Δ is the Laplacian. In more complicated cases, we may think of a nonlocal operator of the form

$$L = \int K(\boldsymbol{x}, \boldsymbol{x}')...d^3x', \tag{15.40}$$

where the dots indicate that the function q occurring after L in (15.38) has to be taken under the integral with coordinates $\boldsymbol{x}'$, then the integration over d^3x' has to be performed and, eventually, a new function of $\boldsymbol{x}$ is generated.

In the following we shall use the simple form (15.39). We shall assume that brain activity is damped without an external signal. This damping may be assumed to be of the form

$$\gamma_1 \dot{q}(\boldsymbol{x}, t), \tag{15.41}$$

as usual in wave propagation phenomena. It is known from a number of biological phenomena that linear damping, as indicated by (15.41), is not sufficient and that nonlinear damping occurs. Such a damping occurred, for instance, in the case of the finger movement model. The corresponding term is of the form

$$\gamma_2 q^2(\boldsymbol{x}, t) \dot{q}(\boldsymbol{x}, t). \tag{15.42}$$

Taking all the terms (15.39)–(15.42) together, we obtain our model of the intrinsic dynamics in the form

$$\ddot{q}(\boldsymbol{x}, t) + \left(\gamma_1 + \gamma_2 q^2\right) \dot{q}(\boldsymbol{x}, t) + Lq(\boldsymbol{x}, t) = 0. \tag{15.43}$$

Our main concern is to introduce the activity of the external signal. Because of the analogy of the present problem to that of the correlation of finger movements, one is led to consider a coupling term between brain activity and external signal that is analogous to terms considered in the case of the finger movement. This leads us back to the list of coupling terms introduced in Sect. 15.3, (15.17)–(15.24), but where we now identify

$$x_1(t) \rightarrow q(\boldsymbol{x}, t), \quad x_2 \rightarrow F(\boldsymbol{x}) \cos \Omega t. \tag{15.44}$$

We repeat that part of this list here where the replacement is made and later used

$$\dot{x}_1 x_1^2 \rightarrow \dot{q} q^2, \tag{15.45}$$

$$\dot{x}_1 x_1 x_2 \rightarrow \dot{q} q F(\boldsymbol{x}) \cos \Omega t, \tag{15.46}$$

$$x_1 x_2 \dot{x}_2 \rightarrow q \left(-\frac{\Omega}{2} F^2(\boldsymbol{x})\right) \sin 2\Omega t. \tag{15.47}$$

Our main goal will be to cover the transition in frequency and phase as described in Sect. 15.2. This requires, in addition to the above list, a further term

$$q \sin \Omega t. \tag{15.48}$$

Taking all terms together, we are led to the equation

$$\ddot{q} + (\gamma_1 + \gamma_2 q^2)\dot{q} + g(\boldsymbol{x}) \sin \Omega t q + f(\boldsymbol{x}) \cos \Omega t q \dot{q}$$
$$+ Lq + L_1 q \sin(2\Omega t) = 0. \tag{15.49}$$

The spatial functions $F(\boldsymbol{x}), g(\boldsymbol{x})$ and $f(\boldsymbol{x})$ represent the spatial distribution of the acoustic stimulus. Why did we choose the terms (15.45)–(15.48)? Again some "secret" of this kind of modeling has to be explained. In order to study the properties of the solution of (15.49), we make the hypothesis

$$q = \xi_0 + \xi_1 e^{i\Omega t} + \xi_2 e^{2i\Omega t} + \text{complex conjugate}, \tag{15.50}$$

because of the parametric driving force. The amplitudes ξ_0, ξ_1, ξ_2 may still depend on space and time. We make use of two approximations that we have already encountered several times in this book, namely the rotating wave approximation and the slowly varying amplitude approximation. This allows us to transform (15.49) into coupled differential equations for the amplitudes in (15.50). We then choose the coefficients in (15.49) in such a way that the observed change of the power spectrum and phase shift at the phase transition results. After building this model, we may forget the individual arguments and solve (15.49) on a computer. To obtain a numerically stable solution, two modifications must be made. We have to replace

$$\gamma_1 \dot{q} \quad \text{by} \quad -\gamma_1 \Delta \dot{q}$$

and the constant L_1 by the operator

$$L_1 = \beta^2 \Delta - \alpha_1^2.$$

When we decompose the numerical solution $q(x, t)$ into Karhunen–Loève modes, quite a good agreement is found between the time-dependent amplitudes and the experimental data (including the occurrence of the transition). On the other hand (15.49) needs some further improvements if we wish to achieve the same accuracy as achieved by the method of Sect. 15.3. It will be an important task for the future to link these results with those of microscopic theories, for instance those formulated by *Nunez* (1995).

15.5 EEG and MEG Analysis Revisited

Let us discuss the results of the preceding two chapters on EEG and MEG analysis in the light of synergetics. As may transpire from Chap. 2 on brain research, a good deal of that research focusses its attention on localized actions of the brain. This is not only true for studies of the actions of the individual neurons, but also of research based on magnetic resonance imaging (MRI) and positron emission spectroscopy (PET scan). Here emphasis is laid on the detection of localized centers, where, in particular, subtraction methods are used. The conclusions we are going to draw on the EEG and MEG analysis presented in the previous chapters are rather different. Here we found that in the cases studied, macroscopic coherent patterns of electric and magnetic activity occur and that these patterns are governed by few order parameters only. This is a conclusive result, because the number of order parameters was much smaller than that of the channels (electrodes or SQUIDs). In the case of the EEG of *petit-mal epilepsy* one might argue that here we are dealing with a pathological case in which many neurons fire coherently because of some hyperactivity. But healthy people also show coherent EEG activity in their α-waves, i.e. when at rest with closed eyes. Here again the macroscopic activity is governed by few order parameters, namely at maximum five, as was shown in previous analyses briefly mentioned in Chap. 14. The results of the Kelso experiment are still more striking, because here a precisely defined task was given to the subject. One might have expected that these tasks were performed in the brain in highly localized areas. But the results are quite different. Again they show that a considerable region of the brain, including parts of its auditory and motor area, is involved in this task in a highly coherent fashion. The experiments clearly show that the coherent obtained patterns are not due to some irrelevant background activity of the brain, but directly connected with the task. This is evident from the power spectrum that has its peaks at the frequency of the acoustic signal and its higher harmonics, or its subharmonics. The dynamics is governed by very few order parameters that are much less numerous than the channels.

These remarks do not exclude the possibility that at smaller temporal and spatial scales additional processes occur, but in the author's opinion one would expect a high correlation between the macroscopic and the microscopic patterns as is expressed by the slaving principle. Quite clearly a considerable amount of work still has to be done to establish this relationship between macroscopic and microscopic activities in detail. As we have shown in these chapters, the dynamics of the order parameters is governed by comparatively simple equations of coupled oscillators or low-dimensional systems that produce chaos. These oscillators or chaotic systems are by no means localized or realized by individual cells; they are rather the property of the whole neural system. Just the same way as the role pattern of a fluid is the outcome of the collective motion of all molecules, or the coherent field of a laser is the outcome of the collective emission acts of the individual atoms. On the other hand, we must be fully aware that brain activity is incredibly more complex than these simple physical phenomena. But nevertheless all these systems share the property of being able to produce phenomena that are highly coherent in space and time. The experimental results of the MEG are particularly clearcut because of the periodic driving. In future work, the study of transitions appears to be of particular interest, because here we might gain more insight into various rate constants.

Part IV **Cognition**

16. Visual Perception

It is my deep conviction that by studying visual perception or, more precisely speaking, *pattern recognition* we can learn a good deal about *cognition*. Thus the considerations of this chapter will serve us later as a metaphor when we deal with some cognitive abilities of humans.

16.1 A Model of Pattern Recognition

In order to devise models for visual perception by humans, we may proceed in at least two ways: The more traditional way might be called *bottom up*. Here we start from model neurons and their connections and then derive the (macroscopic) properties of their network. We shall return to this approach later in this book (Chap. 18). The other line of thought, which follows the spirit of phenomenological synergetics, may be called a *top-down approach*. Here we start with the tasks that a macroscopic system has to fulfill and then only in a later step do we look for realizations by means of networks of elements that we may again call *model neurons*.

So let us consider the task to be carried out by a macroscopic system for visual perception. This system may be the human brain, or an animal brain, or a highly sophisticated computer. Visual perception is a vast field. For instance, whenever we – or an animal – see a bright spot, visual perception is at work. Our goal is, however, more specific and, at the same time, more ambitious. We want to study visual perception at the cognitive level, or in other words, we want to understand how a brain may recognize objects (or patterns) through vision. Therefore, in the following we shall speak of pattern recognition. First of all, what precisely do we understand by pattern recognition? When we see a face, we want to know the name of the person, or at least to establish whether we know this person from the past. Thus we wish to associate a name with a face. Or, in other words, pattern recognition may be interpreted as the action of an *associative memory*. There are other examples of an associative memory, for instance, a telephone book. When we look up the name Alex Miller, the telephone book provides us with his telephone number. If we wish to express the action of an associative memory in abstract terms, we may say that it serves to complete an incomplete set of data to yield the full set. But how do we realize such an associative memory?

In the following we wish to realize an associative memory by some dynamics. Quite often a dynamics can be visualized by means of the motion of a ball in a landscape. In such a landscape, we identify the recognized patterns with the bottoms of the valleys, whereas an incomplete pattern, i.e., a not yet recognized pattern, may be visualized as the position of a ball lying on a slope (Fig. 16.1). During the recognition process, the ball will be pulled into the closest valley and thus the recognition task will be accomplished. The central question, of course, is in which way we can link the features of an individual pattern to the dynamics of that ball. To this end, we invoke an analogy between pattern recognition and pattern formation. In other words, we claim that pattern recognition by a brain or a computer is nothing but pattern formation. To explain this idea in more detail, let us consider the example of a fluid heated from below. As we have shown in Chap. 4, fluids heated from below may form roll patterns. Let us consider a circular vessel that is heated from below and where the critical temperature difference has been established so that, in principle, a roll pattern can evolve. Let us further assume that we prescribe one upwelling roll in a specific direction (Fig. 16.2, left column). Then computer simulation shows that the liquid is able to complement this single roll to a whole pattern in the course of time. When we prescribe a different direction of the initial roll, a new roll pattern

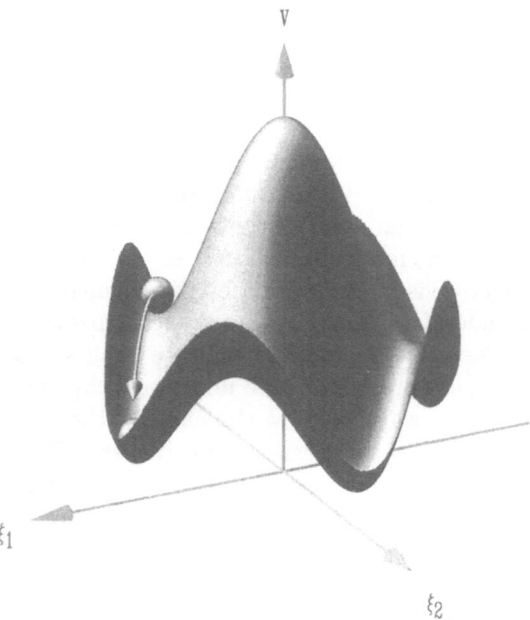

Fig. 16.1. Visualization of pattern recognition by the dynamics of a ball moving in a potential landscape. The minima of the potential correspond to a recognized pattern, a position on a slope corresponds to an incomplete, i.e. not yet recognized, pattern

a)   b)   c)

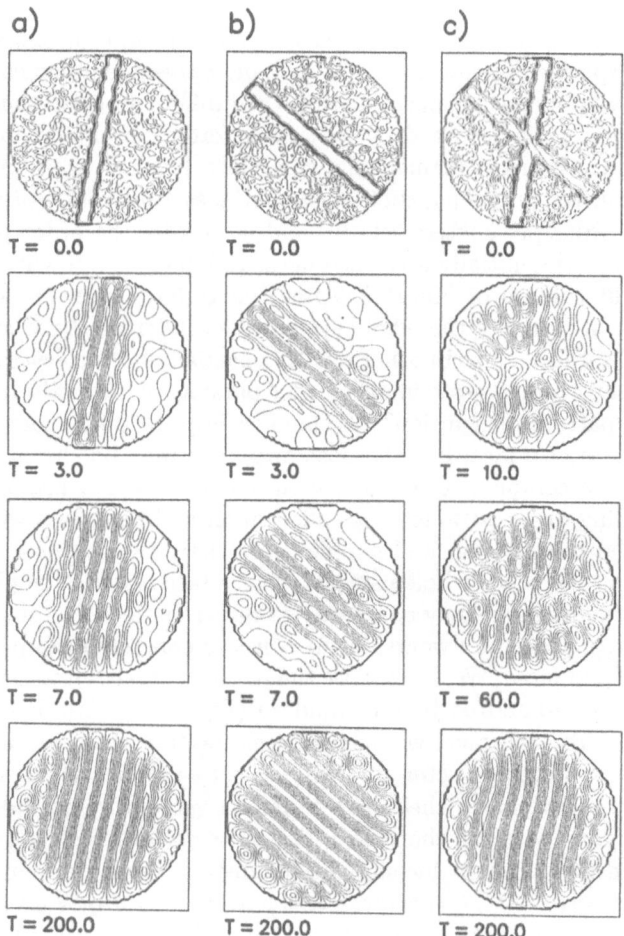

Fig. 16.2a–c. Computer simulation of pattern formation in a liquid in a circular vessel that is heated from below. The heating is beyond the critical point. *Left column*: Initially an upwelling roll is prescribed; in the course of time the fluid complements this pattern to a whole roll pattern lying in the same direction. *Middle column*: Same as left column, but with a different initial state. *Right column*: A conflict situation is constructed by providing the fluid with two initially given rolls, the one roll being somewhat stronger than the other one. In the course of time only one roll and its corresponding pattern win the competition

in that direction will develop (Fig. 16.2, middle column). Finally, we force the liquid into a conflict situation by prescribing two initial rolls in different directions, whereby one roll is somewhat stronger than the other (Fig. 16.2, right column). As the computer calculation reveals, the fluid first seems to seek a compromise, but then the originally stronger roll wins the competition and determines the pattern which finally evolves.

Let us interpret these results in terms of order parameters and the slaving principle. We first prepare an initial state that may be represented by means of a superposition of all possible roll patterns with their different orientations. Each of these roll patterns is governed by its specific order parameter, but one order parameter is strongest, namely the one which belongs to the initially prescribed roll. After the preparation of the initial state, a competition between the various order parameters sets in, and is won by the order parameter that belongs to the initially given strongest roll. This is also clearly exhibited by the right column of Fig. 16.2. After this order parameter has won the competition, it enslaves the whole system, i.e., it forces the whole fluid into its ordered roll-like state. In other words, a partially ordered system is brought into its fully ordered state via the order parameter competition.

What happens in pattern recognition? We claim exactly the same, namely originally some features of a face, like the nose and eyes, may be shown to the human brain or to a computer to define the initial state of that pattern. Then the corresponding order parameter is called upon and competes with all others; it wins the competition and supplements the given features by the other features that define the whole pattern. Thus a face may be complemented and also the family name may be added, if this is required (Fig. 16.3). In order to see how this procedure works, let us consider concrete examples, namely the recognition of faces. We decompose the pictures of faces into their individual pixels, say, one-hundred by one-hundred (Fig. 16.4). We denote a pixel by its index j. To each pixel we may attribute a grey value v_j. The set of all grey values v_j forms a vector $v = (v_1, v_2, ..., v_N)$. When there are several faces given, we distinguish their corresponding vectors by an index k and write v_k. We now consider the recognition process, in which a test pattern vector is prescribed that is incomplete, for instance, one containing only the eyes and the nose. We call this given test pattern vector q. We then wish to devise a dynamics so that in the course of time the vector q develops and, eventually, reaches a vector v_{k_0}, where v_{k_0} is one of the stored prototype pattern vectors. As shown by some analysis, which we do not repeat here, however, we may construct the dynamics in surprisingly close analogy to the

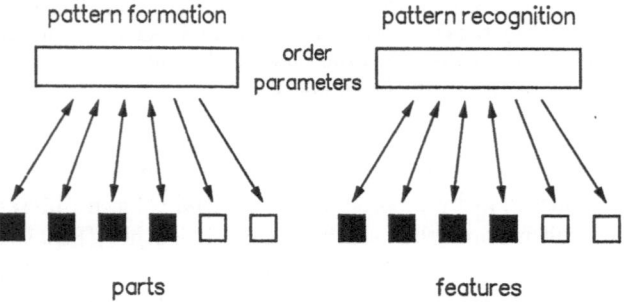

Fig. 16.3. Analogy between pattern formation and pattern recognition (see text)

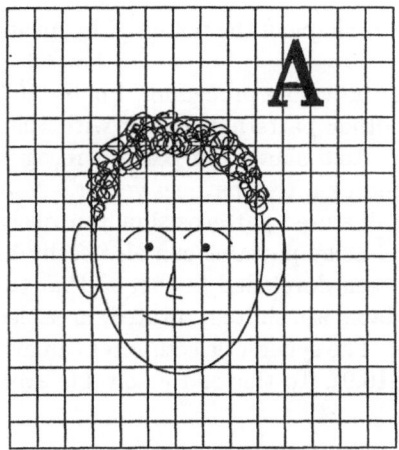

Fig. 16.4. Example of the construction of a prototype vector v. Its components are the grey values of the corresponding pixels j

fluid dynamics example. To this end, we introduce adjoint vectors v_k^+ that are defined by the orthogonality relations

$$(v_k^+ v_{k'}) = \delta_{kk'}. \tag{16.1}$$

The adjoint vectors may be constructed from the original vectors v_k by means of

$$v_k^+ = \sum_{k'=1}^{M} A_{kk'} \overline{v}_{k'}, \tag{16.2}$$

where $\overline{v}_k$ is the transposed vector of v_k, and where the coefficients $A_{kk'}$ can be determined by means of (16.1) (cf. Appendix B). The pattern recognition dynamics is described by the equations

$$\dot{q} = \sum_{k=1}^{M} \lambda_k (v_k^+ q) v_k - B \sum_{k,k'=1}^{M} (v_{k'}^+ q)^2 (v_k^+ q) v_k - C(q^+ q) q, \tag{16.3}$$

where the individual terms on the right-hand side have the following meaning: The first term contains the attention parameters λ_k. We shall elucidate their role and their interpretation below in Sect. 16.2. By decomposing the vectors v_k^+, v_k into their components v_{kj}^+ and v_{kj}, respectively, we may write the vector component j of the term $(v_k^+ q) v_k$ in the form

$$\sum_{j'} v_{kj} v_{kj'}^+ q_{j'}.$$

In this expression, $v_{kj} v_{kj'}^+$ is the learning matrix of pattern k. The third term serves to discriminate between the patterns and the last term limits the growth of the grey values of the pixels. The dynamics of q can be visualized, as so often in our book, by the motion of a ball in a potential landscape. In general, this dynamics takes place in a high-dimensional space, but when we consider a two-dimensional example in which we have only two pixels and two prototype patterns, we may visualize the shape of the potential by means of Fig. 16.1.

A few explicit examples can illustrate the whole procedure. Figure 16.5 represents a number of stored faces (or other patterns) jointly with their family names (encoded by letters). Figure 16.6 shows the pattern recognition process, where part of a face is offered as a test pattern vector. Note that the recognition process makes use of all stored patterns simultaneously. The dynamics is very robust: For instance, noisy patterns can also easily be recognized (Fig. 16.7). Our approach has been developed further so that the recognition process is possible even if the faces are shifted, rotated, or scaled in space, or if they are deformed (Fig. 16.8). The synergetic computer can also learn to recognize facial expressions (Fig. 16.9). To this end a suitable aver-

Fig. 16.5. Examples of stored prototype patterns including the encoded family names

Fig. 16.6. Example of a recognition process where part of a face is prescribed as initial state

Fig. 16.7. Recognition of noisy faces

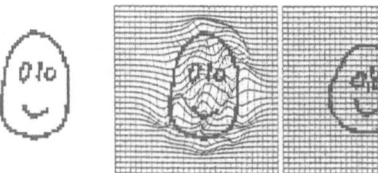  

Fig. 16.8. Recognition of a deformed face

Fig. 16.9. Examples of facial expressions recognized by a synergetic computer that has learnt facial expressions

age over faces of different persons, each with the same facial expression, must be taken and an association between that expression and the corresponding interpretation (e.g., joy) must be established. Depending on the chosen pathways and attention parameters, the computer can recognize a face irrespective of its expression, or a facial expression irrespective of the individual person. (In our experiments, the success rate was about 80 percent.)

We shall not dwell on the details here, however, because we wish to underline another aspect, namely the relationship between a network of neurons and order parameters. The equations (16.3) can be interpreted as describing the activities of a network of model neurons in the following manner: We may attach to each index j a model neuron with activity q_j. We rewrite (16.3) in terms of its components

$$\dot{q}_j = \sum_\ell \lambda_{j\ell} q_\ell + \sum_{\ell mn} \lambda_{j\ell mn} q_\ell q_m q_n, \tag{16.4}$$

where the coefficients $\lambda_{j\ell}, \lambda_{j\ell mn}$ can be derived from (16.3) and are composed of the components of the prototype vectors v_k and their adjoints, v_k^+. For instance, $\lambda_{j\ell}$ is given by

$$\lambda_{j\ell} = \sum_k \lambda_k v_{kj} v_{k\ell}^+, \tag{16.5}$$

where v_{kj} and $v_{k\ell}^+$ are the components of the vectors v_k and v_k^+, respectively. According to equation (16.4), model neuron j changes its activity as a result of inputs from other neurons (Fig. 16.10). According to the first sum on the right-hand side of (16.4), the model neuron j multiplies the signals from neuron ℓ by factors $\lambda_{j\ell}$, which may be interpreted as synaptic strengths, and then sums up over all inputs ℓ. The second term describes inputs from

Fig. 16.10. Visualization of (16.4) by a parallel network (see text)

three neurons at a time. These inputs are multiplied by one another and by a factor $\lambda_{j\ell mn}$. This factor may again be interpreted as some kind of synaptic strength. Finally, the sums are taken over ℓmn. To some extent, these procedures are reminiscent of those ascribed to real neurons, but with two basic differences. Namely:

1) In the present model we use soft nonlinearities, whereas in the conventional model of neurons one uses a threshold function between input and output. We might interpret the soft nonlinearities by the idea that they stem from the action of groups of neurons.

2) We use single *and* triple inputs. When, instead of soft nonlinearities we use a smooth threshold function and expand it into a power series of inputs q_k, we shall find triple inputs as the leading nonlinear term.

In the context of our book it is important to note that (16.3) and (16.4) represent the action of a whole network of individual neurons and that our model allows us to study the connection between the individual activities and the origin of order parameters. The transition from the microscopic neuronal level to the order parameter level is achieved by decomposing q into its mode skeleton quite in analogy to the discussion of pattern formation in Chap. 10

$$q(t) = \sum_{k=1}^{M} \xi_k(t) v_k + w(t). \tag{16.6}$$

ξ_k is the order parameter that belongs to the pattern v_k. w represents a remainder term that vanishes during the development of q in the course of time. Thus the evolving recognized patterns are determined by the order parameters ξ_k. When we insert (16.6) into (16.3), multiply this equation by v_k^+ and form the scalar product, then, because of the orthogonality relation (16.1) and $(v_k^+ w) = 0$, we immediately obtain

$$\dot{\xi}_k = \lambda_k \xi_k - B \sum_{k' \neq k}^{M} \xi_{k'}^2 \xi_k - C \sum_{k'=1}^{M} \xi_{k'}^2 \xi_k. \tag{16.7}$$

This also provides us with a key to determine the initial values of ξ_k, namely by means of

$$\xi_k(0) = \left(v_k^+ q(0) \right). \tag{16.8}$$

Equations (16.6)–(16.8) establish the relation between the microscopic spatial patterns of activities of the model neurons and the order parameter level. As

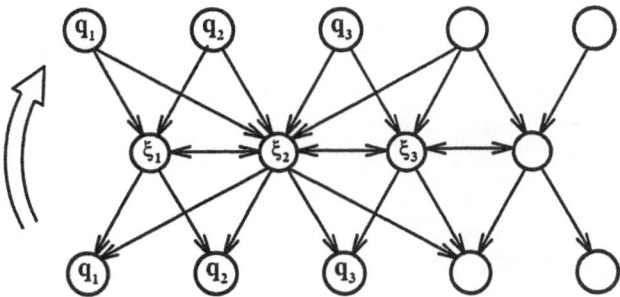

Fig. 16.11. Relationship between the activities of the individual model neurons (*top row*) and the dynamics of the order parameters ξ_k (*middle row*). The order parameters determine the output of the last row with activities $q_j(t)$. (For further details see text)

we see, each pattern k is governed by its order parameter ξ_k. In a way we may state that the order parameter ξ_k represents the "idea of the pattern k".

The relationship between the activities q_i of the individual neurons and the order parameters can be visualized by means of Fig. 16.11. The uppermost layer contains the individual neurons, j, with their activities q_j. The initial activity at time $t = 0$ is then projected to the middle layer of order parameters by means of the rule (16.8). Then the dynamics between the order parameters starts, and only one wins the competition, this one determining the activity pattern of the neurons which are represented by the lower level in Fig. 16.11. In reality, of course, the lower level coincides with the uppermost level. Actually, this visualization of the interrelation between the individual neurons and the order parameters is inherent in Fig. 16.3, where we invoked the analogy between pattern formation and pattern recognition. In that figure we showed only one order parameter. When we note that in Fig. 16.11 the uppermost layer and the lowest layer coincide and when we mirror Fig. 16.11 around the horizontal axis through the order parameter level, we immediately find that these two figures are identical. Figure 16.11 allows another interpretation, namely, we may identify the uppermost layer with that of, say, sensory neurons and the lower layer with motor neurons. Then the order parameters govern motor action.

16.2 The Role of Attention Parameters. Ambiguous Figures

Let us try to elucidate the role of the attention parameters λ_k. When we show a scene, such as that of Fig. 16.12, to the computer, it first recognizes the woman in the foreground. When we set the attention parameter corresponding to the woman equal to zero and show the same scene to the computer again, it recognizes the man in the rear. In this way, the computer was able to

Fig. 16.12. Example of a scene that has been recognized by the synergetic computer

consecutively recognize scenes composed of five partly hidden faces. I believe that this result may indicate how human brains deal with complex scenes. They first recognize a part of it, then the corresponding attention fades away, and attention is focussed on another part of the scene. This interpretation will be substantiated below.

In rare cases the computer failed to recognize the faces of a scene properly, but recognized, say, a third face instead. But as it turns out, humans can also be deceived by pictures. Consider Fig. 16.13: Here most people first recognize Einstein's face, but when they look more carefully, they will recognize three bathing girls. This leads us to the field of ambiguous patterns. A few examples are shown in Figs. 16.14–16.16. The ambiguity may have different sources; for instance, it may be based on different spatial interpretations as in the case of the Necker cube, or it may involve semantic ambiguities, as in the case of the vase or faces, or young woman/old woman. Our brain deals with these ambiguous patterns in a very peculiar way. For instance, in the picture of vase/faces, we recognize a vase for a while, then we recognize two faces, then the vase again, and so on. In other words, our perception oscillates back and forth between these two interpretations. At the beginning of the 20th century, the Gestalt psychologist *Köhler* suggested an explanation of this result. According to his ideas, once a pattern has been recognized the corresponding attention fades away and then new attention can be focussed on the other interpretation. He offered some physical explanation of these processes, but this is nowadays considered obsolete and no longer shared by most neurophysiologists or psychophysicists. On the other hand, in the frame of our order parameter approach, it is quite simple to simulate these processes. To this end, we start from the order parameter equation for two order parameters ξ_1, ξ_2 representing two different percepts, such as vase and face. By specializing (16.7), we obtain

Fig. 16.13. Einstein's face or three bathing girls?

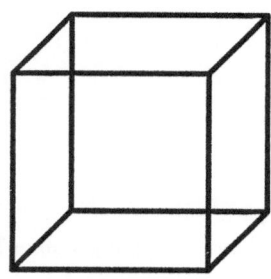

Fig. 16.14. The Necker cube

$$\dot{\xi}_1 = \xi_1 \left[\lambda_1 - C\xi_1^2 - (B + C)\xi_2^2 \right],\tag{16.9}$$

$$\dot{\xi}_2 = \xi_2 \left[\lambda_2 - C\xi_2^2 - (B + C)\xi_1^2 \right].\tag{16.10}$$

Note the completely symmetric role played by ξ_1 and ξ_2.

In addition, we subject the attention parameters to a dynamics that takes into account saturation, i.e., an attention parameter decreases once the corresponding order parameter increases. The equations for the saturation of attention parameters then acquire the following form:

$$\dot{\lambda}_1 = \gamma \left(1 - \lambda_1 - \xi_1^2 \right),\tag{16.11}$$

Fig. 16.15. Vase or faces?

Fig. 16.16. Old woman or young woman?

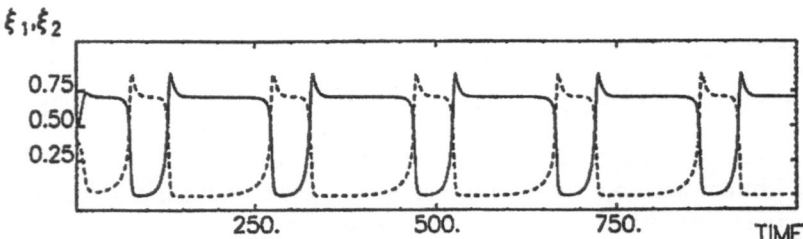

Fig. 16.17. Plot of the order parameters ξ_1, ξ_2 corresponding to two interpretations, for instance vase/faces, versus time

$$\dot{\lambda}_2 = \gamma \left(1 - \lambda_2 - \xi_2^2\right). \tag{16.12}$$

A solution of the coupled equations (16.9)–(16.12) is shown in Fig. 16.17, where we indeed find the observed oscillations of perception.

16.3 Influence of a Bias

In many cases one pattern is more quickly recognized than another. For instance, according to detailed psychological studies in the case of Fig. 16.16, eighty percent of male persons recognized the young woman first and twenty percent the old woman. Thus a certain bias is present in the recognition of ambiguous figures. We shall discuss our model of this phenomenon in some detail, because it sheds light on an interesting relationship between the size of the bias and the duration of the periods during which a pattern is perceived. Let us label the prototype pattern vector of the old woman by 1 and that of the young woman by 2. Then we may reformulate the above statement by saying that a percentage $p_1 = 20$ percent of the persons identify the test pattern vector q, which corresponds *objectively* to the actually presented pattern with v_1, and a percentage $p_2 = 80$ percent with v_2. This leads us to the idea of reconstructing the test pattern vector q from these experimental data by putting

$$q = p_1 v_1 + p_2 v_2. \tag{16.13}$$

In order to model this intrinsic bias, we resort to the fundamental mechanism introduced in the original approach of the previous section (Fig. 16.1). With respect to two dimensions, that approach means that the landscape of the potential V plotted over the plane spanned by ξ_1, ξ_2 is divided by a ridge along the diagonal (Fig. 16.18). In this case, there is no intrinsic bias so that $p_1 = p_2 = 50$ percent and q coincides with the ridge along the diagonal. But with bias, this ridge must be displaced so that the line of *neutral force*, i.e. the vector q, is given by formula (16.13). Some simple mathematical analysis shows that the ridge can be displaced by adding a potential V_b to the potential that is plotted in Fig. 16.1, where V_b has the form

$$V_b = -2B\xi_1^2\xi_2^2\left(1 - 4\alpha\frac{\xi_1^2 - \xi_2^2}{\xi_1^2 + \xi_2^2}\right). \tag{16.14}$$

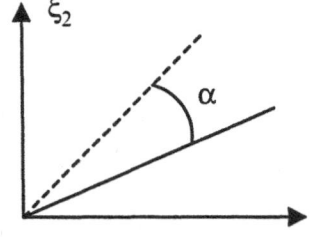

Fig. 16.18. Visualization of the bias angle α in the plane of the order parameters ξ_1, ξ_2

This additional potential contains only the parameter α. Indeed, (16.14) is constructed in such a way that the ridge of the total potential forms the same angle α with the v_1-axis as does q (16.13) (Fig. 16.18). In order to obtain the new equations for the order parameters, we just have to add to the right-hand sides of (16.9) and (16.10) the negative derivatives of V_b with respect to ξ_1 and ξ_2, respectively: We thus find our final equations in the form

$$\dot{\xi}_1 = \xi_1 \left[\lambda_1 - C\xi_1^2 - (B+C)\xi_2^2 \right] - \frac{\partial V_b}{\partial \xi_1}, \tag{16.15}$$

$$\dot{\xi}_2 = \xi_2 \left[\lambda_2 - C\xi_2^2 - (B+C)\xi_1^2 \right] - \frac{\partial V_b}{\partial \xi_2}. \tag{16.16}$$

When we solve these equations together with the equations for the attention parameters, we obtain the results shown in Fig. 16.19. The relationship between the times T_1 and T_2 during which the individual patterns (e.g. young woman or old woman) are identified and the bias parameter α_0 is given by

$$\alpha = \alpha_{\text{crit}} \frac{T_2 - T_1}{T_1 + T_2}, \tag{16.17}$$

where

$$\alpha_{\text{crit}} = \frac{1 - B}{4B}. \tag{16.18}$$

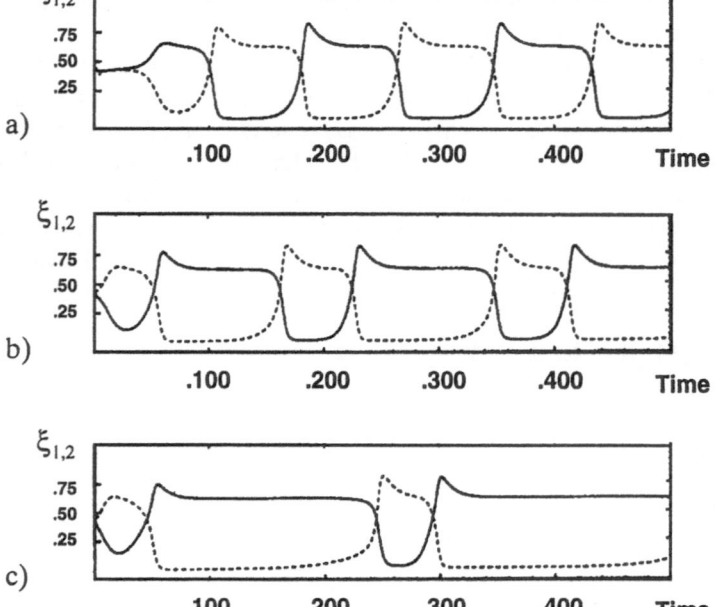

Fig. 16.19a–c. Oscillations with increasing bias

As a detailed analysis of the equations (16.15), (16.16), (16.11), (16.12) reveals, oscillations of perception can take place only if

$$- \alpha_{\text{crit}} < \alpha < \alpha_{\text{crit}}. \tag{16.19}$$

Otherwise the pattern with the bigger bias is recognized all the time. As Fig. 16.19 shows, the time during which the percept with bigger bias is seen by an individual person increases with increasing bias α_0. It is a quite remarkable experimental and then also theoretical finding that there is a relationship between the behavior of a group of observers and of the individual. Namely the group behavior shows up in the percentage of the first recognized pattern, while the individual behavior shows up in the length of the perception time of the preferred pattern.

Let us treat two other striking features that occur in the recognition of ambiguous patterns, namely hysteresis and fluctuations. We came across the phenomenon of hysteresis in Sect. 5.1, where we illustrated this phenomenon in human perception by means of Fig. 5.3. According to our previous considerations, hysteresis means that the state of our brain depends on previous experience. Thus the same pattern may be interpreted in different manners depending on patterns we have seen before. In order to treat this effect within the framework of the present approach, we begin with the following consideration: If we have not seen a pattern before, or at least not for a long time, we will attach a bias parameter (angle α_0) after a certain duration of habituation. When we are looking at a somewhat different picture the original bias will fade away and will relax towards a bias attached to this new picture. This adaptation will take some while, however, so that we now let the bias parameter $\alpha \equiv \alpha_0$ in (16.14), (16.15), (16.16) become a time-dependent function which obeys an equation of the form

$$\dot{\alpha}(t) = -\mu[\alpha(t) - \alpha_{0j}]. \tag{16.20}$$

Here μ is the damping parameter, whereas α_{0j} represents the "objective" bias of the pattern with label j. (The "objective" bias can be measured by means of the reversion times, though this is not our concern at present.) The solution to (16.20) for $t \geq t_j$, where t_j is the initial time, is given by

$$\alpha(t) = \alpha_{0j} + [\alpha(t_j) - \alpha_{0j}]e^{\mu(t-t_j)}. \tag{16.21}$$

We now proceed as follows: We assume that at time $t = t_1$ the observer starts to regard pattern 1 for some time, say, τ. The index j is put equal to 1 and thus $\alpha_{0j} = \alpha_{01}$. The initial value $\alpha(t_1)$ is subjective and may be chosen arbitrarily in our simulation. After time τ, i.e. at time $t = t_1 + \tau \equiv t_2$, the observer begins to look at the second picture $j = 2$. α_{0j} is chosen equal to α_{02}, whereas the initial value of α, $\alpha(t_2)$ is chosen by means of a solution of formula (16.21) at its final time, i.e. $t_2 = t_1 + \tau$.

We may continue this procedure in a piecewise fashion, where each time the final value of α in one interval serves as the initial value in the next interval. In this way we obtain a curve, such as that shown in the lower part

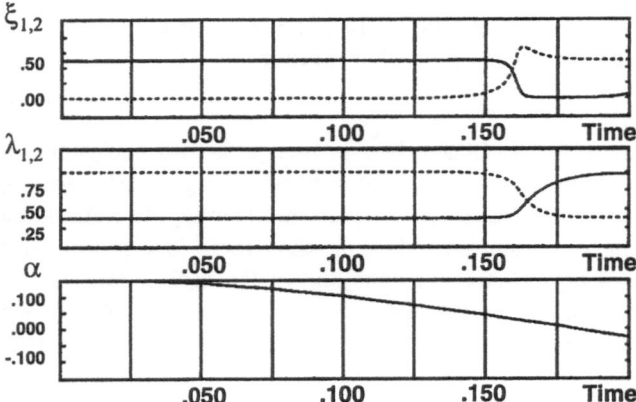

Fig. 16.20. Hysteresis effect in the recognition of face/woman

of Fig. 16.20. Simultaneouly we solve the equations for ξ_k and λ_k, where the time evolution of $\alpha(t)$ is taken into account. The time evolution of ξ_1, ξ_2 and λ_1, λ_2 is shown in the two upper parts of Fig. 16.20. When we perform the same calculation now going backwards in time (Fig. 16.20), we immediately find that the crossing over between the two order parameters ξ_1 and ξ_2 occurs at a different value of the bias. This is the hysteresis effect.

16.4 The Role of Fluctuations of Attention Parameters

In this section we wish to show how our previous results can be refined when fluctuations of the attention parameters are taken into account. In this way good qualitative agreement with psychophysical results obtained by *Borsellino* et al. (1972) are obtained. The agreement can even be made quantitative if the fluctuation parameters and the other parameters of the model system are chosen appropriately. We first briefly remind the reader of the basic experiments of *Borsellino* et al. These authors carefully studied the distribution of the oscillation periods, which are not strictly fixed and fluctuate to some extent (Fig. 16.21). By varying the size of the objects, e.g. Necker cubes, shown to the subjects, *Borsellino* et al. varied the visual angle (angle subtended by the pattern at the eye) and studied its impact on the fluctuations of the lengths of the recognition times. On plotting the frequency (number of specific outcomes) versus duration, they find a distribution function whose shape and center of gravity depend on the visual angle. While for small angles the distribution function has a narrow width and is located at comparatively short times, for increasing angles the width increases and the position of the maximum shifts to longer durations. For different observers the same type of behavior results. *Borsellino* et al. were able to distinguish between fast, slow, and typical observers. In these cases,

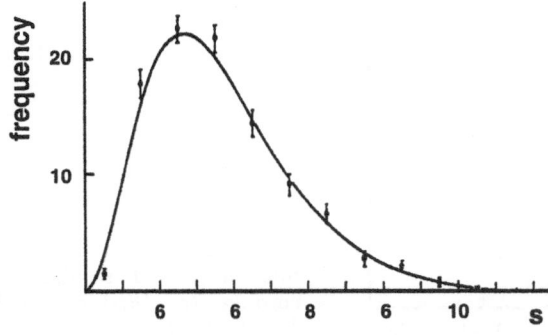

Fig. 16.21. Relative fre-
quency of various periods
T. (Experimental results of
Borsellino et al. (1972))

the maxima of the frequency distribution were shifted to shorter times (fast observer) or to longer times (slow observer).

To describe the time development of the order parameters, we adopt (16.15), (16.16) or, in the absence of bias, (16.9), (16.10). The time-dependent attention parameters λ_1 and λ_2 are now assumed to obey the equations

$$\dot{\lambda}_1 = \gamma \left(1 - \lambda_1 - \xi_1^2\right) + F_1(t), \tag{16.22}$$

$$\dot{\lambda}_2 = \gamma \left(1 - \lambda_2 - \xi_2^2\right) + F_2(t). \tag{16.23}$$

λ_1 and λ_2 are subjected to fluctuating forces F_1 and F_2 as indicated in (16.22), (16.23) above. We assume that the average value of F_j vanishes,

$$\langle F_j(t) \rangle = 0, \tag{16.24}$$

and that the fluctuations have a short memory, or more precisely, that

$$\langle F_j(t) F_k(t') \rangle = \delta_{jk} Q \delta(t - t'), \quad j, k = 1, 2. \tag{16.25}$$

In our simulations we allow the size of the fluctuations, Q, to be varied.

The equations (16.9), (16.10), (16.22), (16.23) were solved by means of an Euler forward procedure. The impact of the fluctuating forces was mimicked by a random number generator. The initial values of ξ_1, ξ_2 were chosen at random in the range $0 - 1$ and the test run was continued long enough to enable us to define a statistics of the frequency of duration, i.e., we could determine a distribution function of frequencies versus duration.

In subsequent test runs, the size of the fluctuating forces was changed and each time the distribution function was determined. The resulting curves are shown in Fig. 16.22. They agree qualitatively quite well with the results obtained by Borsellino et al.

In order to compare our results with their's for fast and slow observers, we changed the relaxation constant γ of the attention parameters. The results are shown in Figs. 16.23 and 16.24. Again a good qualitative agreement is found which can, to some extent, be made quantitative when the parameters are chosen appropriately.

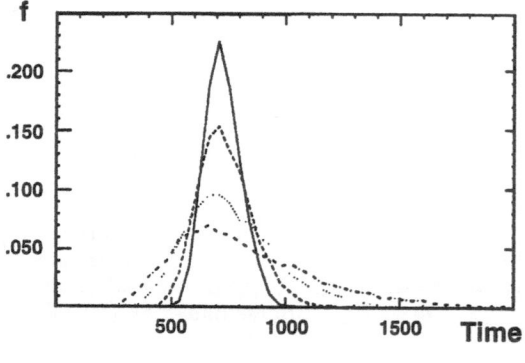

Fig. 16.22. Same as 16.21, but theoretical results according to *Ditzinger* and *Haken* (1990)

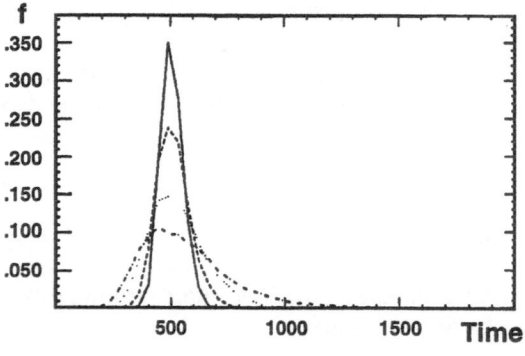

Fig. 16.23. Relative frequency of times T for a fast observer

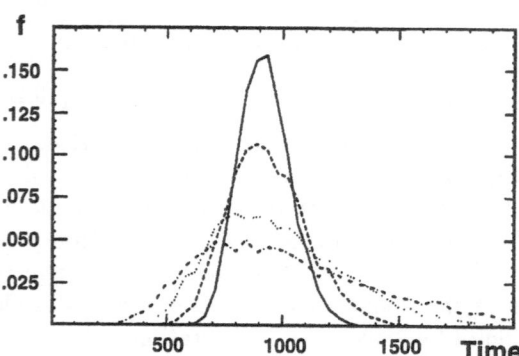

Fig. 16.24. Relative frequencies of observed times for a slow observer

Finally, we studied the impact of fluctuations on the recurrence of pattern interpretation when 3 interpretations are possible. The results are shown in Fig. 16.25. As is evident, the patterns are recognized in a random sequence.

One might ask whether fluctuations act only on the attention parameters or also on the order parameters describing recognition. Our numerical results, which took into account both kinds of fluctuations, showed that meaningful results can be obtained only when we allow for the impact of fluctuations on the *attention parameters* but not on the order parameters. These results

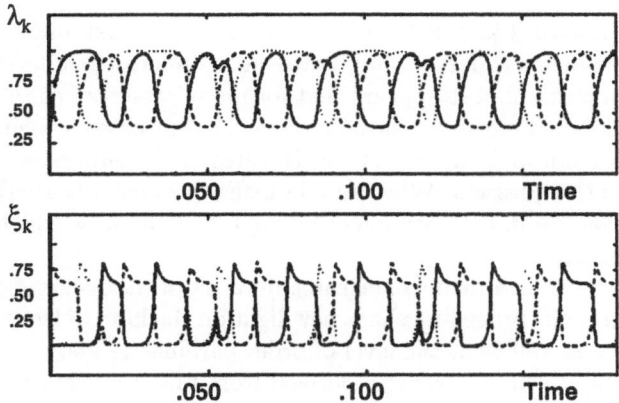

Fig. 16.25. When three interpretations (illustrated by solid, dashed, and dotted curves) are possible and noise on attention parameters is allowed, a random sequence of interpretations occurs

need not necessarily exclude the possibility that fluctuations acting directly on the order parameters play some role. But as a general conclusion, we may definitely state that the impact of fluctuations on the attention parameters is by far the more important .

The good agreement between our results and the experimental results of *Borsellino* et al. is surely quite significant. It supports the idea that the change in shape and position of the distribution function found in *Borsellino's* experiments can be explained by a change in the size of the fluctuations. As mentioned before, the experimentally determined distribution function changed as a function of the vision angle. This function became broader with increasing vision angle (magnification). In our interpretation this means that the attention parameters had to fluctuate more intensely with a bigger vision angle which is certainly a plausible assumption.

16.5 Learning Patterns

Human vision is certainly the field of cognition that has been studied best – experimentally and perhaps also theoretically. In the previous sections of this chapter we have conceived pattern recognition as the emergence of new patterns formed in the human brain or in some kinds of parallel computers, such as the synergetic computer. We believe that the concepts of synergetics shed new light on the relationship between processes at the microscopic level and the emergent macroscopic properties described by order parameters. Pattern recognition appears to be a process of competition between order parameters that is controlled by previously learnt patterns, by attention parameters, and by bias. In addition, fluctuations of attention parameters play an important role. Similar to the case of finger movements, the concepts of synergetics allow us to model cognitive processes in a way that may even be checked quantitatively. Synergetics thus helps to forge a link between the microscopic level and the macroscopic. In particular, it appears that for many processes at the macroscopic level, the detailed nature of the subsystems, in our case the neurons, is not so important and only a few general features of the subsystems are relevant. This raises the question: Why are real neurons so complicated? Most probably the answer lies in the question of learning and memory, which is not as yet well understood.

In conclusion, let us consider a model of learning in terms of the synergetic computer. As we shall see, our procedure sheds new light on the kind of learning at the neuronal level as well as at the level of order parameters and their dynamics. Just recall the learning paradigm derived from finger movements Sect. 8.1), where it was said that the potential landscape of order parameters is deformed by learning. But there it was not at all obvious how learning, changes of the neural network, and changes of the potential landscape are connected with each other. So let us consider this question, which, of course, is of quite a fundamental nature, more closely. To this end, let us start with the dynamics of the recognition process that is described by (16.3). Essential elements here are the stored prototype patterns v_k and their adjoint vectors v_k^+. When we write down the equations for the individual components q_j of the vector q and interpret them as neural actitivies, we obtain the equations (16.4), where the coefficients $\lambda_{j\ell}$ and $\lambda_{j\ell mn}$ are the synaptic strengths. They can be expressed by means of the stored prototype patterns, for instance by (16.5). Quite evidently, our approach has a great advantage over all other conventional neural computers or network models, because it allows us to establish a direct relationship between the synaptic strengths on the one hand and the stored prototype patterns on the other. Thus, instead of studying the changes of the synaptic strengths, we may study the learning of prototype patterns. This is achieved by the following approach: As we have mentioned above, the equations for the dynamics of the pattern recognition process, i.e. (16.3), may be derived from a potential V that depends, on the one hand, on the stored prototype patterns v_k (or their adjoint vectors v_k^+) and, on the other hand, on the test pattern vector q, i.e. $V = V(v_k, q)$. An explicit example of V is provided by

$$V(v_k^+, q) = -\frac{1}{2} \sum_{k=1}^{M} \lambda_k \left(v_k^+ q\right)^2 + \frac{1}{4} \sum_{kk'=1}^{M} B_{kk'} \left(v_k^+ q\right)^2 \left(v_{k'}^+ q\right)^2, \qquad (16.26)$$

where

$$B_{kk'} = \begin{cases} B + C & \text{for } k \neq k' \\ C & \text{for } k = k' \end{cases} \qquad (16.27)$$

Now we invoke a kind of duality principle: To achieve pattern *recognition*, the ball symbolizing the pattern vector q had to run down the potential landscape (Fig. 16.1). Or, to express this in more abstract terms, the potential V had to be *minimized* by a gradient dynamics of q. Here the prototype vectors v_k

Fig. 16.26. Example of the learning of features. In this case three prototype patterns (upper part of the figure) had to be learnt. They were given to the computer in a random sequence. The results of the learning procedure on the vectors and the joint vectors are shown in the lower part of the figure, where learning time increases to the right

are considered fixed. We now turn the argument around and consider the test pattern vectors q as offered but fixed patterns and the prototype patterns v_k and their adjoints as patterns to be learnt, i.e. as patterns to be adjusted. Again, the adjustment must be achieved by a minimization procedure. To this end we formulate the general potential V (16.26) and sum up over all offered test patterns vectors q. Then the vectors v_k^+ must be determined such that $\langle V \rangle = \sum_j V(v_k^+, q_j)$ becomes a minimum, which can again be achieved by a gradient dynamics by the $v_k^+ s$. [1] Thus learning appears as an optimization process. A few examples of this approach are shown in Figs. 16.26 and 16.27. They show convincingly that this procedure leads to an efficient learning of prototype patterns. Once the prototype patterns have been learnt, we can immediately construct the corresponding synaptic strengths $\lambda_{j\ell}, \lambda_{j\ell mn}$ and

---

[1] For practical applications, the conditions $(v_k v_k) = 1$ and (16.2) must be taken care of by additional constraints in the form of potentials V_2 and V_3 that must be added to (16.26). The details are given in Apprendix C.

Fig. 16.27. Learning of partly hidden faces. A random sequence of such patterns, each time with a different position of the squares, was offered to the computer. The learning procedure established clear prototype patterns that coincide with the uncovered faces of Fig. 16.26, top part

thus construct the recognition dynamics according to (16.3). In this way, we have established a connection between the learning procedure on the one hand and the change of the synaptic strengths on the other. At the same time, we see that an abstract potential landscape, namely $\langle V \rangle$, is changed by the learning procedure. It is interesting to note that the learning procedure can be influenced by an appropriate choice of the attention parameters. In this way one may facilitate the formation of classes or categories.

Finally, by the projection (16.8), we may proceed from the microscopic level, described by q, to that of the order parameters, described by ξ_k. When making this transition, we have to transform the potential V depending on q and v_k into a potential V that depends on the order parameters and attention parameters only. In the case of pattern recognition, the resulting potential turns out to be of a universal nature. It simply reads

$$\tilde{V} = -\frac{1}{2} \sum_{k=1}^{M} \lambda_k \xi_k^2 + \frac{1}{4} \sum_{k,k'} B_{kk'} \xi_k^2 \xi_{k'}^2 , \qquad (16.28)$$

with $B_{kk'}$ given by (16.27). Obviously, the *patterns* v_k^+ have disappeared. But the learning effect is hidden in the projection rule (16.8) which contains the *patterns* v_k^+.

16.6 A Model for Stereo Vision

Our world is three-dimensional. To each of us this statement appears self-evident; it is not, however, to the visual system of our brain. Why is this so? Let us briefly consider how visual information is processed by our eyes. As we all know, the eye works like a little camera. The light rays stemming from an object penetrate through the lens and then hit the retina, where they generate some kind of image. Since the retina is two-dimensional, the image generated on it is two-dimensional, too. So the world should appear two-dimensional to us, but it does not. How then does our brain reconstruct the three-dimensional world? A number of mechanisms have been proposed and checked by psychophysical experiments. A simple guess can be by size. For instance we know the average size of people and know that with increasing distance their apparent size decreases. But this may give rise to optical illusions, too (Fig. 16.28). Another guess is based on the size of textures. A further clue is again based on experience and may be called shape from shade. From the shades of an object we guess its form and thus its third dimension. In all these cases, we need some preknowledge to estimate the third dimension. An important and fundamental mechanism was revealed by *Julesz* (1991), who generated random dot stereograms (Fig. 16.29). When we look at these random dot pictures, such that the left eye can see only the left image and the right eye can see only the right image, we immediately recognize the square floating above a background. This recognition of depth without knowing any meaning of the objects is called early vision.

What kind of information is given to the brain to allow it to reconstruct the third dimension? Consider Figs. 16.30 and 16.31. In Fig. 16.30 two points of an object have the same distance from an observer. According to the geometrical reconstruction we easily find that the objects have the same separation on both retinas. Now consider Fig. 16.31, where the objects have a different distance from the observer. Then, as we may easily see from this figure, the distances between the two objects are different on the two retinas. When we compare the two images, we find that the points corresponding to the same object in space are shifted with respect to each other. This shift depends on the positions of the images on the retinas and on the depth

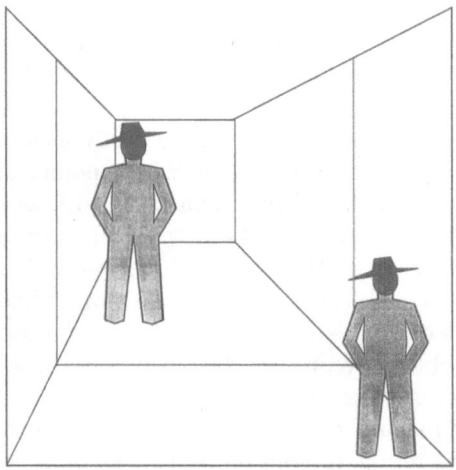

Fig. 16.28. An optical illusion relating to the sizes of people. Both people are actually the same size

Fig. 16.29. Example of a Julesz random dot stereogram. The left and right random dot patterns are different when seen by the left and right eye, respectively. By putting a sheet of paper between the two stereograms, perpendicular to the page of this book, a square floating above a background will be recognized. The square on the right-hand side is the disparity map which encodes the shifts a and b (see text). This map clearly demonstrates that the computer has *recognized* the square floating above a background

objects

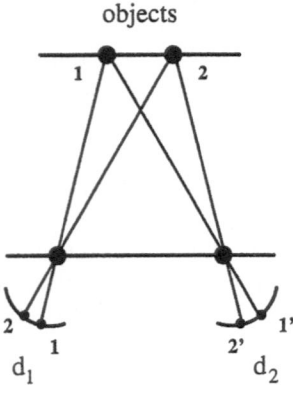

Fig. 16.30. Two objects, 1 and 2, give rise to corresponding object points on the retinas of the two eyes, where the distances d_1 and d_2 are the same

and is called *disparity*. Knowing the disparity we may calculate the distance of objects by means of geometrical considerations. (Actually, there are two further cues, namely vergence and accommodation.)

We shall not enter these considerations here, but wish to study a more fundamental problem, namely: How does the brain know which points on the retina belong to the same object? This is indeed a nontrivial task as is shown by Fig. 16.32. Let us follow the light rays backwards. Each crossing point between two light rays may stem from the same object, but we see that, for instance, we can recognize three objects in front of the plane where the two objects are located. In terms of mathematics, the problem of reconstructing an object is an ill-posed problem. So our first task will be to devise a mechanism by which the recognition process is made unique. Before we do so, we analyse the difficulties a little bit further. Due to the finiteness of the resolution on the image plane, i.e. the retina, a point on that plane does not

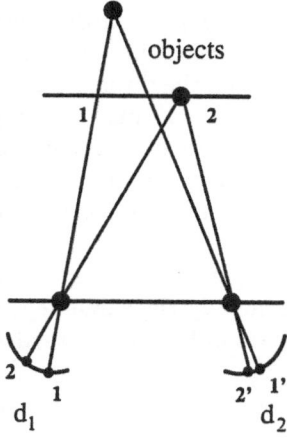

Fig. 16.31. When the distances of the object points are different with respect to the plane of the observer, the distances d_1 and d_2 on the retinas differ

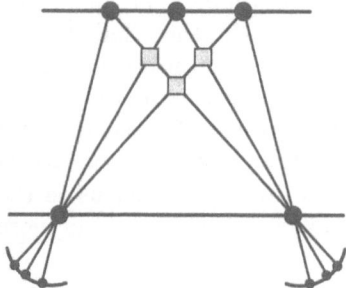

Fig. 16.32. Illustration of the ambiguity in recognition of the distances of objects

correspond to a point but to a segment in space. A point of the 3D space, which is out of focus, corresponds to an area in the image plane. The left eye (camera) may see certain points of the object which are not seen by the right eye (camera) and vice versa. These occluded points, therefore, have no partner. There may be many pixels of the image with the same grey values; therefore, there is often no unique solution to our problem. Because of the different view points, the left and right image planes can show globally different brightness levels. So even when the structure of the images is the same, on the brightness level there are no corresponding points in the images.

Our remarks lead to the following consequences: The finiteness of the resolution of the image planes leads to a finite resolution in 3D space and particularly to a finite resolution in depth, and we cannot use single grey level values as features, but need features which have more structure to be uniquely matched. These should not depend on absolute values of brightness and have to correspond to definite locations on the object in space (*Marr* and *Hildreth* (1980), *Marr* and *Poggio* (1976, 1979)). To overcome these difficulties, in accordance with these authors, we shall use several constraints.

1) *The search area constraint.* We assume a certain range of possible disparity values corresponding to the human visual system. Fusion, i.e. identification of different images on the retina, is only possible if the disparity d lies within a certain range $d < 18'$, called the *Panum fusional area.* Roughly speaking the disparity is mainly in the horizontal, or more precisely speaking, in the so-called *epipolar* direction, but since the lens system may not be perfect, we should also consider the occurrence of a small disparity in the perpendicular direction.

2) *The continuity constraint.* At least in general, in our natural surroundings we always find extended surfaces of approximately the same depth. Therefore, the disparity values should also vary smoothly over extended areas in the image plane.

3) *The uniqueness constraint.* It says that the correspondence problem should have a unique solution.

To summarize what we have said so far: Our task will be to find points on the two images on the retinas which correspond to each other. This will be called the correspondence problem. To solve it, we proceed in several steps. We first introduce the features that we want to match and then introduce a similarity measure between two features. Then we use the experimental fact that there are so-called disparity tuned neurons and we design equations for the temporal evolution of their activities. This will then lead us to the solution of the correspondence problem. Let us follow up these steps in more detail.

In order to define the *matching features* we assign to each image point an appropriate region around it. Each image point is, therefore, related to a certain feature area. Our goal is: Starting from a feature in the left image to find the feature in the right image which is most similar to it. This directly gives us the disparity of the two corresponding image points. To solve the problem of varying brightness of the two images we do not take the grey values in the areas but the structure-relevant differences to the local mean brightness of the areas to be matched. These so-called local features contain all the information we need, namely the local structure of the image. The larger the area, the more unique is the match, but the lower the spatial resolution. A large area in the image plane clearly has to correspond to a connected part of the 3D space; therefore, for such features there exists a one-to-one correspondence of image plane and object.

To show how we introduce the matching features, let us consider a one-dimensional example, where the image points lie on the x-axis. Each image point possesses a grey value $g(x)$ (Fig. 16.33). We define the surrounding of point u, which we take of size $2c$ (Fig. 16.34). Since we are interested in local variations, we subtract the average grey value $\bar{g}$ from $g(x)$. This average value is obviously defined by

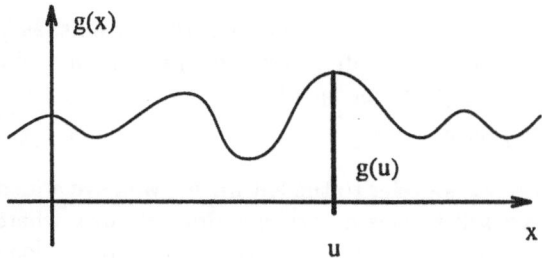

Fig. 16.33. A plot of grey values $g(x)$ versus x

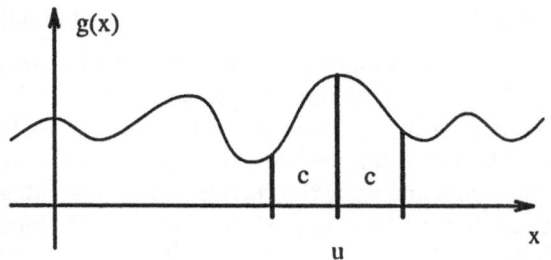

Fig. 16.34. Definition of a surrounding of the point u

$$\frac{1}{2c} \int\limits_{u-c}^{u+c} g(x)dx = \bar{g}(u).$$
(16.29)

In order to cut out the section of length $2c$ around the image point u, we introduce the window function, f, which is defined by

$$f(x - u) = \frac{1}{2c} \quad \text{for} \quad |x - u| \leq c$$
$$= 0 \quad \text{for} \quad |x - u| > c$$
(16.30)

and is illustrated in Fig. 16.35. We shall assume that the window function is normalized

$$\int f(x - u)dx = 1.$$
(16.31)

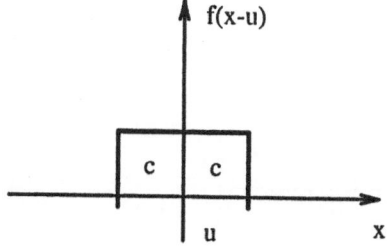

Fig. 16.35. Illustration of the window function f

By means of f, we may cut out a section of $g(x)$ along the whole x-axis, as is shown in Fig. 16.36. As already stated, we shall subtract $\bar{g}(u)$ and thus define the matching feature for the left image by means of

$$L(x; u) = f(x - u)[g_{\mathrm{L}}(x) - \bar{g}_{\mathrm{L}}(u)], \tag{16.32}$$

(cf. Fig. 16.37). Because the grey values refer to the left image, we have added the index L to g. The right eye will receive a corresponding picture, where the features are more or less the same, but are shifted in space. In analogy to (16.32), we define

$$R(x; u) = f(x - u)[g_{\mathrm{R}}(x) - \bar{g}_{\mathrm{R}}(u)]. \tag{16.33}$$

Figure 16.38 shows R with a section taken along the x-axis, where the features correspond to those of Fig. 16.37. The task of the brain is to determine the shift, i.e. the disparity a, by means of a comparison of the features of Fig. 16.38 with those of Fig. 16.36. To enable such a comparison, we shift the matching features of Fig. 16.38 so that they coincide with the corresponding matching feature of Fig. 16.36. This means that we have to perform a shift of R along the x-axis by an amount a. Indeed, when we now introduce $R(x + a, u + a)$ (Fig. 16.39) this shift is accomplished. The problem the brain or the computer have to solve is the following: They have to determine the shift a by means of the form of the matching features so that the matching features coincide as much as possible. We must bear in mind that the two features need not fully coincide, for instance, because of occlusions, etc. Before we go on, we generalize our considerations to two dimensions in the following obvious way:

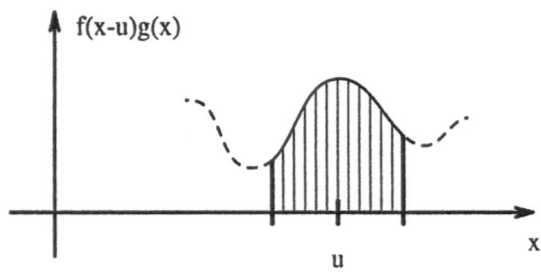

f(x-u)g(x)

u

x

Fig. 16.36. The window function multiplied with the grey value distribution versus space coordinate x

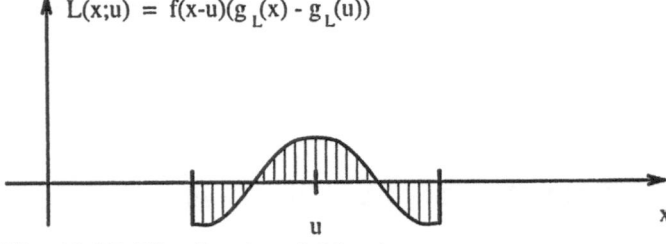

$L(x;u) = f(x-u)(g_{\mathrm{L}}(x) - g_{\mathrm{L}}(u))$

u

x

Fig. 16.37. Visualization of $L(x; u)$

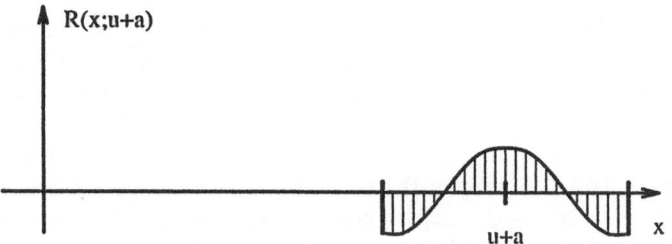

Fig. 16.38. An example of $R(x; u + a)$. In this case R is shifted against L

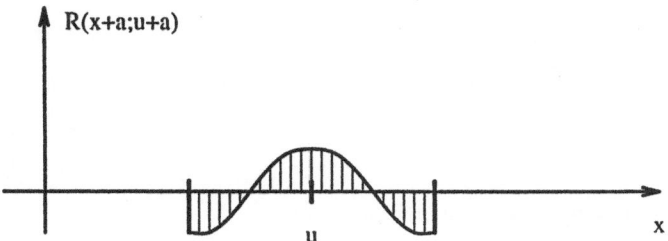

Fig. 16.39. A displacement of R along the x-axis may lead to coincidence between R and L (Fig. 16.37)

$$\left. \begin{array}{c} x \to x, y \\ u \to u, v \\ a \to a, b \end{array} \right\} .$$
(16.34)

We establish a *similarity* between the left image point u, v and the right image point $u + a, v + b$ by considering the distance between the features defined by L and R and integrate over the surroundings of u and v and $u + a, v + b$, respectively. This difference becomes a minimum if the matching is achieved in the optimal way by adjusting a and b.

In the following we introduce a positive measure of the similarity so that we define the following *similarity measure*:

$$S(u, v; a, b) = E - \iint |L(x, y; u, v) - R(x + a, y + b; u + a, v + b)|\, dx dy.$$
(16.35)

In it E is a positive constant that is chosen big enough so that S remains positive. Quite clearly, when the similarity between the features around u, v and $u + a, v + b$ is best, S has a maximum value. Our goal will be to invent a dynamics that assures us, by an adjustment of a and b, that (16.35) becomes a maximum in the course of time. To this end we introduce the variables $\xi_{ab}^{uv}(t)$ and take the relation

$$S(u, v; a, b) = \xi_{ab}^{uv}(t = 0)$$
(16.36)

as the initial values for these variables. We identify ξ_{ab}^{uv} as the activity of the "disparity neuron" which corresponds to the image point u, v and the disparity a, b. We assume that in the course of time ξ_{ab}^{uv} changes its activity because of a process in which the activities of the disparity neurons compete. Finally just one configuration wins which corresponds to the correct distributions of disparities according to the three-dimensional positions of the objects.

We borrow the basic idea for this process from the synergetic computer model studied in Sect. 16.1. We subject the variables ξ_{ab}^{uv} to a set of differential equations in which the various disparities compete with each other and the greatest similarity wins. In generalization of that model we introduce an additional term in which points $u'v'$ that are neighbors of uv and have the same disparity ab, help the variable ξ_{ab}^{uv} to win. These considerations lead us to the equations

$$\dot{\xi}_{ab}^{uv} = \xi_{ab}^{uv} \left[\lambda_{ab}^{uv} - (B + C) {\sum_{a',b'}}' (\xi_{a'b'}^{uv})^2 - C\, (\xi_{ab}^{uv})^2 + D {\sum_{u',v'}}' \left(\xi_{ab}^{u'v'} \right) \right],$$

$$(16.37)$$

where we use the following analogies with (16.7):

$$\xi_k \rightarrow \xi_{ab}^{uv}(t), \tag{16.38}$$

$$\lambda_k \rightarrow \lambda_{ab}^{uv}. \tag{16.39}$$

The primes on the sums indicate that in the first sum $a', b' \neq a, b$ and in the second sum $u', v' \neq u, v$. Let us discuss the individual terms in this equation.

We choose λ_{ab}^{uv} as

$$\lambda_{ab}^{uv} = S(u, v; a, b), \tag{16.40}$$

i.e. we attribute the biggest growth rate to the configuration a, b that is initially biggest. There are three reasons why we do not simply choose this highest value to find the correct disparity a, b of a given image point uv:

1) There may be no unique or significant maximum of $S(u, v; a, b)$.
2) We have not yet implemented the continuity constraint and
3) Choosing is a serial process and therefore slow. In analogy to the human brain only parallel operating algorithms are fast enough to process large amounts of data. The terms containing B and C in (16.37) serve for a competition between the variables ξ_{ab}^{uv} with different indices ab, while the term containing D supports the variables ξ_{ab}^{uv} with the same index ab by means of its neighbors u', v'.

As we have seen in Sect. 16.1, equations of the form (16.37) can be easily interpreted as the action of a parallel network. At the present state of computer technology, we had to solve the set of equations (16.37) on a serial computer instead of using a parallel network.

Note that the number of equations (16.37) is rather large because we used about 100×100 pixels (u, v) and displacements (a, b) of the order of 10. At the end of a calculation we find that some of the ξ_{ab}^{uv}'s are zero, and others are not equal to zero. From the latter ones we can read off the position u, v of an object and the disparity a, b. For instance in Fig. 16.29, right-hand side, the black parts show that at these points the disparity has a certain constant value, the white parts show that here the disparity has another constant value – corresponding to a floating of the square above a background. Note that in this example the disparity in the y-direction (vertical) was chosen equal to zero. Figures 16.40–16.42 show that the depth recognition process described here is quite robust, as is known from human subjects, too. These findings have been substantiated by further computer experiments (*Reimann* and *Haken* (1994)).

To conclude this section on stereopsis a remark should be made about the recognition of 3-D objects in 2-D pictures. (Such pictures have become very popular and can be found in books, on wallpaper, advertisements, etc.) The principle is shown in Fig. 16.43. It is actually related to the ambiguity of depth recognition as was demonstrated by Fig. 16.32. When we construct extended objects that are superimposed on each other in an appropriate way (for which computer programs are available), the scheme of Fig. 16.43 can be extended so that whole objects are recognizable in three dimensions. If a

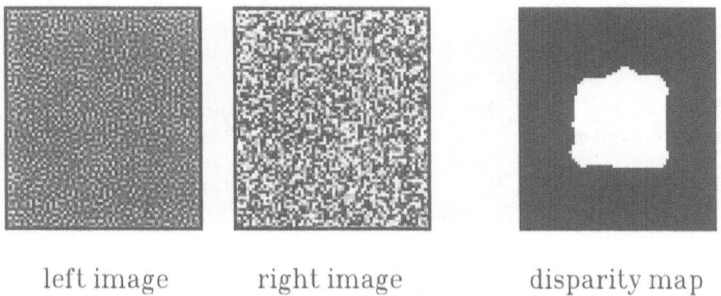

left image   right image   disparity map

Fig. 16.40. Same as Fig. 16.29, but with differently coarse-grained random dot stereograms. The computer – like humans – still recognizes the square floating above the background (after *Reimann* and *Haken* (1995), as are Figs. 16.41, 16.42, 16.45–16.48)

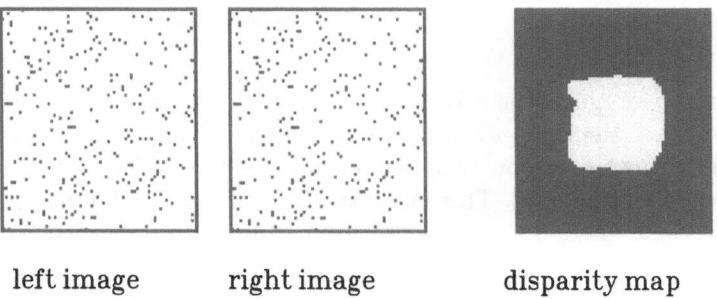

left image   right image   disparity map

Fig. 16.41. Same as Fig. 16.40, but with sparse random dot patterns

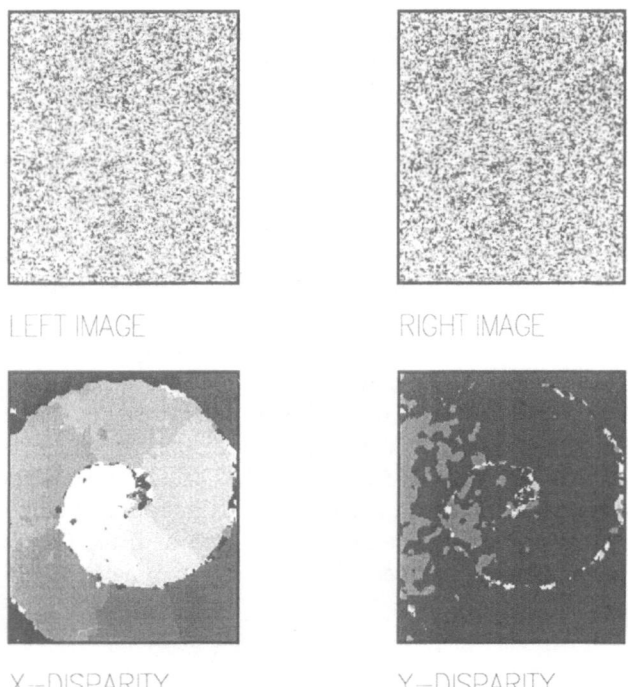

Fig. 16.42. Same as Fig. 16.41, but with a spiral. Note how the disparity map reveals the different depths

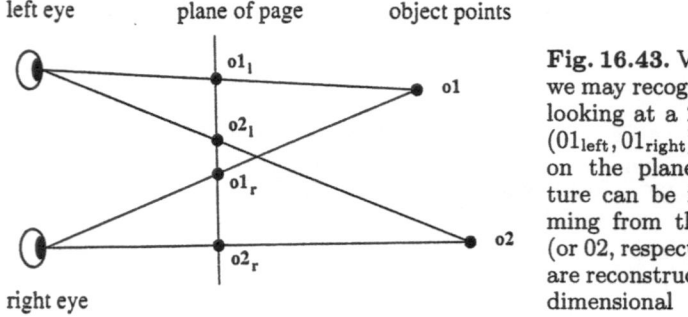

Fig. 16.43. Visualization of how we may recognize 3-D objects by looking at a 2-D picture. Pixels $(01_{\text{left}}, 01_{\text{right}})$, or $(02_{\text{left}}, 02_{\text{right}})$ on the plane of the 2-D picture can be identified as stemming from the same object 01 (or 02, respectively). The objects are reconstructed as being three-dimensional

person wants to recognize the 3-D object, he or she must direct his/her eyes appropriately, i.e. the vergence must be controlled. This vergence control can be achieved by the computer by minimizing a potential function V with respect to the vergence shift. This function can be chosen in different ways. A simple procedure is to set

$$V = -S, \tag{16.41}$$

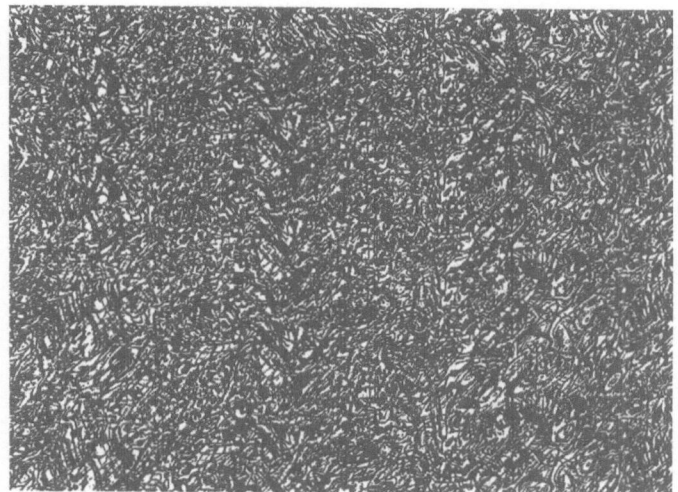

Fig. 16.44. Section of a commercial 3-D picture

where S is defined by (16.35), and $a \equiv s, b = 0$ if we consider only a horizontal shift. We must observe two basis differences between our previous application of S and the present one.

1) In contrast to our previous procedure, we now choose

   $$R(u, v) \equiv L(u, v)$$

2) R and L refer to larger features or whole objects, whereas above we were concerned with a "fine-tuning".

Figure 16.44 shows a section of a 3-D picture in which a unicorn is hidden. The minima of V (Fig. 16.45) determine the gross disparity values – then the computer can do the fine tuning. Figures 16.46 and 16.47 show how the disparity maps reveal a single unicorn or triple unicorns.

In order to recognize depth, a human or a computer has to compare two images in space (on the retinas). But we may also compare the change of images in space *and* time. As *Reimann* (1995) has shown, the procedure described above can easily be generalized to enable the computer discover optical flow as shown in Fig. 16.48.

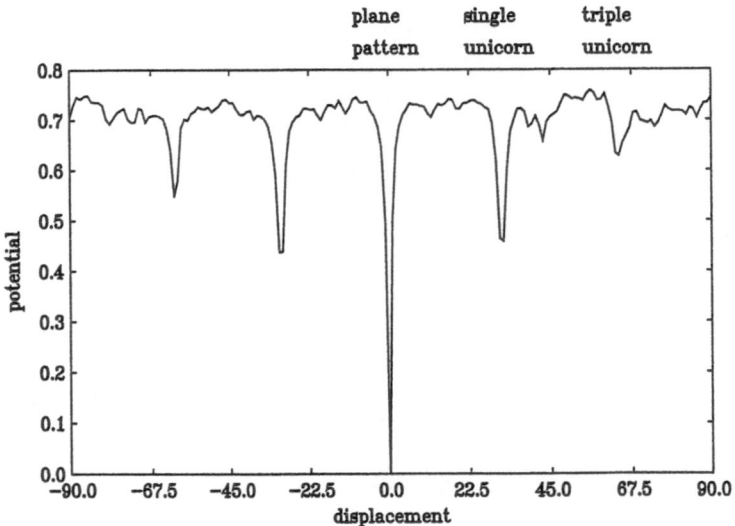

Fig. 16.45. The potential (16.41) plotted versus displacement

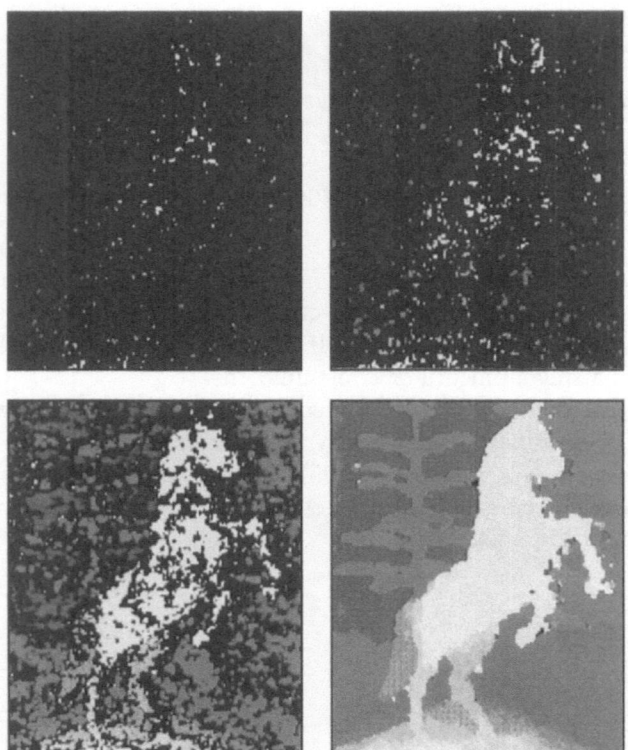

Fig. 16.46. Temporal evolution of the disparity map belonging to a picture of which Fig. 16.44 shows a segment

Fig. 16.47. Disparity map of the triple unicorns using the corresponding minimum of the potential of Fig. 16.45

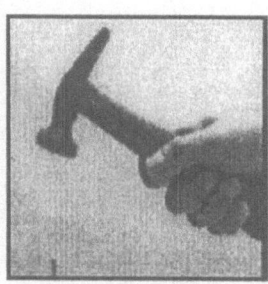

time t

time t+τ

optical flow

Fig. 16.48. Determination of optical flow connected with a movement of which two snapshots are shown. From *Reimann* (1995)

17. Decision Making as Pattern Recognition

In this chapter we will follow up our idea that visual pattern recognition can serve as a metaphor for the understanding of human cognitive abilities. One typical problem confronting humans is decision making. This has to be done in our personal daily life, but also in economy and companies, especially by managers, and it is an important task in politics, and so on. When we analyse the problem of decision making more closely, we quickly find that there are a number of intrinsic difficulties, of which we mention a few. In general, the information we have about a problem on which we have to make a decision is incomplete. In mathematical terms, the problem is often ill-posed in a way that may be reminiscent of the problems we discussed in stereopsis. Quite often decision making has to be done in conflict situations. Each specific decision bears its own risks. The problem of decision making implies that, in general, there are multiple choices and a repertoire of actions. In studying these problems, both quantitative and qualitative methods have been applied and there is, of course, a considerable literature on decision making.

In this chapter we wish to shed new light on this problem by invoking an analogy between decision making and pattern recognition. In general, there is a discrepancy between the known data and the required data needed to decide upon a specific action (Fig. 17.1). In the ideal case the known data coincide with the required data. In general, however, the known data are insufficient, i.e., there are a certain number of unknown data. How do humans fill the gap of unknown data? This is what we want to analyse in the following. A simple though nontrivial example is provided by a tennis player. Some analysis shows that his or her time for the necessary reaction is too short to allow an analysis of the sensory input data before starting motor action. Thus tennis players have to act on specific clues that are based on their experience and training. But, in general, the problem is still more complicated. Consider to this end Fig. 17.2. It indicates that, at least in general, the known data can be complemented in a variety of ways to fill in the gap of the unknown data. Depending on how we fill in the unknown data, different decisions or actions may be taken. This figure is in a sense oversimplified, because even if all the data are known there may still be several decisions that are compatible with all the known data.

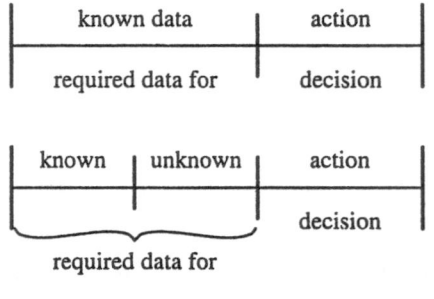

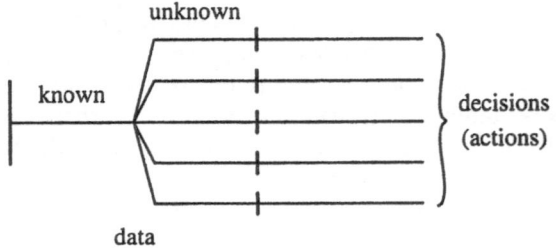

Fig. 17.1. *Upper part:* The known data coincide with the required data for taking action or making a decision. *Lower part:* The known data are insufficient

Fig. 17.2. The known data may be complemented in a variety of ways. Depending on these different ways, different decisions or actions may be taken

How do we fill in the unknown data? Our main theme will be that we often rely on a similarity between a given situation and a previous situation. When we want to cast this similarity into a mathematical frame, we have to look for similarity measures (see below). Of course, in a nonmathematical way we may rely on analogies or metaphors. A number of psychological factors are of importance, such as awareness, attention, bias, and beliefs. When we take seriously the analogy, which we shall discuss below, with pattern recognition, we can expect time-dependent choices as in the case of ambiguous figures. This implies that we make decisions that show oscillations, or in the course of time there may be random choices. In more detail, we propose to draw the following analogies between pattern recognition and decision making (cf. Table 17.1). In decision making the data correspond to patterns treated in pattern recognition. The data may be quantitative or they may consist of specific rules, laws, or regulations. They may be in the form of algorithms, or when we think of computers, in the form of programs or flow charts. Diagrams may also be considered as constituting such data. In pattern recognition the patterns may consist of pictures or of the arrangements of objects. The patterns may be visual or acoustic signals. Quite often these patterns are encoded as vectors, which may be constant or time-dependent. Of course, in decision making the data may be multi-dimensional.

So far we have been discussing the analogy between the objects dealt with in decision making and in pattern recognition. In both cases the prototype patterns or the sets of known complete *data* may be learned or given. Incomplete data in decision making have their analog in pattern recognition in the form of incomplete test patterns. How can we exploit this analogy to

Table 17.1. Correspondence between the elements and processes of pattern recognition and those of decision making

Pattern Recognition	Decision Making
patterns pictures arrangement of objects visual, acoustic signals movement patterns actions (often encoded as vectors)	data, quantitative, qualitative, yes/no rules, laws, regulations algorithms, programs flow charts, diagrams orders multi-dimensional in short: "data"
prototype patterns learned or given	sets of known complete "data" learned or given
test patterns	incomplete data in particular "action" lacking
similarity measure dynamics bias attention, awareness	$\rightarrow$
unique identification or oscillations between two or more percepts hysteresis complex scenes saturation of attention	unique decision or oscillations between two or more decisions do what was done last time even under changed circumstances multiple choices failure, new attempt based on new decisions "heuristics"

study decision making? In analogy to pattern recognition we may introduce a similarity measure, for instance, the overlap of prototype patterns and the test pattern. We can then establish a dynamics that is based on the similarity measure and may also include bias, attention parameters, or awareness. So, from a formal point of view, the whole procedure that we encountered in Chap. 16 in pattern recognition may be transferred to a scheme describing decision making.

What will be the consequences? They are listed up in Table 17.1. In pattern recognition and in decision making, we may find a unique identification and a unique decision, respectively. But in a number of cases we may be confronted with oscillations between two or more percepts, or between two or more decisions. These oscillations are not unusual in our daily life as everybody knows. Here we can trace them back to a fundamental mechanism of the human cognitive abilities. A very important analogy arises when we remember the hysteresis effect that we came across in pattern recognition (cf. Fig. 16.20). Translating this effect into decision making means the following: A person does what he or she did last time even under changed circumstances. The analogy between pattern recognition and decision making can be carried further. In pattern recognition we dealt with complex scenes, where we saw that the computer and probably the human brain analyses such a scene by means of a saturation of attention. Once part of a scene has been recognized, we focus our attention on the other objects. In our analysis of decision making, multiple choices correspond to complex scenes and the saturation of attention, we met in pattern recognition, can now be translated as follows: Based on our attention we make a first choice. When we encounter a failure, the *attention* parameter for that endeavor is put equal to zero. We then make a new attempt based again on our attention for a new kind of endeavor, and so on. Depending on our previous experience there may be a hierarchy of attention parameters through which we work starting with the highest attention parameter. This interpretation is related to *Wagenaar's* (1993) notion of *heuristics*.

Summarizing these ideas we can state: The mechanisms discussed in the case of pattern recognition can be translated into those of decision making. This can be done not only at a qualitative level but also quantitatively at the level of computer algorithms in analogy to the synergetic computer. Quite obviously, our analysis is by no means complete and other strategies may be of equal importance. Artificial intelligence and here especially the approach by expert systems must be mentioned. A problem encountered here is that of branching, where the various branches become extremely numerous and decision making eventually becomes very difficult. We believe that this branching problem can be circumvented by the approach we have outlined above, because, as in pattern recognition, the various possibilities are taken care of in a parallel fashion.

18. The Brain as a Computer or Can Computers Think?

18.1 An Excursion: *What is Thinking?*

Everybody knows, of course, what thinking is about. But when we are forced to define it more rigorously, this turns out to be a hard task. First of all it is fairly simple to define actions that are not based on thinking, according to our common understanding. Such actions may be reflexes, for instance, when we close our eyes if an object is quickly approaching our face, or when we move our leg if somebody hits our knee with a hammer. Instincts are not supposed to be related to thinking and even the rather complex behavior of insects is often attributed to instincts. (It appears, however, that a new kind of understanding of insect behavior might emerge.) Thinking is not involved in unconscious actions, such as walking or swimming, actions that are controlled by the cerebellum and the spinal cord.

We may properly speak of thinking when higher brain activities are involved, such as in recognition and production of language, analysis of scenes, pictures, etc., i.e. in all cases where meaning is attributed to perceived visual or acoustic patterns. Thinking is involved when we analyse situations, which may be the diagnosis by a doctor or a housewife studying the special offers in a supermarket. A good deal of thinking is based on drawing analogies, on associations, and on classifications. Thinking is involved in planning and in making decisions. Actually, we studied a model for decision making in the previous chapter. We are all convinced that playing chess or solving mathematical problems involves a good deal of thinking. Thinking lies at the basis of the development of theories about physical, chemical, social and many other processes, or when we solve engineering tasks. Quite generally, while thinking we devise mental pictures of the world or mental representations. A number of experiments have been done to substantiate these concepts. For instance, when people are asked to mention a flower, most of them immediately mention a rose, for a tool a hammer, and for a wild animal a lion. In the study of mental representations, scientists speak of internal maps that we devise, for instance, of cities or of our surrounding. Thinking is connected with consciousness and introspection, both vast fields that remain largely unexplored despite many efforts. An issue of great current interest is the question of whether animals can think or have consciousness, and numerous experiments have been reported which indicate that quite diverse species can indeed *think*.

In conclusion of this section let me mention that there are whole books devoted to the problem of intelligence and how to establish measures of intelligence. A widely known example is the intelligence quotient or IQ, where by well-defined questions to be tackled by test persons a number for their IQ can be calculated. In more recent times these procedures have increasingly been questioned. In the present author's opinion, the question of intelligence and how to measure it is more subtle, because intelligence may be highly task related. Some people are gifted for some specific kinds of tasks, other people for other kinds. In addition there might be some kind of hierarchy of intelligence when we compare intelligence in the animal kingdom with that of humans. In the context of this chapter, present-day computers are still far away from fulfilling tasks that one might call intelligent so that no IQ has yet been established for computers.

18.2 Computers

The mechanization of mental processes has a long history that dates back to *Leibniz* and others who devised mechanical calculators or mechanical computing machines. These machines were constructed to perform additions, subtractions, multiplications, and divisions. A big step forward was made by the development of electronic computers and the invention of effective computer architectures, notably by *Zuse* and *von Neumann*. According to the architecture devised by *von Neumann*, a computer is composed of a processor in which all the operations are performed and a memory. Actually, we find the same architecture in the Turing machine, which we shall discuss below. While the first electronic computers were composed of electronic valves, modern computers are based on semiconductor devices, namely silicon chips, on which myriads of individual functional elements are integrated. Future computers will probably be based not only on semiconductors but also on lasers and molecular electronics. All these computers have in common that they perform their operations sequentially, i.e., one after another. Quite often loops are built in, for instance, when a solution to a problem is found by approximations and the approximations can be made better and better by repeating the individual steps. Irrespective of the material substrate, the individual components of a computer are logical elements that can realize the mathematical operations of a so-called Boolean algebra. It is interesting to note that these mathematical operations are not multiplications of numbers, say, but logical operations, such as 'AND' and 'OR'. An example of how such operations can be realized by a simple mechanical device based on water pipes is shown in Fig. 18.1. A connection between these logical operations and the operations of multiplication, addition, subtraction, and division can be easily established when one uses a binary system of numbers. In this system any

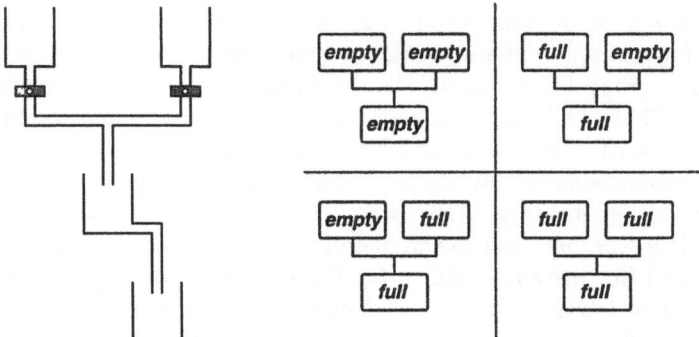

Fig. 18.1. A simple example of the realization of the logical function 'or' by means of interconnected vessels that can be filled by water. The right-hand side of this figure shows the individual cases

number is represented by zeros and ones only. Let us illustrate the relationship between logical relations and multiplication by means of the following example: As beginners at school know, the multiplication rules include $0 \times 0 = 0, 0 \times 1 = 0, 1 \times 0 = 0$ and $1 \times 1 = 1$. These rules are completely mimicked by the *water computer* of Fig. 18.1, if we identify zero with *empty* and one with *full*. At the same time, the *water computer* realizes the logical operation 'OR'. The lower vessels become full when the left *or* the right upper vessels (or both) have been full. Similary, other algebraic and logical rules, such as addition, can be mimicked.

An important concept in the theory of computation, and, as it nowadays turns out, in some brain theories, is that of the *Turing machine* or, more precisely speaking, of the *universal Turing machine*. Such a machine was first devised by the famous British mathematician *Turing* (1936). It consists of a tape that may be, at least in principle, infinitely long. On it the digits *zero* and *one* can be printed or erased by a head that moves along that tape. The head contains a program. According to that program and the digit it just reads off the tape, the head may replace the digit by another one or leave it unchanged and may move on in either direction, where it again erases and prints, or leaves a digit and then moves on again. The program to, can be encoded by means of digits. Before the Turing machine begins operation, a series of digits are imprinted on the tape. These numbers are both the program to be read by the head and the initial values on which the program operates. Then the machine is supposed to operate for some time, until it comes to a halt and the final result can be read off from the tape. In the sense of the Turing machine, a mathematical problem is called computable if the machine comes to a halt after a finite time. In this way, the Turing machine plays an important role in discussions and proofs relating to computability.

To illustrate the issue of computability, let us think of a number of symbols, A, B, C, for example. The operations connecting these symbols are assumed to form a group (cf. Chap. 9). We can form all sorts of expressions like AA, AB, AC, BCA, i.e., any combinations of these letters. Each combination is called a 'word'. One then defines specific relations, such as AB = 1, i.e., special combinations are put equal to unity. Quite clearly, because of AB = 1, we obtain, for instance, ABC = C. The *word problem* consists in the task of deciding whether, due to the defining relations such as AB = 1, two words are equal (for instance ABC = C). Consider now the general case of an arbitrary set of letters and a set of defining relations. Can we devise a general procedure (an algorithm) by which we can decide whether or not two words are equal. Such a problem can be translated into a problem on the Turing machine and it can be shown that we cannot determine whether the Turing machine comes to a halt or not. In other words, there is no general procedure possible by which we may solve the *word problem*. Actually, the halting problem is closely connected with *Gödel's* incompleteness theorem, but we shall not dwell on this relationship here.

Since in some discussions on the nature of brain activities the concepts of the Turing machine and Gödel's theorem play a role, we add a few comments on the halting problem. There are, indeed, intricacies with respect to the halting problem that were already mentioned by *Wiener* (1948) in his famous book on cybernetics, though in a different context. In the context of our book, we can, for instance, put the problem of the recognition of ambiguous figures to the Turing machine. In Chap. 16 we saw how the corresponding equations can be established. We can then ask the Turing machine to decide whether it '*sees*' a vase or two faces. Quite clearly, the solution is oscillatory, i.e., the Turing machine will never come to a halt. So at this level, the problem *vase or faces* is, indeed, *undecidable*. But nature teaches us quite clearly how it solves this problem, namely by means of oscillations. So when we transform the problem to one of a higher level, namely by asking "Are there oscillations?", the Turing machine will immediately answer in the affirmative and come to a halt. Though this example is rather simple, it allows us to draw some interesting conclusions: Whether or not a problem can be solved does not only depend on the process of solution, but also on the kind of question we ask and/or, in particular, on our insight concerning what we accept as a solution. Indeed, great discoveries, for instance in mathematics, are based on this kind of insight. Just remember the discovery of the imaginary unit $i = \sqrt{-1}$. Here we recognize that 'i' is an acceptable solution to the equation $i^2 = -1$ provided we enlarge the concept of numbers.

If we take these comments seriously, we may certainly say that a brain does not act like a Turing machine. Put in other terms, when a Turing machine produces a set of numbers as the solution to a mathematical problem, it is the human being who attaches meaning to these numbers, or, if the Turing machine finds a problem undecidable, the human being may find out *what* is really happening! We shall come back to the question of meaning in a later chapter.

18.3 Artificial Intelligence

Following the advent of the electronic computer, it became clear that this machine cannot only manipulate numbers but also work on symbols. This has become a common place when we simply think of a PC as word processor. On the other hand, we must be aware that the operations of a word processor are still rather primitive and are certainly not connected with intelligent behavior by the computer. The goal of artificial intelligence is indeed far removed from this. It rests on the idea that objects and actions are represented in our brain by means of symbols. According to artificial intelligence, thinking means processing these symbols according to specific rules. According to this idea, programs were developed to demonstrate that a number of problems can be solved in this way. An example is that of arranging little colored building blocks, such as cubes, cylinders, and pyramids. It is possible, for instance, to put a pyramid on a cube but not a cube on a pyramid. In this way *Winograd* (1972) developed programs so that a computer could execute commands, such as: Put the red pyramid on the blue cylinder and the whole arrangement on the green cube, etc. Another example of the application of artificial intelligence is the design of chess computers which indeed, are now quite powerful and occasionally even beat grand masters. The *secret* of the success of the computer lies in its speed: It can check a large number of moves in a very short time. It definitely works in a completely different manner from the way a skilled chess player acts – the computer works by brute force rather than by intelligence.

Another now famous example is the computer program Eliza written by *Weizenbaum* (1966). He devised a program that could somehow mimic a psychoanalyst. This program posed questions and acted upon the responses of patients. Actually, it was based on some tricks that were taken from typical questions of analysts, for instance, when a patient mentioned his or her mother, the computer program asked: Tell me more about your mother. *Weizenbaum* himself was rather sceptical about this approach and was fully aware of the limits of computers and their kind of '*thinking*'.

In specific well-defined areas computers can be used to perform tasks that are otherwise done by humans, for instance, bookings in banks, hotels, or travel terminals, provided the users follow specific rules. Another use of computers that is envisaged is as expert systems. For instance, some doctors are interested in devising programs that may replace the diagnosis made by humans. Practitioners know, however, that this approach is beset with enormous difficulties because of the complexity of diagnosis. The same symptoms may be caused be quite different diseases and if one wishes to explore all possibilities or to eliminate the very unlikely possibilities, the number of questions and decisions to be made becomes enormous. In addition, one must realize that in practically all cases the efficiency or capability of the computer is not based on its own ability to think, but rather on the cleverness of the programmer.

From the experience gathered with all these systems, a general conclusion can be drawn: A prerequisite for any present-day computer work is a finite number of well-defined objects and well-defined rules. Computers cannot (or can hardly) handle vagueness, ambiguities, and suchlike. (This must not be confused with fuzzy logic, which is actually based on well-defined rules!) The brain seems to use quite different strategies, some of which were discussed in Chap. 17.

18.4 Neurocomputers and Connectionism

When we compare the speed with which typical tasks of daily human life are performed by computers and by the human brain, we quickly find a big discrepancy. For instance, when we perceive a scene, our brains (of which the eyes are a part!) must process billions of bits. On the other hand, it is well-known that the individual constituents of the brain, namely the neurons, work at a slow pace in the range of milliseconds. In spite of the slowness of its elements, the brain can process an enormously high amount of information in a fraction of a second. By contrast, computer elements are very fast, but the conventional computer is slow (or not even capable) of recognizing scenes. For these reasons the brain cannot act like a serial computer; rather it works in parallel. Since in the seminal work by *McCulloch* and *Pitts* in 1943 concepts for parallel computers were developed, we just remind the reader of that model. In it one assumes that a network is composed of individual model neurons, each of which possesses two states, a resting state and an active state. When a neuron in its resting state receives signals from other neurons, it remains in the resting state if the sum of the signals is smaller than a certain critical value, the so-called threshold. If the sum of the signals exceeds that threshold, the neuron switches to its active state and emits a signal itself. These authors demonstrated that a network of interconnected neurons of this type can perform all the logical processes of a Boolean algebra provided the links between the neurons are chosen appropriately. *Rosenblatt* (1962) was the first to realize such a network by means of his perceptron. The story is told that subsequently this field was not followed up further because of a harsh criticism by *Minsky*, who had shown that the perceptron could not solve the XOR-problem, the *'exclusive or'* operation. Experts tell me, however, that others were fully aware of this and similar obstacles. At any rate, for a while this field was followed up by very little research until it was eventually revived in the beginning of the eighties.

Progress was made by the introduction of three-level adaptive filters as shown in Fig. 18.2. The first layer contains the input layer, from which in-

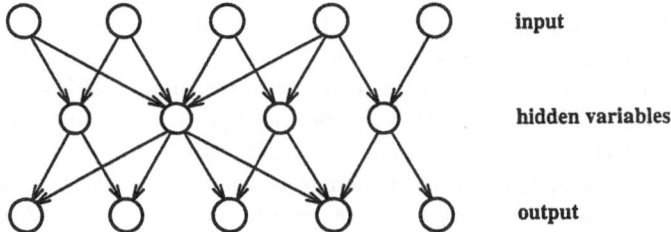

Fig. 18.2. An adaptive filter. The arrows indicate the synaptic strengths. Note that not all connections are shown here. In contrast to more common representations, here the input layer is the uppermost layer, and the output layer the lower layer, in agreement with the corresponding representations of Chap. 16. There is a decisive difference between the synergetic computer (Fig. 16.11) and an adaptive filter: In the former the middle layer nodes are connected to one another, in the latter they are not

formation is transferred to a second layer containing the so-called hidden variables. (In contrast to the network of Fig. 16.11, the hidden variables are not connected with each other.) From the middle layer signals are emitted to a third, output layer. The main problem consists in fixing the projections from one layer to the next. These projections depend, on the one hand, on the inputs emerging from each individual model neuron, but also on the weights according which these signals are transmitted. The major problem consists in the determination of these weights or *synaptic strengths*. (The word *synaptic* is chosen in analogy to processes between real neurons; Chap. 2). A widely used concept to fix these synaptic weights is back propagation.

Let us explain this concept in more detail. According to a widespread philosophy in this field, a neural network learns by means of examples. So a number of problems are presented to the network, whose answers are known. The network is first equipped with more or less randomly chosen synaptic strengths. Then the results calculated by the network are compared with the known exact results. The idea is to minimize the error by consecutively adjusting the synaptic strengths. Because this procedure is done backwards starting from the output layer to the middle layer and then from the middle layer to the input layer, it is called back propagation.

A famous example was developed by *Sejnowski* and *Rosenberg* (1987), who taught a network how to read written English language aloud. As is known, the pronunciation of nouns and vowels depends on the word in which they appear. For instance, a in bar is voiced differently to the a in bad, and an s is read differently depending on whether or not it occurs in connection with h. *Sejnowski* and *Rosenberg* succeeded in teaching such a network and the results were comparable with the performance of children in the first one or two years of elementary school. Note that in this way no semantics was learnt by the computer, simply the rules governing how to pronounce words.

The comparison between the performance of a neural network and a human brain in learning tasks is still flourishing. But there are at least two difficulties: Up to now there exists no general theory of learning or of how to devise a neural network, say with respect to the number of components in its various layers, and there is also no general understanding of how such a network really acts. We can follow up the change of the synaptic strengths and can also visualize them, but this does not give us any insight into the general principles of how the overall state of the network is established and how it can be adequately described.

We conclude this section with the remark that for the world of physicists interest in this field was greatly stimulated by a paper of *Hopfield* (1982), who demonstrated a formal equivalence between the McCulloch and Pitts model and spin glasses in solid state physics. Spin glasses may be visualized as being composed of individual magnets, each having a north and a south pole. We know that the interaction energy between such magnets is higher when they lie parallel and lower when they point in antiparallel directions. Assuming that we fix the interaction strengths between magnets at different sites in some lattice, we can attribute a total energy to the whole system depending on the orientation of the individual magnets. The orientations 'up' and 'down' correspond to the two states in the two-level neuron as introduced by *McCulloch* and *Pitts*. The Hopfield model has the advantage that one can calculate a characteristic energy of the system that is connected with the individual states of the elementary magnets. On the other hand, the system is plagued by so-called ghost states, i.e., by states which do not correspond to stored patterns, and the system can be trapped in such states.

To conclude this section, let me compare neurocomputers with the synergetic computer of Sect. 16.1. Despite some formal analogies (parallel processing, three-level realizations) there are a number of fundamental differences between neurocomputers (including spin glass models and adaptive filters) and the synergetic computer. The main differences are, besides the detailed properties of the neurons that we discussed in Sect. 16.1, the following:

1) In a number of cases, such as in the Hopfield model, there are so-called ghost states, i.e., the recognition process gets trapped in states that do not correspond to actually stored or learned patterns. In order to circumvent the difficulty of trapping, complicated and time-consuming procedures have been developed, in particular *simulated annealing*. In this process random pushes are exerted on the ball in a potential landscape to push it out of the unwanted trap and to move it into deeper-lying states that correspond to the actually learned patterns. The synergetic computer totally avoids this difficulty, because there are no ghost states or other unwanted states. It is highly unlikely that the brain uses simulated annealing or anything similar.

2) In the case of the synergetic computer, the concepts of order parameters and enslavement allow us to characterize the total states of the network

and even to attribute *meaning* to them. This possibility of interpretation is lacking in the case of neurocomputers.

3) In actual realizations of neurocomputers, it has so far not been possible to include more than a few hundred neurons.

4) In the case of the synergetic computer we have a learning theory that allows us to determine the synaptic strengths and their connection with the prototype patterns. Such a learning theory is absent in the case of neurocomputers.

Though the development of neurocomputers undoubtedly has been a very important step in modeling neuronal activities, I believe that the synergetic computer based on general synergetic concepts comes closer to the goal of understanding brain activity.

18.5 Can Computers Think?

How can we test whether a computer can think? One proposal for such a test is due to *Turing* (1950). Let us consider the following arrangement: A person is sitting on one side of a wall and on the other side is either another human or a computer. The two sides may communicate, say, by means of typewriters or by fax. The human can ask questions to establish how the other side answers. Does the other side answer like another human, or do the answers indicate that the other side is a computer? Today real contests are hold in which, in all cases, humans have eventually found out whether the other side is a computer. It appears that the ability of the computers to act in a human-like fashion depends on their programmers rather than their own *intelligence*. Amusingly enough, such a test might work the other way around, namely as a proof that the other side is a computer. For instance, there are some tasks that are very difficult to solve for humans in a short time, for instance, the multiplication of large numbers. A computer may be able to do this very quickly and thus reveal that it is a computer and not a human being, unless, of course, it is programmed to decline such types of task. But, here again, it is the foresight of the programmer rather than the insight of the computer that is relevant.

When we summarize our present experience with all kinds of computers, we can state the following: What originally seemed to be difficult for a computer is now rather simple, for instance, playing chess, and what initially seemed easy is difficult, namely the recognition of patterns, faces, or facial expressions, analysis of scenes, perception of language. In the opinion of some authors, such as *Dreyfus* (1972), there are strong indications that there is some kind of nonformalizable experience. In human life, our behavior in certain situations is based on intuition rather than on algorithms. Difficulties arise in other seemingly simple tasks, such as translations. Quite often, words or sentences have a double meaning. Can a computer recognize such a double meaning or the irony underlying some sentences?

To me it seems to be too early to give a definite answer to these questions, but my basic attitude is more optimistic with respect to algorithms. For instance, we saw in the chapters on ambiguous figures and on decision making, how algorithms can deal with problems having no unique solution. Similarly, we hardly or never see the same face twice. A face changes all the time – due to different facial expressions, illumination, age, etc. But as we mentioned in Chap. 16, we can subject faces to deformations, etc., by algorithms. Thus the vagueness can be resolved, though occasionally not uniquely. In other words, there is still a huge potential for transforming unformalizable experience into formalizable experience. Or, to turn the argument around: there are still considerable possibilities for improving computers including making them more error tolerant in the sense of allowing, for instance, deformations in an abstract sense.

On the other hand, present-day computers have a long way to go before they can genuinely think and brain research may provide us with insights presently not dreamt of. We shall come back to some of these questions in Chap. 20.

19. Networks of Brains

The central theme of synergetics is the cooperation of parts of a system. So far, more or less indirectly, we have been studying the cooperation of neurons within a brain. This book would be incomplete without at least some discussion of the cooperation of brains. Indeed, it is because of the cooperation of human brains that the human race has become the dominant species in the animal kingdom. First of all, there is the new quality 'language' that is made possible only through the cooperation of brains. The origin of language is still a mystery, but a few speculations do not seem to be too far fetched. In the early stages, language may have started from chance events, where utterances were initially randomly connected with certain objects or events. When these utterances were taken up by a group and attached to the same objects and events, primitive language could have been formed. Strangely enough, the situation resembles what happens in a laser, where first a few photons are emitted randomly, eventually leading to a coherent wave – the order parameter of the laser. Quite evidently, at the same time it was necessary that the brain could store these utterances and associations with objects and events and that it could recall them in context with objects and events. Thus the attachment of meaning to words could occur only by means of a context. This can also be easily observed when we watch how young children learn basic elements of a language. After individual words denoting simple objects, later combinations of two words are used and suddenly more complex combinations appear. Language has evolved over long periods of time. There were cruel 'experiments' in the middle ages in which children were allowed to grow up without verbal contact with their surrounding. They never developed a language. Thus quite evidently the acquisition of language is passed on from generation to generation.

The transfer of knowledge in the animal kingdom is very limited. This is not so between humans because of language in the spoken or written form. Thus the enormous advantage that humans have over animals is due to this collective tradition of knowledge that has so far been laid down, in particular, in libraries. At present we are witnessing a new revolution due to the computer. It not only allows us to store information, but also to process it in a variety of ways. The potential impact of this new revolution on the further development of human kind can hardly be overestimated.

In addition to the internal storage and processing of information in each individual brain, we now have to take into account the external storage and processing of information. This has led *Portugali* (1994) to coin the concept of inter-representation networks (IRN). (For a relationship between these concepts and cognitive maps, etc. see references.) Based on work by *Haken* and *Portugali* (1994), we wish to explain some of these concepts in the remainder of this chapter. Because of the interplay *internal – external*, new order parameters evolve both in the individual and also within the collective. Incidentally, the cognitive system must be seen not as an internal network representing the external environment, but as an internal–external network, part of whose elements are internally represented or stored in the mind/brain, and some of which exist, stored or externally represented, in the outer environment. *Portugali* suggested that some of the nicest experimental examples of the operation of IRNs are the so-called Bartlett scenarios devised by *Bartlett* (1932) as part of his studies on *remembering*. The general structure of the Bartlett scenarios will help to convey the notion of IRN.

A typical Bartlett scenario evolves like this: A subject is given a text or shown a figure and is asked to memorize it. He/she is then further asked to externally reproduce it out of memory (i.e., to rewrite the text or redraw the figure, etc.). This external representation is given to another subject to memorize and reproduce, and so on. The usual result of such experiments is that, after several strong fluctuations in the reproduction, the text or the figure are stabilized and do not change much from iteration to iteration. The interpretation offered is that what we have here is (i) a cognitive network composed of internal and external elements and representations, (ii) a sequential interplay between the internal and external elements of the system, and (iii) a typical synergetic process: this sequential interplay first exhibits strong fluctuations between competing configurations of texts/figures, which then lead to the emergence of a certain order parameter which enslaves both the external and the internal elements/representations of the system. Thus, instead of the usual process of pattern formation in which the order parameter(s) enslave(s) some external subsystems, and the usual process of pattern recognition in which the order parameter(s) enslave(s) some internal features, we have here an integrated process – the order parameter(s) enslave(s) both the externally represented subsystems and the internally represented features.

19.1 A General Model of IRN in Terms of Synergetics

In order to cast this integrative view into a graphic and, subsequently, mathematical form, we must remind the reader of the synergetic network model of pattern recognition as presented in Chap. 16. According to these concepts, the synergetic computer can be realized by a three layer network (Fig. 19.1). These layers are:

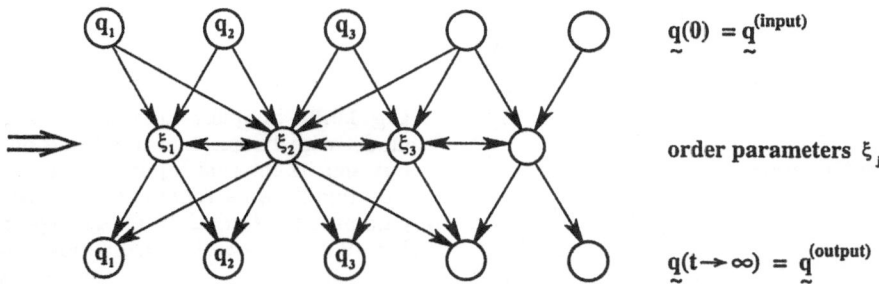

$q(0) = q^{(input)}$

order parameters ξ_j

$q(t \to \infty) = q^{(output)}$

Fig. 19.1. Three-level network of the synergetic computer. The first (upper) layer consists of model neurons that receive the input. This first layer projects onto the second layer that represents the order parameters. The third layer represents the output from the order parameter layer. Though formally similar to a neural computer arrangement, the algorithm of the synergetic computer is quite different, e.g. the model neurons interact by means of soft nonlinearities. Note that learnt patterns are encoded in the connections between the first and second and between the second and third layers. In the case of static patterns, the connections between the order parameters are of the same universal form, whereas in the case of dynamical patterns the order parameter connections may depend on the movement patterns to be generated. These remarks also hold for the remaining figures in this chapter

1) The input layer with (model) neurons labeled by k, where $q_k(0)$ represents the initially given input activity of neuron k;
2) the middle layer representing the order parameters ξ_j;
3) the output layer with neurons labeled by ℓ where $q_\ell(\infty)$ represents the final acticity of neuron ℓ.

For what follows it will be convenient to look at the network of Fig. 19.1 from the side as indicated by the arrow. We then arrive at Fig. 19.2. Now we are in a position to cast our integrative view into a graphic representation (Fig. 19.3). Here we have two kinds of inputs, $q^{(i)}$ (i: internal) and $q^{(e)}$ (e: external) and two kinds of outputs, again internal and external. The inputs are taken at time $t = 0$, whereas the outputs are taken at $t > 0$, and in most cases for $t \to \infty$ (see below). The middle node symbolizes the brain, in

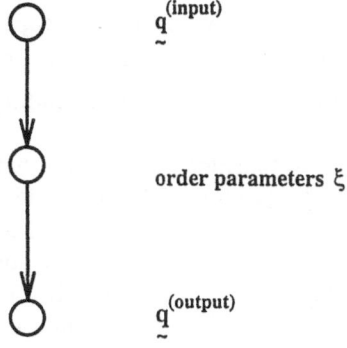

$q^{(input)}$

order parameters ξ

$q^{(output)}$

Fig. 19.2. The network of Fig. 19.1 seen from the side as indicated by the arrow in Fig. 19.1

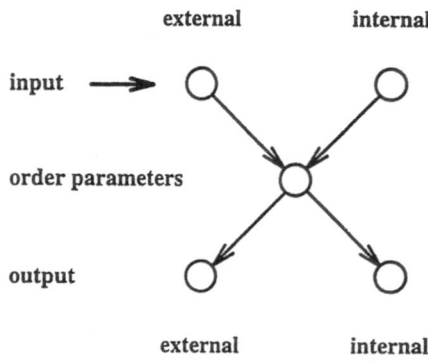

external   internal

input →

order parameters

output

external   internal

Fig. 19.3. Simplest case of an IRN model with its external input and output and its internal input and output. The middle area represents the order parameters. Note that in analogy to Fig. 19.2, a network corresponding to Fig. 19.1 is seen from the side so that each circle represents a whole set of model neurons

which one or several order parameters ξ_j have been established. The index j distinguishes between different order parameters. In the context of this book it is important to note that the same order parameters ξ_j may govern quite different external outputs. For instance, the order parameter ξ_j may be connected with a specific output pattern $v_j^{(e)}$. Such an output pattern may be a text or a drawing as in the Bartlett scenarios, or some other action, such as movements or writings, etc. that may lead to an external storage, say, in handwriting or in computers. The total set of possible drawings is then represented by the output vector $q^{(e)}(t)$

$$q^{(e)}(t) = \sum_j \xi_j(t) v_j^{(e)}. \tag{19.1}$$

This output vector develops in the course of time. In general we will consider it for a large time so that the temporal change of the order parameters has finished. In an analogous fashion the order parameters ξ_j may govern the formation of *internal* patterns, such as internally stored images or learned patterns. These patterns are denoted by $v_j^{(i)}$. The total set of possibilities is represented by

$$q^{(i)}(t) = \sum_j \xi_j(t) v_j^{(i)}. \tag{19.2}$$

The reader should not be deceived by this notation in which patterns, both external and internal, are denoted by the same letter, v. Note that the upper index may indicate quite different patterns, for instance, $v_j^{(e)}$ may refer to spoken words while $v_j^{(i)}$ may refer to an internally stored image corresponding to that word.

In the next step of our analysis, we have to consider the causes that generate the order parameters. To this end, according to Fig. 19.3, we consider two different inputs to the order parameter level, namely an *external* and an *internal* input. The external input is denoted by $q^{(e)}(0)$, the internal one by $q^{(i)}(0)$. The zero in the bracket indicates that these inputs are taken at time t equal to zero. Again the inputs may have quite different modalities.

The vectors q represent sets of data in different modalities and may have different dimensions. $q^{(e)}$ is externally given via the sensory system, such as visual, auditory or tactile. $q^{(i)}$ is internally given, for instance, by vague ideas, fantasies, dreams, thoughts, etc. An important step from the upper level, namely the input level, to the middle level dealing with order parameters is by means of preprocessing. For instance, the patterns given may not be complete or might be distorted or displaced in space, rotated or differently scaled or deformed. The patterns given must be checked internally against stored prototype patterns that we shall denote by u_j. We shall assume that this preprocessing is performed whenever necessary. Quite evidently, at this stage, the concept of Gestalt enters. This preprocessing may be performed either on q or on u_j or on both.

By means of the prototype patterns u_j, we may decompose the *externally given* data vector $q^{(e)}$ according to

$$q^{(e)} = \sum_j \xi_j^{(e)} u_j^{(e)} + w^{(e)}, \tag{19.3}$$

where $w^{(e)}$ is a remainder term that need not be considered (cf. Chap. 16). Again, we use adjoint vectors $u_j^{(e)+}$ which obey the orthogonality relation

$$\left(u_j^{(e)+} u_k^{(e)} \right) = \delta_{jk}. \tag{19.4}$$

When we multiply (19.3) by the adjoint vector, we immediately obtain

$$\xi_j^{(e)}(0) = \left(u_j^{(e)+} q^{(e)}(0) \right), \tag{19.5}$$

where the argument 'zero' on both sides indicates that we take these values at time $t = 0$, i.e. at the beginning. In complete analogy to the externally given signal, the *internally given* signal also may be processed according to

$$q^{(i)} = \sum_j \xi_j^{(i)} u_j^{(i)} + w^{(i)}, \tag{19.6}$$

where, in general, we may assume that different criteria for the prototype patterns apply or that the prototype patterns may have different modalities. This is why we distinguish the prototype patterns by upper indices e or i, respectively. In analogy to (19.5), we may form

$$\xi_j^{(i)}(0) = \left(u_j^{(i)+} q^{(i)}(0) \right). \tag{19.7}$$

Now the question arises of how new order parameters for the total system, external and internal, are determined. To study this we introduce the weighted superpositions of the order parameters (19.5) and (19.7) according to

$$\xi_j(0) = \alpha_j |\xi_j^{(e)}(0)| + \beta_j |\xi_j^{(i)}(0)|. \tag{19.8}$$

We then subject the order parameters ξ_j to a competition process that is well-known from pattern recognition by the synergetic computer of Chap. 16.

It means that in our approach we are dealing with the recognition of an internally or externally given pattern in which different order parameters ξ_j with indices j, compete and where one order parameter eventually wins the competition, namely the one that had obtained the highest value (19.8) at the beginning. This competition is described by the equations

$$\dot{\xi}_j = \left[\lambda_j - (B + C)D + B\xi_j^2\right]\xi_j,\tag{19.9}$$

where D is given by

$$D = \sum_{j'}\xi_{j'}^2 \quad , \quad B, C > 0\tag{19.10}$$

and B and C are positive constants. Equation (19.9) has the property that only one order parameter wins, or in other words, the "winner takes all" strategy applies. We may also conceive of mechanisms in which order parameters cooperate, but we shall not be concerned with this possibility here.

Note that all the steps indicated above, including preprocessing, can be performed – and have been performed – by a computer, so that our approach is entirely operational. This remark holds also for the rest of this chapter.

19.2 Collective Cognitive Processes

One of the most important properties of inter-representation networks (IRN) concerns the collective potential of the externally represented elements of the network. Once a person's cognitive construction becomes an external element in the person's IRN, it at the same time also becomes public. That is to say, other people might use it for various purposes, including it as an element in their own personal IRN. The external element thus enters a collective social or cultural process. To see how it works, we shall refer to the classical case of the Bartlett scenario. As noted, a typical experiment starts with a given input, such as a story, a drawing, etc., and proceeds with a sequence in which one person's memory of the previous input becomes an input to the next person, etc. Figure 19.4 illustrates an example. Similar experiments with Bartlett scenarios were performed by *Stadler* and *Kruse* (1990). *Bartlett, Stadler* and *Kruse* designed their scenarios as means to illustrate what the mind/brain remembers and what not. We use it as an illustration of the dynamics of collective, interpersonal, cognitive processes and of the role of external representation in it.

The interesting result of the various Bartlett scenarios is that after several steps the story or the figure stabilizes and does not change any more. In terms of synergetics this implies that a certain order parameter has enslaved the system and brought it to a steady state. In terms of the present discussion we should emphasise that the results show how, as a consequence of the interplay between internal and external representations of individuals, a collective agreement has been reached among many individuals, without

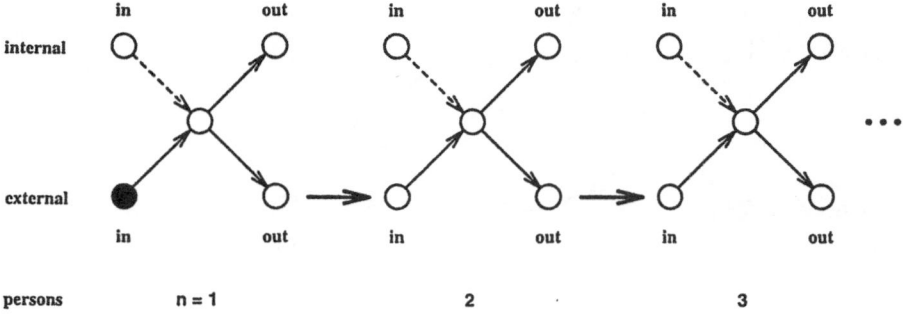

in   out   in   out   in   out

internal

external

in   out   in   out   in   out

persons   n = 1   2   3

Fig. 19.4. A sequence of communications between persons (see text)

these individuals being aware that they are engaged in a collective agreement. The order parameter which eventually enslaves the system is thus a
collective order parameter.

To sum up, let us consider the iteration steps that we discussed above in
their mathematical form.

19.3 Iterations

The processes we described in the preceding section can be cast into a mathematical form which we shall now describe.

1) *Intrapersonal.* The externally produced output of a person may be first
laid down, say, by writing, and then reading this again can serve as an input
to the same person. This leads also to a description of Bartlett's experiments with respect to a single person, in which a pattern is given as input,
is then reproduced as an output by a drawing, is then again seen, and so on.
(Fig. 19.5). The iteration can be formally described by

$$q_{n+1}^{(e)}(t = 0) = q_n^{(e)}(t \to \infty), \quad n = 1, 2....$$ (19.11)

The index n refers to the number of iterations. Note that this process as
indicated by $q_n^{(e)}(t \to \infty)$ may be rather complicated because of preprocessing. The reproduction of the pattern, for instance by a drawing, may be not
perfect; thus one has to consider a convergence as shown experimentally by
Bartlett. Following the advent of PCs, the scenario of Fig. 19.5 must be replaced by Fig. 19.6, where the computer may actively change the transfer of
information.

2) *Interpersonal.* A second class of iterations can be achieved interperson-
ally. The coupling between persons can then be represented in the form of
Fig. 19.4, or by means of the formula

$$q_{n+1}^{(e)}(t = 0) = q_n^{(e)}(t \to \infty), n = 1, 2, ...$$ (19.12)

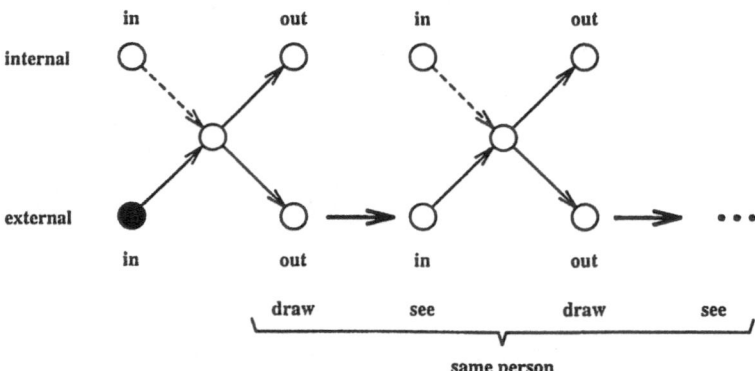

Fig. 19.5. The same as Fig. 19.4, but with respect to the same person

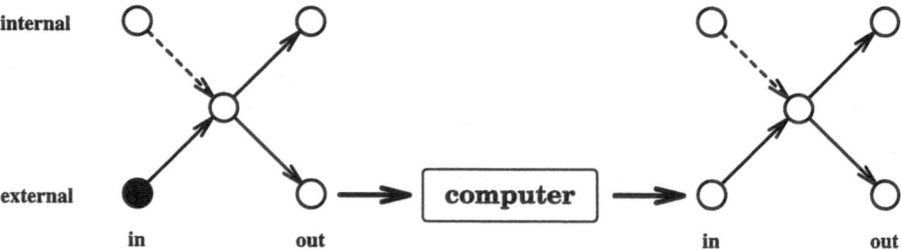

Fig. 19.6. Communication via a computer

The index n refers to the person involved in the corresponding step before the transfer takes place. An example is provided again by Bartlett's experiments and can be verified by computer calculations that are currently being performed. From a formal point of view, we may state that the process of Fig. 19.4 is indistinguishable from the process of Fig. 19.5, provided the preprocessing is governed by the same explicit laws for each person.

3) *Interpersonal connection via a common reservoir.* Another interpersonal connection is achieved by means of Fig. 19.7, where each individual gives his or her output to a common reservoir (a library, a data highway, or a public computer) and receives inputs from that reservoir. From a formal point of view, we have the relations

$$q_k^{(e)}(0) = W_k q_{common}, \quad k = 1, ...n \tag{19.13}$$

and

$$q_{common} = \sum_k c_k q_k^{(e)}(t \to \infty). \tag{19.14}$$

The index k enumerates the individuals. W_k is a *personal window* operator that selects part of the information that is stored in q_{common}.

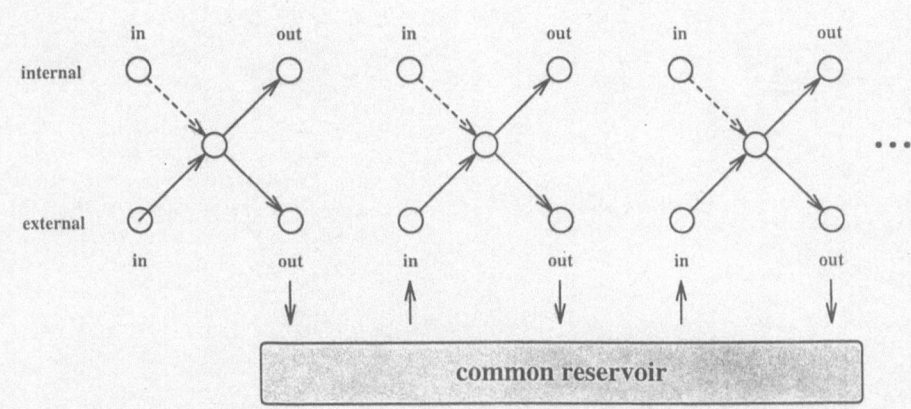

Fig. 19.7. Communication between persons and a common reservoir, such as a library or a public computer

19.4 Concluding Remarks

In this chapter we have seen how the concepts of IRN may be cast into a graphical and mathematical form. The graphical representation allows a simple visualization and provides us with a prescription for the mathematical procedure. According to the latter, input data encoded in vectors q are pre-processed and transformed into order parameters and prototype patterns. The order parameters then govern new external and internal output patterns that themselves may occur in different modalities.

Lack of space forbids me from presenting more details; in fact, this could fill another whole book.

20. Synergetics of the Brain: Where Do We Stand? Where Do We Go from Here?

20.1 Looking Back

The present book may be viewed as an attempt to develop a coherent theory of brain activities at the macroscopic level. We conceived the brain as a giant complex system that obeys the laws of synergetics, i.e., it operates close to instability points, where the macroscopic patterns are determined by order parameters. The bridge between the macroscopic and microscopic level is provided by the slaving principle. In the past, because of the complexity of brain functioning, verbal descriptions of brain activity have certainly dominated the field of brain theory. Things are presently changing because of two main lines of research: One may be characterized as connectionism and may be traced back to the models of *McCulloch* and *Pitts* which we encountered briefly in Chap. 18. The other coherent trend of mathematical modeling of the brain rests on synergetics as outlined in this book. This does not mean that still other approaches are not available, but it would seem that such approaches are much narrower than the one treated here. Quite often verbal descriptions have the advantage of being more flexible as a consequence of ambiguities inherent in language. By contrast, mathematical approaches are operational, i.e. they allow a rigorous check of statements made. Probably the most adequate treatment should lie in between, namely it should not be so rigid as present mathematical treatments, and it should be more quantitative than usual verbal descriptions.

Let us recapitulate what is at the heart of this book. We have emphasized the development of concrete and explicit mathematical models that are as close as possible to experimental findings. We started with experiments and models on movement coordination that allowed us to show how well basic concepts of synergetics may be applied to the phenomena of coordination. A central theme was the description of behavior by means of one (or few) order parameter(s) and to demonstrate how, by means of learning, the brain reduces the number of order parameters.

Another concrete example was that of EEG analysis, in particular in the case of epileptic seizures, where a concrete model was developed. In the case of sensory and motor coordination, as was exemplified by *Kelso's* MEG experiments, the difficulties with respect to modeling became more evident, namely the difficulty in finding *unique* models.

We then treated various aspects of vision including stereo vision and the interpretation of ambiguous figures. The modeling was again done at the macroscopic phenomenological level though it appeared that direct microscopic models may become derivable from the phenomenological approach, which allows one to make contact with concepts of neural computers. Again the dominant role played by the concept of order parameters became evident.

We then made a bold step to cognitive abilities by claiming that processes of vision can be considered as metaphors for other cognitive abilities and we elaborated this idea for the case of decision making, which was put in close analogy to pattern recognition. Finally we showed how we may develop models of networks of brains, where a hierarchy of order parameters is established. In a number of cases brain activities were modelled by means of coupled nonlinear oscillators, such as in several of finger movement experiments.

Before we start a discussion of more fundamental issues, a word should be said on the kind of models we used and on their visualization. At a first glance, the concept of a potential landscape, be it in movements or in vision, seemed to play an important role. Its advantage lies in the fact that it provides ready intuition about important processes leading to qualitative macroscopic changes in behavior and cognition. I found these landscape descriptions very useful, for instance when I predicted the dramatic change of the properties of laser light (*Haken* (1964)). (As an aside for the physicist: Potentials in mechanics and for systems in thermal equilibrium are, of course, well known; their existence in nonequilibrium systems, such as the laser, is by no means obvious, however.)

In spite of these successes we must not overlook the limitations of the potential landscape models. Even in the relatively simple case of the laser, this model holds only under very special conditions. This is even more true in connection with brain activity. Indeed, the landscape of the Haken–Kelso–Bunz model discussed in Sect. 6.2 is only an approximation of the oscillator model, again due to *Haken–Kelso–Bunz* (1984), which I discussed in Sect. 7.2. To study multifrequency tapping tasks or oscillations in perception, we had to go beyond potential landscapes. Taking all considerations together, it appears that the brain operates close to instability points, whereby it continuously changes its states. For a specific task a potential landscape for order parameters may appear, then disappears, is replaced by another landscape, chaos of some kind sets in, oscillations of various kinds occur, and so on. For instance, when we are sitting quietly with our eyes closed, our EEG shows α-waves. Their analysis (*Friedrich, Fuchs, Haken* (1990)) shows low-dimensional chaos.

In looking back at this book we might be inclined to say that the brain acts as if it solves differential equations, especially those of nonlinear oscillators. But this is clearly just our interpretation. Another possible trap of interpretation of our approach must be avoided: The occurrence of oscillations does not imply that there are hard wired oscillators in the brain. The "wiring" (which may be traced back to synaptic strengths) appears fairly (though not entirely) flexible. These remarks lead me into a discussion of more fundamental aspects.

20.2 Mind and Matter – An Eternal Question

The approaches presented here bring out an essential idea of synergetics, namely that self-organization of a system is indirectly steered by setting control parameters to appropriate values. This fixing of control parameters is by no means a trivial task. Whenever we had to fix these parameters in model equations, be it for the finger tapping equations or for the MEG analysis, the solution of the equation depended sensitively on the values of the parameters. Thus here a very deep problem arises, namely the question: Who sets the control parameters in the brain? Does the idea of *Eccles* come in here: That the brain is a machine or a computer, and its program, or in terms of self-organization, its control parameters, are set by the mind? I am deeply convinced that the control parameters are set by the brain via other processes of self-organization at a different level than the one treated in the equations determining, for instance, specific movements. There are a number of hints at how such a setting can be achieved: One is by learning, i.e., by a change of synaptic strengths. An indirect indication for the setting of control parameters is given by means of the so-called *Bereitschaftspotentiale* discovered by *Kornhuber* and *Deecke* (1965). In the corresponding experiments a subject is asked, say, to move his or her index finger whenever he or she wishes to do so. Then, at some time, the finger is lifted. But, and this is the decisive discovery, some 60 milliseconds before this action, specific electric potentials in the brain can be observed in the EEG. It is as if the brain prepares the action. In my opinion, the establishment of *Bereitschaftspotentiale* is again an act of self-organization which precedes other actions of self-organization by suitably setting control parameters. The obvious difficulty is: What triggers the self-organization of the *Bereitschaftspotentiale*? I believe it is here that microscopic events are transformed into macroscopic manifestations in the form of electric potentials. It is my conviction that all actions of the brain that are nowadays perhaps believed to be immaterial are connected with material processes. For instance, the command (transmitted by material pathways) is materially stored by neurons (or synapses, etc.) and then (perhaps spontaneously) activated (perhaps by a fluctuation). The experimental proof of my argument is difficult, however, at least at present, because we know too little about the material basis of memory.

Not, however, that I do not claim that properties of the mind are a mere outcome of material brain activity, rather my point of view is based on the order parameter concept and the slaving principle including that of circular causality. In other words, my interpretation is that the abstract processes governed by order parameters (and their changes) and the material processes described by the individual variables of the system condition each other. It may well be that these statements are of an untestable or "philosophical" nature. The reason lies in the fact that the brain is extremely complex and the emergence of new qualities may happen at a variety of different levels from the microscopic to the macroscopic, and it may be difficult to establish all the correlations that are necessary for proving that a new quality has emerged.

As we have outlined at several occasions in this book, the occurrence of order parameters and the action of the slaving principle generally imply an enormous information compression. Specific complex microscopic configurations are governed by one or a few order parameters. A lucid example for the action of information compression is language itself. A simple word like dog comprises a huge variety of different races, colors, shapes, postures, etc. Communication only becomes possible because of this and similar kinds of information compression. At the same time, information compression gives rise to ambiguities, and the efficiency of a language lies in the balance between uniqueness and ambiguity.

It is interesting to note that information compression can also be found in motor control. As we have shown in the case of the pedalo experiment, this motion is eventually, after learning, governed by a single complex order parameter that obeys a rather universal order parameter equation, namely the Van der Pol oscillator equation. On the other hand, the individual order parameters must be made efficient by means of a translation into the many degrees of freedom, for example, of muscle cells. This process can be viewed as information inflation. So in a way the slaving principle has two aspects: Seen from one side it serves to compress information; seen from the other, it generates information inflation.

Another aspect worthy of discussion is the nature of the order parameters. With few exceptions, order parameters are immaterial, for instance they may represent a phase angle as in the finger movement example. This, of course, immediately leads us to the mind–body problem, namely, how can an immaterial quantity, such as an order parameter, steer the behavior of a material system, such as muscles. From a purely mathematical viewpoint there is certainly no difficulty. The phase angle and the contraction of muscle cells can be described by mathematical variables and their equations of motion. According to the results of synergetics, the individual parts with their variables q of a system give rise to order parameters and, in turn via the slaving principle, the order parameters ξ govern the behavior of the parts, which is expressed by

$$q = f(\xi), \tag{20.1}$$

i.e., q is a function of ξ. But in science and still more in philosophy, we wish to interpret relationships or, in other words, to provide them with meaning. For instance, Newton's law

$$ma = F, \tag{20.2}$$

i.e., the mass of a particle times its acceleration is equal to the force F acting on the particle, is interpreted by saying: The force F is the *cause* of the particle's acceleration. What then would be the corresponding interpretation of (20.1): q represents the variables of the material constituents, e.g. muscle cells, while ξ is an immaterial quantity (the mind?). By analogy between (20.1) and (20.2) we would say: The mind determines the behavior of matter.

On the other hand, as mentioned above, q gives rise to ξ: or now interpreted again: matter determines the mind. (*Mind from Matter* is the title of a famous book by *Delbrück*). Finally let us invoke circular causality: Mind and matter condition each other, or in other words, mind and matter are two sides of the same coin. This is my point of view, but it is not new. As I learned from *Atlan*, it is that of *Spinoza*. I am afraid it is here where quite different points of view may be expressed and disputed. In my opinion in the present case the difficulty starts when we go from mathematics to the ontology of brain and mind.

Whatever the outcome of such disputes, I like the order parameter concept and the slaving principle at least as a *metaphor* for the mind–body problem, and perhaps more.

20.3 Some Open Problems

It is well-known in science that the solution of one problem often raises a dozen new questions. Of course, this also applies to the approach presented in this book. The brain is an extremely complex system, and, as I stated in the beginning, it has many facets. Indeed, there are numerous questions that haven't been answered in this book, or whose answers are not known at all. I list a few of them: One is the question of where memory is located. Is it located in synapses or, more specifically, in receptors? Or is memory connected with microtubuli as is suspected by some scientists, such as *Hameroff* (1987).

A problem that I didn't discuss at all is that of brain growth and development. This is quite fundamental, because structure and function condition each other. The topic is so immense, however, that it deserves another book.

A further problem which I intentionally left out of my discussion is consciousness. As *Freeman* (1995) remarked in a recent book of his, this question is brought up again and again and at least every fifty years. In my own experience the closer scientists are to brain research in their own work, the less they speak about consciousness. This is, of course, only a generalization and there are certainly exceptions. But it appears that those people who are concerned with brain activity are rather reluctant to discuss this problem. Prominent counter-examples may be *Crick* and *Koch* (1990), as well as *Edelman* (1992). They present scientific approaches to this problem, but personally I prefer to omit it. The same is true for qualia, such as the perception of color or of pain, which, in my opinion, are not – or not yet – accessible to mathematical modeling along the lines outlined here.

So what is the future of the approach that I have I presented? Clearly we may endeavor to develop more complicated mathematical models in the frame of synergetics, and we could treat more complicated movements or kinds of behavior. A large field of modeling that is beginning to be developed is that of coupled nonlinear oscillators to treat specific experiments in vision, as mentioned in Chap. 2 (see, for instance, *Tass* and *Haken* (1995)).

By way of conclusion let us discuss a few general issues:

1) *Is the brain a machine?* In order to discuss this issue, we must bear in mind that the concept of a machine has been changing considerably over the centuries. Originally a machine was a simple device, such as a lever or a hammer to perform mechanical work. Nowadays we speak of a computer as a machine. Furthermore, a variety of concepts that are borrowed from biology are now applied to machines. We find concepts, such as self-organization, self-repair, self-assembly, self-steering, etc. in the context of the construction of machines. Note how the "Self" creeps into machines! So when we compare the brain with a machine, we must carefully discuss what kind of machine we have in mind. The brain is certainly not a machine in the original sense of the word, namely a man-made device that performs specific tasks. But when we equip a machine with more and more biological aspects, then perhaps, eventually, there is not much difference between the brain and a machine. It is as if there is an ambitious race between the human brain and the human brain (this is *not* a misprint!). On the one hand, the human brain wants to construct a machine whose prowess equals that of the brain, and, on the other hand, it wants to prove that it is superior to any such machine. (A related issue is the comparison between brain and computer. We discussed this in Chap. 18 so it will not be repeated here.)

2) *The brain and chips – or brain protheses.* An interesting task will be the physical linking of neurons and chips, work being pursued, for instance, by *Fromherz* (1994). Here we are only at the beginning and it is certainly too premature to make any definite statements about further developments, for instance about chip-implants in damaged brains, or improving the capacity of a brain (brain protheses). Only the future can tell us whether this is science fiction or not. But from the abstract point of view of synergetics, cooperative effects can lead to the same macroscopic behavior of systems with quite different microscopic components. All that counts are the order parameters.

3) *Creativity.* Finally a few words about creativity may be in order. So far, I have omitted this problem completely. Indeed, to me creativity appears to be the deepest at all puzzles concerning the brain. It means the birth of thoughts that have never been generated before and, in particular, whose generation was extremely unlikely. One may compare the creation of a new

idea to a jigsaw puzzle. It may even be relatively easy to characterize the act of creation in simple verbal terms; for instance, as the competition and cooperation of different ideas in the form of order parameters. But it is here that my criticism of verbal descriptions applies: Making such statements is rather idle and does not give us any operational approach or a *recipe* for how to solve a puzzle or how to find a new basic idea. Maybe it is good that the nature of genius is still shrouded by mystery.

Appendices

A. Analysis of Time Series

As we have seen in Chap. 2, EEGs show, in general, a rather complicated structure in their time series when we plot the electric potential versus time (see Fig. 2.5). Thus it seems that the electric activity of the brain, or more precisely speaking, the electric dynamics of the brain is very complicated and stems from many degrees of freedom. It therefore came as a surprise to many scientists when *Babloyantz* (see References of Chap. 14) showed that the dimension of the underlying dynamics is comparatively small, say, of the order of 6 or 7, depending on the state of the person, who may, for instance, be in deep sleep or in some other state of sleep. Similar results hold for epileptic seizures. It turns out that the dimensions here are not the usual integer dimensions that we are used to for a line or a square, but so-called fractal dimensions. In the following we shall show how fractal dimensions are defined and how they can be determined from measured data or from mathematical models of dynamical processes. At the end of this appendix, in A.4, we shall also discuss some of the difficulties this method is confronted with.

We first consider the problem of analyzing a time series by taking a simple example.

A.1 Time Series Analysis

Let us consider the pendulum, which we treated in Sect. 5.2. Its motion can be visualized in the phase-plane, where we plot the trajectory versus the coordinate x and the velocity v (Fig. A.1). In view of later generalisations, we shall also denote the x-axis by the coordinate q_1 and the v-axis by the coordinate q_2. The velocity is, of course, connected with the position x by

$$v = \dot{x}, \tag{A.1}$$

where the dot means as usual the time-derivative. In the new notation we may also write (A.1) in the form of

$$q_2 = \dot{q}_1. \tag{A.2}$$

Furthermore we note that the coordinate x, or equivalently q_1, obeys the oscillator equation

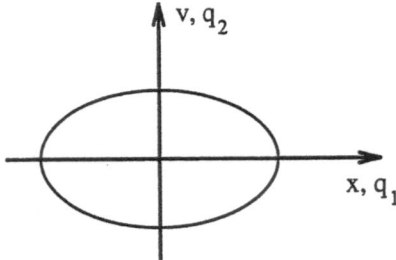

Fig. A.1. A trajectory described by $q_1 = A \sin \omega t$, $\quad q_2 = \dot{q}_1 = \omega A \cos \omega t$

$$\ddot{q}_1 = -k q_1, \tag{A.3}$$

where k is a positive constant.

Let us consider the case where we measure only the q_1 coordinate, which we plot as a function of time according to Fig. A.2. In it q_1 is given by

$$q_1 = A \sin \omega t. \tag{A.4}$$

Then we ask ourselves: Can we reconstruct from this time series the trajectory or the attractor in the phase-plane of Fig. A.1. In the present case, this is quite simple, because of the relation (A.2), i.e.

$$q_2 = \dot{q}_1. \tag{A.5}$$

From this we can immediately deduce

$$q_2 = \omega A \cos \omega t. \tag{A.6}$$

By plotting (A.4) and (A.6) in the phase-plane, as time t proceeds, we obtain, of course, the trajectory of that figure.

These relations can be cast into a more general and abstract form. Let us assume that a system described by the variable q_1 obeys an equation of the form

$$\ddot{q}_1 = f(q_1, \dot{q}_1), \tag{A.7}$$

where f is a given function of q_1 and $\dot{q}_1$. We may deduce the new variable q_2 by

$$q_2 = \dot{q}_1 \tag{A.8}$$

and may now replace the system of equations (A.7) and (A.8) by

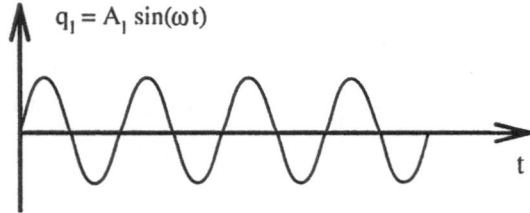

Fig. A.2. Example of a time series, $q_1 = A \sin \omega t$

$$\dot{q}_1 = q_2 \tag{A.9}$$

and

$$\dot{q}_2 = f(q_1, q_2). \tag{A.10}$$

This means that we can represent the trajectory in the phase-plane by these two equations. Quite evidently, to deduce a trajectory in the phase-plane from a time series is a trivial task.

These considerations can easily be generalized to the case of n dimensions. The basic assumption, however, is that we are dealing with a dynamical system and that we know the dimension of the phase space, i.e. the number of independent variables. We first introduce an abbreviation for the j th derivative by

$$\frac{d^j q_1}{dt^j} \equiv q_1^{(j)}. \tag{A.11}$$

We assume that the dynamics is described by an equation of the form

$$q_1^{(n)} = f\left(q_1, q_1^{(1)}, ..., q_1^{(n-1)}\right). \tag{A.12}$$

We may introduce the new coordinates

$$q_1^{(0)} \equiv q_1 \tag{A.13}$$

and

$$q_j = q_1^{(j-1)}. \tag{A.14}$$

Equation (A.12) can then be immediately replaced by the following set of equations:

$$\dot{q}_1 = q_2, \tag{A.15}$$

$$\ddot{q}_1 \equiv \dot{q}_2 = q_3, \tag{A.16}$$

$$\dot{q}_n = f(q_1, q_2, ..., q_n). \tag{A.17}$$

These define trajectories in a phase space with n dimensions. Following the sequence of time, we may immediately construct the trajectory simply by taking higher and higher derivatives of q_1 according to (A.14). In this way one can easily calculate the attractor.

There is, however, a basic difficulty when one wishes to apply this procedure to actually measured time series. Because any time series can consist only of discrete points and is, in reality, neither continuous nor even differentiable, the evaluation of derivatives introduces considerable numerical errors. Therefore, another method for constructing an attractor was introduced by *Takens* (1980) and others and has become very useful. Again our goal is to reconstruct the trajectory of Fig. A.1 from a time series as given in Fig. A.2. The basic idea is to introduce a time-shift T so that the coordinate q_1

$$q_1 = A \sin \omega t \tag{A.18}$$

is transformed into

$$q_2(t) = Cq_1(t+T)$$
$$= CA \sin[\omega(t+T)], \tag{A.19}$$

which is actually the time-sequence of the variable q_2, provided $\omega T = \pi/2$. The quantity C in (A.19) is a scaling constant. This example shows that we can reconstruct the trajectory of Fig. A.1 by means of a suitable time-shift T. However, a difficulty arises in practical applications, because we do not know a priori from the experimental data what time-shift T must be used. Thus we consider the case where T is different from that used in (A.19). From the second row in (A.19) we readily obtain

$$q_2 = CA(\sin \omega t \cos \omega T + \cos \omega t \sin \omega T) \tag{A.20}$$

With help of (A.20), we may show that q_1 and q_2 now obey an equation of the form

$$(q_2 - \alpha q_1)^2 + \beta q_1^2 = (CA)^2 \tag{A.21}$$

with

$$\alpha = C \cos \omega T$$
$$\beta = C \sin \omega T. \tag{A.22}$$

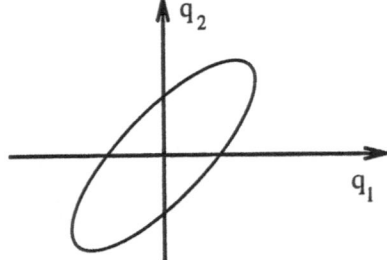

Fig. A.3. Reconstructed trajectory (see text)

The resulting new trajectory is shown in Fig. A.3. From a comparison between Figs. A.3 and A.1 we may conclude that by using an arbitrary time-shift or time-delay T, we obtain essentially the same *attractor* but now rotated and deformed in phase-space. A pathological case occurs if $\beta = 0$ which happens for $T = 2\pi/\omega$. In this case, (A.21) can be replaced by

$$q_2 - Cq_1 = \pm CA \tag{A.23}$$

and the attractor consists of two individual lines only, i.e., the attractor is no longer resolved. This example shows clearly that the reconstruction of an attractor can, at least in some cases, depend sensitively on the choice of T. Below we shall show how to reconstruct attractors for more complicated systems. But before we do so, we discuss how to determine the dimensions of an attractor, in particular of a chaotic attractor.

A.2 Definition of Dimensions

We first consider possible definitions of dimensions starting with the quite obvious cases of a line, a square, and a cube. But our goal will be to introduce the concept of fractal dimension, i.e., a dimension that is non-integer.

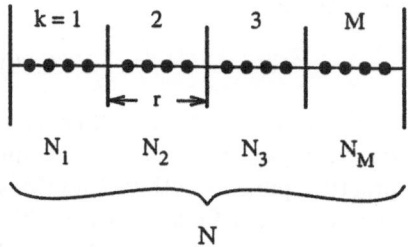

Fig. A.4. Illustration of the definitions of N_k, N, and M used in (A.24)

Let us start with one dimension and consider a single line with n points that all have the same separation (cf. Fig. A.4). The total number of points shall be N. $N_k(r)$ is the number of points in the interval k of length r. For the moment all cells are assumed of equal size so that we put $N_k(r) = N(r)$. The number of cells necessary to cover all points will be denoted by $M(r)$ and obeys the relation

$$M(r) \times N(r) = N. \tag{A.24}$$

For later purposes we write (A.24) in the form

$$\frac{N(r)}{N} = \frac{1}{M(r)}. \tag{A.25}$$

Quite evidently, in one dimension

$$N(r) \propto r. \tag{A.26}$$

Similarly in two dimensions we obtain

$$N(r) \propto r^2, \tag{A.27}$$

and in the case of D dimensions

$$N(r) \propto r^D, \tag{A.28}$$

or more precisely

$$N(r) = Cr^D, \tag{A.29}$$

where C is a constant that depends on the shape of the individual coverings, for instance, cubes or spheres, etc. Taking the logarithm of (A.29), we obtain

$$\ln N(r) = \ln C + D \ln r \tag{A.30}$$

and hence

$$D = \frac{\ln N(r)}{\ln r},$$
$$(A.31)$$

at least in the case $r \to 0$, where

$$\frac{\ln C}{|\ln r|} \to 0.$$
$$(A.32)$$

On account of (A.25) we may cast (A.31) into the form

$$D = -\lim_{r \to 0} \frac{\ln M(r)}{\ln r}.$$
$$(A.33)$$

Equation (A.31) is our first central result; it tells us how to define a dimension D. This definition actually holds not only for the objects we treated here, but also for fractals, as we shall see below. For practical purposes other definitions of D are used quite often. We therefore show how these definitions arise. First of all, the density of points need not be uniform. We thus introduce the local density of a cell k in the form

$$P_k = \frac{N_k(r)}{N}.$$
$$(A.34)$$

This expression may also be interpreted as the probability of finding a point in cell k.

To arrive at a precise mathematical definition of P_k, we take the limit

$$P_k = \lim_{N \to \infty} \frac{N_k(r)}{N}.$$
$$(A.35)$$

This means in practical applications that we have to take N sufficiently large.

It may seem somewhat surprising that instead of the simple definitions (A.31) or (A.33) more complicated definitions of a dimension are used. One of the reasons lies in the fact that the density of points need not be homogeneous and that one may apply weights in different ways. Let us start with the general definition of the dimension D_q, namely

$$D_q = \frac{1}{q-1} \lim_{r \to 0} \frac{\ln \left(\sum_{k=1}^{M(r)} P_k^q \right)}{\ln r},$$
$$(A.36)$$

where $q \geq 0$ but not necessarily integer. To elucidate the relationship between (A.36) and the elementary definitions (A.31) or (A.33), we consider the special case of a homogeneous distribution and $q \neq 1$,

$$P_k = P \equiv \frac{N(r)}{N}.$$
$$(A.37)$$

Then the sum occurring under the logarithm in formula (A.36) can immediately be evaluated to yield

$$\sum \dots = M(r) \left(\frac{N(r)}{N} \right)^q,$$
$$(A.38)$$

or because of (A.25)

$$\sum \cdots = \left(\frac{N(r)}{N}\right)^{q-1}.$$
(A.39)

Inserting this result into (A.36), we obtain

$$D_q = \lim_{r \to 0} \frac{\ln\left(\frac{N(r)}{N}\right)}{\ln r},$$
(A.40)

where $q-1$ has cancelled. Now as we have seen, $N(r)$ or $\frac{N(r)}{N}$ is proportional to r^D

$$\frac{N(r)}{N} \propto r^D.$$
(A.41)

Thus we immediately obtain

$$D_q = \frac{\ln r^D}{\ln r} = D\frac{\ln r}{\ln r} = D,$$
(A.42)

i.e., the usual definition of a dimension. We now consider some simple special cases of (A.36). In the case $q = 0$, the sum under the logarithm reduces to

$$\sum_{k=1}^{M(r)} 1 = M(r)$$
(A.43)

so that we obtain instead of (A.36)

$$D_0 = -\lim_{r \to 0} \frac{\ln M(r)}{\ln r}.$$
(A.44)

This is the so-called *Hausdorff* dimension. The case

$$q = 1 \qquad D_1 = \lim_{q \to 1} D_q$$
(A.45)

is somewhat tricky, because the denominator and the numerator in (A.36) vanish. However, we may define D_1 by (A.45). Some standard mathematics shows that D_1 can be written in the form

$$D_1 = \lim_{r \to 0} \left(\frac{-\sum_{k=1}^{M(r)} P_k \ln P_k}{\ln r} \right).$$
(A.46)

This is called the information dimension, because the information of a probability distribution P_k is defined by

$$i = -\sum P_k \ln P_k$$
(A.47)

with the condition

$$\sum P_k = 1.$$
(A.48)

Finally the most important definition in our context arises from $q = 2$, where we introduce the correlation dimension ν by identifying

$$\nu = D_2. \tag{A.49}$$

The different definitions of dimensions for $q = 0, 1, 2...$ need not to give the same numerical results, but we may show, quite generally, that

$$D_2 \leq D_1 \leq D_0. \tag{A.50}$$

After these preparations, we may now deal with the dimension of attractors.

A.3 Dimension of Attractors

First we recall how we obtain attractors. Nowadays they may be obtained by the numerical solution of dynamical equations (A.12) or (A.15)–(A.17) on a digital computer. It is important to know that in any case we shall obtain only a discrete set of points corresponding to the discrete set of time-steps used. Similarly when we are dealing with experimental data, because of the limited time resolution of the measuring operations, we obtain only discrete time series from which to construct an attractor according to the rules elucidated in A.1. When we are dealing with experimentally given data in the form of a time series of a single variable, we do not know a priori the dimension of the underlying phase space. Thus we have to try different dimensions of phase spaces. We call the procedure of choosing a dimension n of phase space *embedding*, i.e., we embed the trajectory in an n-dimensional phase space by constructing an adequate number n of variables according to the laws given in Sect. A.1.

We thus construct n-dimensional vectors at discrete time points t_i

$$q_i \equiv q(t_i), \tag{A.51}$$

where $i = 1, ..., N$. In this way we obtain N points in phase space. We cover this ensemble of points with cubes of side-length r. Remember that $M(r)$ was defined as the minimum number of cubes necessary to cover all points. By practical applications it has turned out that the dimension D_2 is most appropriate. However, a number of transformations will be necessary. We remember that from (A.34), we immediately obtain

$$P_k^2 = \frac{N_k(r)^2}{N^2}. \tag{A.52}$$

This may be interpreted as the probability of finding two points in P_k, or, in other words, the probability that the distance between these two points is less than r. If we form

$$\sum_{k=1}^{M(r)} P_k^2, \tag{A.53}$$

i.e., sum up over all cubes, we obtain the probability that the distance between any pairs is less than r. But this may also be defined as the number of pairs for which

$$| q_i - q_j | < r, \quad i \neq j \tag{A.54}$$

divided by N^2, i.e., by

$$\lim_{N \to \infty} \frac{1}{N^2} \quad \text{(number of pairs for which } | q_i - q_j | < r, i \neq j\text{)}.$$

Equation (A.53) can be expressed differently once again, namely as the so-called correlation integral in the form

$$C(r) = \lim_{N \to \infty} \frac{1}{N^2} \sum_{i=1}^{N} \sum_{\substack{j=1 \\ j \neq i}}^{N} H \left(r - | q_i - q_j | \right). \tag{A.55}$$

In it the double sum counts all the points for which (A.54) holds. In (A.55) we have made use of the so-called *Heaviside* function, which is defined by

$$H(x) = \begin{cases} 1 & \text{for } x \geq 0 \\ 0 & \text{for } x < 0 \end{cases} \tag{A.56}$$

According to *Grassberger* and *Procaccia* (1983), the dimension ν is now defined by

$$\nu = \lim_{r \to 0} \frac{\ln C(r)}{\ln r}. \tag{A.57}$$

Equation (A.57) must be used with some care. First of all taking away the $\lim_{r \to 0}$, (A.57) is equivalent to the statement

$$C(r) \propto r^{\nu}, \tag{A.58}$$

whereby

$$\nu \leq n \tag{A.59}$$

must hold. The $\lim_{r \to 0}$ means that r must be taken small compared to the extension of the attractor, namely if r becomes of that size the probability of finding two points in this attractor with practically an arbitrary distance clearly becomes 1

$$C(r) \to 1. \tag{A.60}$$

On the other hand, if r becomes too small in any numerical simulation, there are not enough pairs present. Thus the art of the whole approach consists in finding the appropriate range for r. A practical approach is provided by the Rapp plot (*Rapp* (1985)). In it one uses

$$\frac{d \ln C(r)}{d \ln r} \tag{A.61}$$

and plots this as a function of

$$\ln C(r). \tag{A.62}$$

The derivative occurring in (A.61) must be taken numerically. Let us see what this definition means by means of the example $C(r) \propto r^{\nu}$. Writing r^{ν} in the form

$$C(r) \propto r^{\nu} = e^{\nu \ln r}, \tag{A.63}$$

we may readily evaluate (A.61) to obtain

$$\frac{d \ln e^{\nu \ln r}}{d \ln r} = \nu. \tag{A.64}$$

In addition we obtain

$$\ln C = \nu \ln r. \tag{A.65}$$

Thus when we plot (A.61) versus (A.62), we obtain a horizontal line at the value ν, which is simply the dimension. Therefore, in practical applications, we have to see where the Rapp plot shows such a horizontal line.

A.4 Some Conclusions

The determination of fractal dimensions may provide us with some clue about how many order parameters to expect. It should be mentioned that an exact determination of fractal dimensions requires long enough time series under steady state conditions. Unfortunately these conditions are practically never fulfilled by physiological (including EEG and MEG) measurements. Therefore, fractal dimensions in these cases can give us only some hints, and must be taken with a grain of salt.

B. Determination of Adjoint Vectors

We multiply both sides of (16.2) from the left by $v_{k''}$ and obtain, on account of (16.1),

$$\delta_{kk''} = \sum_{k'=1}^{M} A_{kk'} (\overline{v}_{k'} v_{k''}), \quad k = 1, ..., M. \tag{B.1}$$

Introducing the abbreviations

$$I = (\delta_{kk''}), \quad A = (A_{kk'}) \quad \text{and} \quad V = (\overline{v}_{k'} v_{k''}), \tag{B.2}$$

we may recast (B.1) as a matrix equation

$$I = AV, \tag{B.3}$$

whose solution can be written

$$A = V^{-1}. \tag{B.4}$$

C. The Potentials Occurring in Sect. 16.5

The potentials V_2 and V_3 quoted in the footnote in Chap. 16.5 read:

$$V_2 = \gamma_2 \frac{1}{2} \left\{ |\,(1 - \boldsymbol{P}_1)\,\boldsymbol{q}\,|^2 + |\,(1 - \boldsymbol{P}_2)\,\boldsymbol{q}\,|^2 \right\}, \tag{C.1}$$

where $\boldsymbol{P}_1, \boldsymbol{P}_2$ are projection operators, whose components are defined by

$$(P_1)_{ij} = (P_2)_{ji} = \sum_{k=1}^{M} v_{ki}^+ v_{kj}^+ \, ; $$

and

$$V_3 = \gamma_3 \sum_{k=1}^{M} (1 - (\boldsymbol{v}_k \boldsymbol{v}_k))^2 . \tag{C.2}$$

γ_2 and γ_3 are positive constants. The minimum of $V_{\text{tot}} = V + V_2 + V_3$ can be found by a gradient dynamics

$$\dot{v}_{ki} = -\partial V_{\text{tot}} / \partial v_{ki}, \tag{C.3}$$

$$\dot{v}_{ki}^+ = -\partial V_{\text{tot}} / \partial v_{ki}^+ . \tag{C.4}$$

References and Further Reading

The literature on brain, behavior, and cognition is enormous and a complete reference list would fill books. Therefore I cite only those references, whose results I use in this book. In addition, I have tried to add some further references that might be relevant. I am aware of the fact that I had to omit important publications. Among more recent books on brain, mind, behavior and cognition are the following:

Aertsen, A. (ed.) (1993): *Brain theory*, Elsevier, Amsterdam

Arbib, A. (ed.) (1995): *The handbook of brain theory and neural networks*, MIT Press, Cambridge, MA

Bock, P. (1993): *The emergence of artificial cognition*, World Scientific, Singapore

Calvin, W.H. (1990): *The cerebral symphony*, Bantam Doubleday Dell, New York

Carpenter, G.A., Grossberg, S. (eds.) (1991): *Pattern recognition by self-organizing neural networks*, MIT Press, Cambridge, MA

Crowley, J.L., Christensen, H.I. (1995): *Vision as process*, Springer, Berlin

Donalds, M. (1991): *Origins of the modern mind*, Harvard University Press, Cambridge, MA

Duke, S.W., Pritchard, W.S. (eds.) (1991): *Measuring chaos in the human brain*, World Scientific, Singapore

Edelman, G.M. (1992): *Bright air, brilliant fire*, Penguin Books, London

Eggermont, J.J. (1990): *The correlative brain. Theory and experiment in neural interaction*, Springer, Berlin

Finke, R.A., Ward, T.B., Smith, M. (1992): *Creative cognition*, MIT Press, Cambridge, MA

Gardner, H. (1985): *The mind's new science. A history of the cognitive revolution*, Basic Books, New York

Gazzaniga, M.S. (1994): *The cognitive neurosciences*, MIT Press, Cambridge, MA

Haken, H., Haken-Krell, M. (1992): *Erfolgsgeheimnisse der Wahrnehmung*, Deutsche Verlags-Anstalt, Stuttgart

Hanson, S.J., Olson, C.R. (eds.) (1990): *Connectionist modeling and brain function: The Developing Interface*, MIT Press, Cambridge, MA

Impedovo, S. (ed.) (1994): *Fundamentals in handwriting recognition*, Springer, Berlin

Kohonen, T., Fogelmann-Soulie, F. (1991): *Cognitiva 90, At the crossroads of artificial intelligence, Cognitive Science and Neuroscience*, North-Holland, Amsterdam

Levy, W.B., Anderson J.A., Lehmkuhle, S. (eds.) (1985): *Synaptic modification, neuron sectivity, and nervous system organization*, Lawrence Erlbaum Associates, London

Logie, R.H., Denis, M. (1991): *Mental images in human cognition*, North-Holland, Amsterdam

Moser, U., v. Zeppelin, I. (eds.) (1991): *Cognitive-affective processes*, Springer, Berlin

Omidvar, O.M. (ed.) (1995): *Progress in neural networks*, Vol. **3**, Ablex Publishing Corporation, Norwood, New Jersey

Petry, S., Meyer, G.E. (eds.) (1987): *The perception of illusory contours*, Springer, Berlin

Pöppel, E. (ed.) (1989): *Gehirn und Bewußtsein*, VCH Verlagsgesellschaft, Weinheim

Purpura, D.P. (ed.) (1993): *Cognitive brain research*, Elsevier Science Publishers, Amsterdam

Reichardt, W.E., Poggio, T. (eds.) (1981): *Theoretical approaches in neurobiology*, MIT Press, Cambridge, MA

Rentschler, I., Herzbacher, B., Epstein, D. (eds.) (1988): *Beauty and the brain*, Birkhäuser Verlag, Basel

Schmitt, F.O., Worden, F., Adelman, G., Dennis, S.G. (eds.) (1981): *The organization of the cerebral cortex*, MIT Press, Cambridge, MA

Stillings, N.A., Feinstein, M.H., Garfield, J.L., Rissland, E.L., Rosenbaum, D., Weisler, S.E., Baker-Ward, L. (1987): *Cognitive science, An Introduction*, MIT Press, Cambridge, MA

Trevarthen, C. (1990): *Brains, circuits, and functions of the mind*, Cambridge University Press, Cambridge

Prologue

Haken, H. (1979): Pattern formation and pattern recognition – An attempt at a synthesis. In Haken, H. (ed.), *Pattern formation by dynamic systems and pattern recognition*, Springer, Berlin

Haken, H. (1977): *Synergetics, An Introduction*, Springer, Berlin

Haken, H. (1983): Synopsis and Introduction, in Başar, E., Flohr, H., Haken, H., Mandell, A.J. (eds.), *Synergetics of the brain*, Springer, Berlin, 3–25

Kelso, J.A.S. (1981): On the oscillatory basis of movements, Bulletin of Psychonomic Society **18**, 63

Kelso, J.A.S. (1984): Phase transitions and critical behavior in human bimanual coordination, American Journal of Physiology: Regulatory, Integrative and Comparative Physiology **15**, R 1000–R 1004

Haken, H., Kelso, J.A.S., Bunz, H. (1985): A theoretical model of phase transitions in human hand movements, Biol. Cybern. **51**, 347–356

Schöner, G., Haken, H., Kelso, J.A.S. (1986): A stochastic theory of phase transitions in human movement, Biol. Cybern. **53**, 247–257

Schmidt, R.C., Carello, C., Turvey, M.T. (1990): Phase transitions and critical fluctuations in the visual coordination of rhythmic movements between people, Journal of Experimental Psychology: Human Perception and Performance **16**, 227–247

Beek, P.I., Peper, C.E., van Wieringen, P.C.W. (1992): Frequency locking, frequency modulation, and bifurcations in dynamic movement systems. In: *Tutorials in motor neuroscience*, Stelmach, G.E., Requin, J. (eds.), North Holland, Amsterdam, 599–622

Körndle, H. (1992): Private communication

Friedrich, R., Uhl, C. (1995): Spatio-temporal analysis of human electroencephalograms: petit-mal epilepsy, unpublished

Fuchs, A., Kelso, J.A.S., Haken, H. (1992): Phase transitions in the human brain: Spatial mode dynamics, International Journal of Bifurcation and Chaos **2**, 917–939

Kelso, J.A.S., Bressler, S.L., Buchanan, S., de Guzman, G.C., Ding, M., Fuchs, A., Holroyd, T. (1992): A phase transition in human brain and behavior, Phys. Lett. A **169**, 134–144

Haken, H. (1987): Synergetic computers for pattern recognition and associative memory, in *Computational systems, natural and artificial*, Haken, H. (ed.), Springer, Berlin, 2–22

Ditzinger, T., Haken, H. (1989): Oscillations in the perception of ambiguous patterns, Biol. Cybern. **61**, 279–287

Ditzinger, T., Haken, H. (1990): The impact of fluctuations on the recognition of ambiguous patterns, Biol. Cybern. **63**, 453–456

Two excellent recent popularizations that emphasize the role of synergetics are:

Freeman, W.J. (1995): *Societies of brains*, Lawrence Erlbaum Associates, Hillsdale, N.J.

and in particular:

Kelso, S.J.A. (1995): *Dynamic patterns: The self-organization of brain and behavior*, MIT Press, Boston

These books contain numerous references.

1. Introduction

1.1 Biological Systems are Complex Systems

Sherrington, C.S. (1906): *The integrative action of the nervous system*, Constable, London

Popper, K.R., Eccles, J.C. (1977): *The self and its brain*, Springer, Berlin

Later, Eccles introduced, instead of the programmer, the quantum mechanical probability field. Cf. Eccles, J.C. (1985): *New light on the mind–brain problem: How mental events could influence neural events*. In: Haken, H. (ed.), Complex systems – operational approaches, Springer, Berlin, 81–106

1.2 Goals of Synergetics

For a popularization see

Haken, H. (1984): *The science of structure, Synergetics*, Van Nostrand Reinhold, New York

2. Exploring the Brain

2.1 The Black Box Approach

For Skinner's work see Catania, C.A., Harnad, S. (eds.) (1988): *The selection of behavior*, Cambridge University Press, Cambridge

For an extension of his earlier work see

Skinner, B.F. (1957): *Verbal Behavior*, Appleton–Century–Crofts, New York

2.2 Opening the Black Box

2.3 Structure and Function at the Macroscopic Level

Brocca, P. (1861): Remarques sur la siège de la faculté du language articulé. Bulletin de la société d'anthropologie, Paris 6
Wernicke, C. (1874): Der aphasische Symptomenkomplex, Breslau

An excellent account of Sperry's and further work is given by Springer, S.P., Deutsch, G. (1993): *Left Brain, Right Brain*, 4th ed., W.H. Freeman and Co., New York

2.4 Noninvasive Methods

Daffertshofer, M., Schwartz, A. (1994): private communication

Computer tomography:

Dann, R., Hoford, J., Kovačič, S., Reivich, M., Bajcsy, R. (1989): Evaluation of elastic matching system for anatomic (CT, MR) and functional (PET) cerebral images, J. Comput. Assist. Tomogr. **13** (4), 603–611
Radü, E.W., Kendall, B.E., Moseley, I.F. (1987): *Computertomographie des Kopfes. Technische Grundlagen – Interpretation – Klinik*, Thieme, Stuttgart, New York

EEG:

Berger, H. (1929): Über das Elektroenkephalogramm des Menschen, Arch. Psychiatr. Nervenkr. **87**, 527–570
Cooper, R., Osselton, J.W., Shaw, J.C. (1984): *Elektroenzephalographie: Technik und Methoden*, 3rd. rev. ed., Gustav Fischer Verlag, Stuttgart
Freeman, W.J. (1975): *Mass action in the nervous system: Examination of the neurophysiological basis of adaptive behavior through the EEG*, Academic Press, London
Lehmann, D. (1971): Multichannel topography of human alpha EEG fields, Electroenceph. Clin. Neurophysiol. **31**, 439–449
Lehmann, D. (1972): Human scalp EEG fields: Evoked, alpha, sleep, and spike-wave patterns. In: *Synchronization of EEG activity in epilepsies*, Petsche, H., Brazier, M.A.B. (ed.), Springer, Berlin
Jansen, B.H., Brandt, M.E. (1993): *Nonlinear dynamical analysis of the EEG*, World Scientific, Singapore
Nunez, P.L. (1981): *Electric fields of the brain: The neurophysics of EEG*, Oxford University Press, Oxford
Nunez, P.L. (1995): *Neocortical dynamics and human EEG rhythms*, Oxford University Press, Oxford

MEG:

Kelso, J.A.S. et al. for references see Chap.15
Hämäläinen, M., Hari, R., Ilmoniemi, R.J. Knuutila, J., Lounasmaa, O.V. (1993): Magnetoencephalography–theory, instrumentation, and applications to noninva-

sive studies of the working human brain, Rev. Mod. Phys. **65**, No. 2, April 1993, 413–497

PET scan:

Raichle, M.E. (1994): Visualizing the mind, Scientific American, April 1994, **270**, No. 4, 36–42
Dann, R., Hoford, J., Kovačič, S., Reivich, M., and Bajcsy, R. (1989): Evaluation of elastic matching system for anatomic (CT,MR) and functional (PET) cerebral images, J. Comput. Assist. Tomogr. **13** (4), 603–611
Martin, W.R.W., Grochowski, E., Palmer, M., and Pate, B.D. (1987): Correlation of structural and functional image in the same patient, J. Nucl. Med. **28**, 634
McIntosh, A.R., Grady, C.L., Ungerleider, L.G., Haxby, J.V., Rapoport, S.I., Horwitz, B. (1994): Network analysis of cortical visual pathways mapped with PET, J. Neuroscience **14** (2), 655–666

MRI:

Damadian, R. (1971): Tumor detection by nuclear magnetic resonance, Science **171**, 1151–1153
Huk, W.J., Gademann, G., Friedmann, G. (1990): *MRI of central nervous system diseases*, Springer, Berlin
Morris, P.G. (1986): *Nuclear resonance imaging in medicine and biology*, Clarendon Press, Oxford

2.5 Structure and Function: Microscopic

For surveys see

Bullock, T.H., Orkland, R., Grinnell, A. (1977): *Introduction to nervous systems*, W.H. Freeman and Co., San Francisco
Thompson, R.F. (1993): *The brain, A neuroscience primer*, 2nd ed., W.H. Freeman and Co., San Francisco
Shaw, G.L., Mc.Gaugh, J.L., Rose, S.P.R. (eds.) (1990): *Neurobiology of learning and memory*, reprint volume, World Scientific, Singapore
Hameroff, S.R. (1987): *Ultimate computing. Biomolecular consciousness and nano-technology*, North–Holland, Amsterdam
Hubel, D.H., Wiesel, T.N. (1962): Receptive fields, binocular interaction and functional architecture in the cat's visual cortex, J. Physiol. **160**, 106–154

see also

Hubel, D. (1988): *Eye, brain and vision*, Scientific American Books, New York
Freeman, W. J. (1991): The physiology of perception, Scientific American **264**, 78–85
Gray, C. M., König, P., Engel, A. K., Singer, W. (1990): Synchronization of oscillatory responses in visual cortex: A plausible mechanism for scene segmentation. In: *Synergetics of Cognition*, Haken, H., Stadler, M. (eds.), Springer, Berlin
Eckhorn, R., Reitböck, H. J. (1990): Stimulus-specific synchronization in cat visual cortex and its possible role in visual pattern recognition. In: *Synergetics of Cognition*, Haken, H., Stadler, M. (eds.), Springer, Berlin

Reference for figures:

Fuchs, A., Friedrich, R., Haken, H., Lehmann, D. (1987): Spatio-temporal analysis of multichannel α-EEG map series. In: *Computational systems – natural and artificial*, Haken, H. (ed.), Springer, Berlin

2.6 Learning and Memory

Hebb, D.O. (1949): *The organization of behavior*, Wiley, New York

Kandel, E.R., Schwartz, J.H. (1985): *Principles of neural science*, 2nd ed., Elsevier, North–Holland, New York

Kandel, E.R. (1979): *Behavioral biology of Aplysia: A contribution to the comparative study of opisthobranch molluscs*, W.H. Freeman and Co., San Francisco

Kuffler, S.W., Nichols, J.G., Martin, A.R. (1984): *From neuron to brain*, Sinauer, Sunderland, MA

3. Modeling the Brain. A First Attempt: The Brain as a Dynamical System

3.1 What Are Dynamical Systems

For some recent books on dynamical systems see

Arnold, V.A. (1983): *Geometrical methods in the theory of ordinary differential equations*, Springer, Berlin

Anosov, D.V., Arnold, V.I. (eds.) (1988): *Dynamical systems I*, Encyclopaedia of Math. Sciences, Vol. 1, Springer, Berlin

Arnold, V.I. (1993): *Dynamical systems V*, Encyclopaedia of Math. Sciences, Vol. 5, Springer, Berlin

Guckenheimer, J., Holmes, P.J. (1983): *Nonlinear oscillations, dynamical systems, and bifurcations of vector fields*, Springer, Berlin

Wiggins, J. (1991): *Global bifurcations and chaos*, Springer, Berlin

4. Basic Concepts of Synergetics I: Order Parameters and the Slaving Principle

Haken, H. (1983): *Synergetics, An Introduction*, 3rd. ed., Springer, Berlin

Haken, H. (1993): *Advanced Synergetics*, 3rd. ed., Springer, Berlin

An important extension of the slaving principle to delay differential equations is given in

Wischert, W., Wunderlin, A., Pelster, A., Olivier, M., Groslambert, J. (1994): Delay-induced instabilities in nonlinear feedback systems, Phys. Rev. E **49**, 203–219

4.2.2 The Laser Paradigm or Boats on a Lake

4.3 Self-Organization and the Second Law of Thermodynamics

von Bertalanffy, L. (1976): *General system theory, foundations, development, applications*, rev. ed. George Braziller, New York

Turing, A.M. (1952): The chemical basis of morphogenesis, Philosophical Transactions of the Royal Society, B **237**, 37–52 see also:

Meinhardt, H. (1982): *Models of biological pattern formation*, Academic Press, London

Wolpert, L. (1991): *The triumph of the embryo*, Oxford University Press, Oxford

Bestehorn, M., Fantz (Neufeld), M., Friedrich, R., Haken, H. (1993): Hexagonal and spiral patterns of thermal convection, Phys. Lett. A, **174**, 48–52

5. Dynamics of Order Parameters

Haken, H. (1983): *Synergetics, An Introduction*, 3rd. ed., Springer, Berlin

Haken, H. (1993): *Advanced synergetics*, 3rd. ed., Springer, Berlin

Graham, R., Haken, H. (1968): Quantum theory of light propagation in a fluctuating laser-active medium, Z. Physik, **213**, 420–450

Graham, R., Haken, H. (1970): Laser light - first example of a second-order phase transition far away from thermal equilibrium, Z. Physik, **237**, 31–46

5.4 Order Parameters and Normal Forms

Catastrophe theory:

Thom, R. (1975): *Structural stability and morphogenesis*, W.A. Benjamin, Reading, MA

Arnold, V.I. (1986): *Catastrophe theory*, Springer, Berlin

Poston, T., Stewart, I.N. (1978): *Catastrophe theory and its applications*, Pitman, London

Normal forms:

Arnold, V.A. (1983): *Geometrical methods in the theory of ordinary differential equations*, Springer, Berlin

Bruno, A.D. (1989): *Local methods in nonlinear differential equations*, Springer, Berlin

6. Movement Coordination – Movement Patterns

6.1 Coordination Problem

Bernstein, N. (1967): *The coordination and regulation of movements*, Pergamon Press, London

Sherrington, C.S. (1906): *The integrative action of the nervous system*, Constable, London

Gibson, J.J. (1979): *The ecological approach to visual perception*, Boston: Houghton-Mifflin

von Holst, E. (1939): Die relative Koordination als Phänomen und als Methode zentralnervöser Funktionsanalysen, Erg. Physiol. **42**, 228–306

von Holst, E. (1935): Über den Prozess der zentralnervösen Koordination, Pflügers Arch. **236**, 149–158

von Holst, E. (1943): Über relative Koordination bei Anthropoden, Pflügers Arch. **246**, 847–865

There is a rich literature on motor control and the generation of motor patterns. For some recent references see for instance:

Kupfermann, I. (1993): The generation of motor patterns, Current Directions in Psychological Science **2**, 126

Cohen, A.H., Rossignol, S., Grillner, S. (1988): *Neural control of rhythmic movements*, Vestebrotes, Wiley, New York

Turvey, M.T. (1990): Coordination, American Psychologist **45**, 938–953

6.2 Phase Transitions in Finger Movement. Experiments and a Simple Model

Experiments:

Kelso, J.A.S. (1981): On the oscillatory basis of movements. Bulletin of Psychonomic Society **18**, 63

Kelso, J.A.S. (1984): Phase transitions and critical behavior in human bimanual coordination. American Journal of Physiology: Regulatory, Integrative and Comparative Physiology **15**, R 1000–R1004

Theory:

Haken, H., Kelso, J.A.S., Bunz, H. (1985): A theoretical model of phase transitions in human hand movements. Biol. Cybernetics **51**, 347–356

6.3 An Alternative Model?

This model is proposed by

Schmidt, R.C., Shaw, B.K., Turvey, M.T. (1993): Coupling dynamics in interlimb coordination, Journal of Experimental Psychology: Human Perception and Performance **19**, 397–415

For a criticism similar to ours see

Fuchs, A., Kelso, J.A.S. (1994): A theoretical note on models of interlimb coordination, Journal of Experimental Psychology: Human Perception and Performance, **20**, No. 5, 1088-1097

6.4 Fluctuations in Finger Movement: Theory

Schöner, G., Haken, H., Kelso, J.A.S. (1986): A stochastic theory of phase transitions in human hand movement. Biol. Cybernetics **53**, 247–257

Haken, H. (1983): *Synergetics, An Introduction*, 3rd. ed., Springer, Berlin

6.5 Critical Fluctuations in Finger Movements: Experiments

Kelso, J.A.S., Scholz, J.P., Schöner, G. (1986): Non-equilibrium phase transitions in coordinated biological motion: Critical fluctuations, Physics Letters A **118**, 279–284

Kelso, J.A.S., Schöner, G., Scholz, J.P., Haken, H. (1987): Phase-locked modes, phase transitions and component oscillators in biological motion, Physica Scripta **35**, 79–87

Scholz, J.P., Kelso, J.A.S., Schöner, G. (1987): Non-equilibrium phase transitions in coordinated biological motion: critical slowing down and switching time, Physics Letters A **123**, 390–394

6.6 Some Important Conclusions

Experiments:

Jeca, J.J., Kelso, J.A.S., Kiemel, T. (1993): Pattern switching in human multilimb coordination dynamics. Bulletin of Mathematical Biology **55**, 829–845

von Holst, E. (1969): Zur Verhaltensphysiologie bei Tieren und Menschen. Gesammelte Abhandlungen Band I, München (Teil I: Relative Koordination)

Schmidt, R.C., Carello, C., Turvey, M.T. (1990): Phase transitions and critical fluctuations in the visual coordination of rhythmic movements between people. Journal of Experimental Psychology: Human Perception and Performance **16**, 227–247

Shik, M.L., Severin, F.V., Orlovskii, G.N. (1966): Control of walking and running by means of electrical stimulation of the mid-brain, Theor. Biol. **142**, 359–391

7. More on Finger Movements

7.1 Movement of a Single Index Finger

Haken, H., Kelso, J.A.S., Bunz, H. (1985): A theoretical model of phase transitions in human hand movements. Biol. Cybernetics **51**, 347–356

Kay, B.A., Kelso, J.A.S., Saltzman, E.L., Schöner, G. (1987): The space-time behavior of single and bimanual movements: Data and model, Journal Experimental Psychology: Human Perception and Performance **13**, 178-192

Freund, H.I. (1983): Motor unit and muscle activity in voluntary motor control, Psychological Reviews **63**, 387–436

Feldmann, A.G. (1980): Superposition of motor programs: I. Rhythmic forearm movements in man, Neuroscience **5**, 81–90

Asatryan, D.G., Feldmann, A.G. (1965): Functional tuning of the nervous system with control of movement or maintenance of a steady posture: 1. Mechanographic analysis on the work of the joint on execution of a postural task, Biophysics **10**, 925–935

Davis, W.E., Kelso, J.A.S. (1982): Analysis of invariant characteristics in the motor control of Down's synchrome and normal subjects, Journal of Motor Behavior **14**, 194–212

Viviani, P., Soechting, J.F., Terzuolo, M. (1976): Influence of mechanical properties on the relation between EMG activity and torque, Journal of Physiology (Paris) **72**, 45–52

7.2 Coupled Movement of Index Fingers

Haken, H., Kelso, J.A.S., Bunz, H. (1985): A theoretical model of phase transitions in human hand movements. Biol. Cybernetics **51**, 347–356

7.3 Phase Transitions in Human Hand Movements During Multifrequency Tapping Tasks

In this chapter I closely follow the experimental and theoretical work of:

Haken, H., Beek, P.J., Peper, C.E., Daffertshofer, A. (1995): A model for phase transitions in human hand movements during multifrequency tapping (unpublished)

7.3.1 Experiment: Transitions in Multifrequency Tapping

We describe the experiments by Beek, P.J., Peper, C.E. (private communication)

For other or related finger tapping experiments, see for example:

Deutsch, D. (1983): The generation of two isochronous sequences in parallel, Percept. Psychophys. **34**, 331–337

Summers, J.J., Rosenbaum, D.A., Burns, B.D., Ford, S.K. (1993): Production of polyrhythms, J. Exp. Psychol. Hum. Percept. Perf. **19**, 416–428

Kelso, J.A.S., de Guzman, C.G. (1988): Order in time: How cooperation beetween the hands informs the design of the brain. In: *Neural and Synergetic Computers*, Haken, H. (ed.), Springer, Berlin, 180–196

Treffner, P.J., Turvey, M.T. (1993): Resonance constraints on rhythmic movements, J. Exp. Psychol. Hum. Percept. Perf. **19**, 1221–1237

Beek, P.J., Peper, C.E., van Wieringen, P.C.W. (1992): Frequency locking, frequency modulation, and bifurcations in dynamic movement systems. In: *Tutorials in motor behavior II*, Stelmach, G.E., Requin, J. (eds.), North-Holland, Amsterdam, 599–622

The phase attractive circle map is treated in connection with finger tapping by

de Guzman, C.G., Kelso, J.A.S. (1991): Multifrequency behavioral patterns and the phase attractive circle map, Biol. Cybern. **64**, 485–495

and further related work:

deGuzman G.C., Kelso, J.A.S., Holroyd, T. (1991): The self-organized phase attractive dynamics of coordination. In: A. Babloyantz, ed., Self-organization, emerging properties and learning, Series B: **260**, 41–62. Plenum, New York

deGuzman G.C., Kelso, J.A.S., Holroyd, T. (1991): Synergetic dynamics of biological coordination with special reference to phase attraction and intermittency. In: Haken, H., Koepchen, H.P. (eds.), *Rhythms in physiological systems*, 195–213, Springer, Berlin

Kelso, J.A.S. (1990): Phase transitions: Foundations of behavior. In: Haken, H., Stadler, M. (eds.), *Synergetics of Cognition*, 249–295, Springer, Berlin

7.4 A Model for Multifrequency Behavior

7.5 The Basic Locking Equation and their Solutions

This theory is based on unpublished work by Haken, H., Beek, P.J., Peper, C.E., Daffertshofer, A.: A model for phase transitions in human hand movements during multifrequency tapping

7.6 Summary of the Main Theoretical Results

Summers, J.J., Ford, S.K., Todd, J.A. (1993): Practice effects on the coordination of the two hands in a bimanual tapping task, Human Movement Science **12**, 111–133
Summers, J.J., Kennedy, T.M. (1992): Strategies in the production of a 5:3 polyrhythm, Human Movement Science **11**, 101–112
Kelso, J.A.S., Bressler, S.L., Buchanan, S., de Guzman, G.C., Ding, M., Fuchs, A., Holroyd, T. (1991): Cooperative and critical phenomena in the human brain revealed by multiple SQUIDs. In: *Measuring chaos in the human brain*, Duke, D., Pritchard, W. (eds.), World Scientific, Singapore

8. Learning

For a recent book see

Savelsbergh, G.J.P. (ed.) (1993): The development of coordination in infancy, Elsevier, Amsterdam

8.1 Learning Changes Order Parameter Landscapes

Schöner, G., Kelso, J.A.S. (1988): A dynamic pattern theory of behavioral change. J. Theoretical Biology **135**, 501–524

A similar mechanism of a changed intentionality potential landscape to model intentionality is invoked by:

Kelso, J.A.S., Scholz, J.P., Schöner, G. (1988): Dynamics governs switching among patterns of coordination in biological movement, Phys. Lett. A **134**, 8–12
Scholz, J.P., Kelso, J.A.S. (1990): Intentional switching between patterns of bimanual coordination depends on the intrinsic dynamics of the patterns, J. of Motor Behavior **22**, 98–124
Zanone, P.G., Kelso, J.A.S. (1992): The evolution of behavioral attractors with learning: Nonequilibrium phase transitions, J. of Experimental Psychology: Human Perception and Performance **18**, 403–421
Schöner, G.S. (1989): Learning and recall in a dynamic theory of coordination patterns, Biological Cybernetics **62**, 39–54
Schöner, G.S., Zanone, P.G., Kelso, J.A.S. (1992): Learning as change in coordination dynamics: Theory and experiment, J. of Motor Behavior **24**, 29–48
Yamanishi, J., Kawamoto, M., Suzuki, R. (1980): Two coupled oscillators as a model for the coordinated finger tapping by both hands, Biol. Cybernectis **37**, 219–225
Yamanishi, J., Kawamoto, M., Suzuki, R. (1979): Studies on human finger tapping, neural networks by phase transition curves, Biol. Cybernetics **33**, 199–208
Tuller, B., Kelso, J.A.S. (1989): Environmentally specified patterns of movement coordination in normal and split-brain subjects, Experimental Brain Research **74**, 306–316

8.2 Learning Changes Number of Order Parameters

see references to Chap. 12

8.3 Learning Gives Rise to New Order Parameters

see references to Chap. 16 and also

Hebb, D.O. (1949): *The organization of behavior*, Wiley, New York

9. Animal Gaits and Their Transitions

Schöner, G., Yiang, W.Y., Kelso, J.A.S. (1990): A synergetic theory of quadrupedal gaits and gait transitions, J. Theor. Biol. **142**, 359–391
Lorenz, W. (1987): Nichtgleichgewichtsphasenübergänge bei Bewegungskoordinationen, Diplom Thesis, University of Stuttgart
Collins, J.J., Stewart, I.N. (1993): Coupled nonlinear oscillators and the symmetries of animal gaits, J. Nonlinear Sci. **3**, 349–392
Hoyt, D.F., Taylor, C.R. (1981): Gait and the energetics of locomotion in horses, Nature **292**, 239
Gambaryan, P. (1974): *How mammals run: Anatomical adaptations*. Distributed by John Wiley and Sons, New York
von Holst, E. (1993): Die relative Koordination als Phänomen und als Methode zentralnervöser Funktionsanalysen, Erg. Physiol. **42**, 228–306

10. Basic Concepts of Synergetics II: Formation of Spatio-Temporal Patterns

Haken, H. (1983): *Synergetics, An Introduction*, Springer, Berlin
Haken, H. (1993): *Advanced Synergetics*, 3rd. ed., Springer, Berlin
Mikhailov, A.S. (1990): *Foundations of Synergetics I*, Springer, Berlin
Mikhailov, A.S., Loskutov, A.Yu. (1991): *Foundations of Synergetics II*, Springer, Berlin

11. Analysis of Spatio-Temporal Patterns

11.1 Karhunen-Loève Expansion, Singular Value Decomposition, Principal Component Analysis – Three Names for the Same Method

Watanabe, S. (1985): *Pattern recognition: Human and mechanical*, Wiley, New York

11.2 A Geometric Approach Based on Order Parameters. The Haken–Friedrich–Uhl Method

We follow essentially

Haken, H., Friedrich, R., Uhl, C., unpublished

For related work see

Friedrich, R., Uhl, C., Haken, H. (1993): Reconstruction of spatio-temporal signals of complex systems, Zeitschrift für Physik B **92**, 211–219

Uhl, C., Friedrich, R., Haken, H. (1995): Analysis of spatio-temporal signals of complex systems, Phys. Rev. E **51**, 3890–3900

For the relationship (11.36) and the notation of stable and unstable modes etc. see Haken, H. *Advanced Synergetics*, 3rd. ed. (1995), Springer, Berlin

12. Movements on a Pedalo

The experiments were performed by Körndle, H.

Körndle, H. (1992): private communication

Karhunen-Loève method, see Chap. 11

The theoretical analysis was performed by

Haas, R. (1995): Ph.D. Thesis, University of Stuttgart
Haas, R., Haken, H., Körndle, H. (1995): Movements on a pedalo: An analysis based on synergetics, unpublished manuscript

13. Chaos, Chaos, Chaos

Poincaré, H. (1892–1899): *Les methodes nouvelles de la mecanique celeste*, Vols. 1-3, Gauthier–Villars, Paris, reprint (1957), Dover, New York
Lorenz, E.N. (1963): Deterministic nonperiodic flow, J. Atmos. Sci. **20**, 130–141
Rössler, O.E. (1977): Continuous chaos. In: *Synergetics, A Workshop*, Haken, H. (ed.), Springer, Berlin

For some books on chaos see

Ott, E. (1993): *Chaos in dynamical systems*, Cambridge University Press, Cambridge
Devaney, R.L. (1993): *A first course in chaotic dynamical systems*, 2nd print., Addison-Wesley, Reading, MA
McCauley, J.L. (1988): *An introduction to nonlinear dynamics and chaos theory*, Royal Swedish Academy of Sciences, Stockholm
Tabor, M. (1989): *Chaos and integrability in nonlinear dynamics*, Wiley, New York
Ott, E., Sauer, T., Yorke, J.A. (1994): *Coping with chaos*, Wiley, New York
Hoppensteadt, F.C. (1993): *Analysis and simulation of chaotic systems*, Springer, Berlin
Peitgen, H.-O., Jürgens, H., Saupe, D. (1993): *Chaos and fractals*, corrected 2nd print., Springer, Berlin
Lasota, A., Mackey, M.C. (1994): *Chaos, fractals and noise*, 2nd ed., Springer, Berlin
Klimontovich, Yu. L. (1991): *Turbulent motion and the structure of chaos*, Kluwer, Dordrecht
Scott, S.K. (1991): *Chemical chaos*, Clarendon Press, Oxford
Schuster, H.G. (1988): *Deterministic chaos: An introduction*, 2nd revised ed., VCH Verlagsgesellschaft, Weinheim
von Koch, H. (1904): Sur une courbe continue sans tangente, obtenue par une construction géométric élémentaire, Arkiv för Matematik, Astronomi och Fysik **1**, 681–704

Shilnikov, L.P. (1965): A case of the existence of a countable number of periodic motions, Sov. Math. Dok. **6**, 163–166; (1970): Math. USSR Sbornik **10**, 91

Glendinning, P., Sparrow, C. (1984): Local and global behavior near homoclinic orbits, J. Stat. Phys. **35**, 645–696

May, R.M. (1976): Simple mathematical models with very complicated dynamics. Nature **261**, 459–467

14. Analysis of Electroencephalograms

For a general treatment of electric fields in the brain see

Nunez, P.L. (1981): *Electric fields in the brain*, Oxford University Press, Oxford, New York

The main results of this chapter are based on

Friedrich, R., Uhl, C. (1995): Spatio-temporal analysis of human electroencephalograms: petit-mal epilepsy, unpublished

For related work see

Lehmann, D. (1971): Multichannel topography of human alpha EEG fields, Electroenceph. Clin. Neurophysiol. **31**, 439–449; (1972): Human scalp EEG fields: evoked, alpha, sleep and spike-wave patterns. In: *Synchronisation of EEG activity in epilepsies*, Petsche, H., Brazier, M.A.B. (eds.), Springer, Berlin

Babloyantz, A. (1985): Strange attractors in the dynamics of brain activity. In: *Complex systems – operational approaches*, Haken, H. (ed.), Springer, Berlin

Babloyantz, A., Nicolis, C., Salazar, M. (1985): Evidence of chaotic dynamics of brain activity during the sleep cycle, Phys. Lett. A **111**, 152

Babloyantz, A., Destexhe, A. (1986): Low dimensional chaos in an instance of epilepsy, Proc. Natl. Acad. Sci. **83**, 3513–3517

Layne, S.P., Mayer-Kress, G., Holzfuss, J. (1986): Problems associated with dimensional analysis of electroencephalogram data. In: *Dimensions and entropies in chaotic systems*, Mayer-Kress, G. (ed.), Springer, Berlin

Fuchs, A., Friedrich, R., Haken, H., Lehmann, D. (1987): Spatio-temporal analysis of multichannel α-EEG map series. In: *Computational systems – natural and artificial*, Haken, H. (ed.), Springer, Berlin

Friedrich, R., Fuchs, A., Haken, H. (1991): Synergetic analysis of spatio-temporal EEG-patterns. In: *Nonlinear wave processes in excitable media*, Holden, A.V., Markus, M., Othmer, H.G. (eds.), Plenum, New York

Friedrich, R., Fuchs, A., Haken, H. (1992): Spatio-temporal EEG-patterns. In: *Rhythms in biological systems*, Haken, H., Köpchen, H.P. (eds.), Springer, Berlin

Friedrich, R., Uhl, C. (1992): Synergetic analysis of human electroencephalograms: Petit-mal epilepsy. In: *Evolution of dynamical structures in complex systems*, Friedrich, R., Wunderlin, A. (eds.), Springer, Berlin

15. Analysis of MEG Patterns

Experiments:

Kelso, J.A.S., Bressler, S.L., Buchanan, S., deGuzman, G.C., Ding, M., Fuchs, A., Holroyd, T. (1991): Cooperative and critical phenomena in the human brain revealed by multiple SQUIDs. In: *Measuring chaos in the human brain*, Duke, D., Pritchard, W. (eds.), World Scientific, Singapore

Kelso, J.A.S., Bressler, S.L., Buchanan, S., DeGuzman, G.C., Ding, M., Fuchs, A., Holroyd, T. (1992): A phase transition in human brain and behavior, Physics Letters A **169**, 134–144

Theory:

Fuchs, A., Kelso, J.A.S., Haken, H. (1992): Phase transitions in the human brain: Spatial mode dynamics, International Journal of Bifurcation and Chaos **2**, 917–939

Jirsa, V.K., Friedrich, R., Haken, H., Kelso, J.A.S. (1994): A theoretical model of phase transitions in the human brain, Biol. Cybern. **71**, 27–35

Jirsa, V.K., Friedrich, R., Haken, H.: Reconstruction of the spatio-temporal dynamics of a human magnetoencephalogram (unpublished)

Bestehorn, M., Haken, H. (1995): unpublished results (Sect.15.4)

Friedrich, R., Fuchs, A., Haken, H. (1992): see Chap.14

16. Visual Perception

16.1 A Model of Pattern Recognition

For a comprehensive survey see

Haken, H. (1991): *Synergetic computers and cognition*, Springer, Berlin

A classic book on associative memory is

Kohonen, T. (1987): *Associative memory and self-organization*, 2nd ed., Springer, Berlin

Analogy between pattern recognition and pattern formation:

Haken, H. (1979): Pattern formation and pattern recognition – an attempt at a synthesis. In: *Pattern formation by dynamic systems and pattern recognition*, Haken, H. (ed.), Springer, Berlin, 2–13

Standard model of synergetic computer:

Haken, H. (1987): Synergetic computers for pattern recognition and associative memory. In: *Computational systems, natural and artificial*, Haken, H. (ed.), Springer, Berlin, 2–22

Haken, H. (1988): Synergetics in pattern recognition and associative action. In: *Neural and synergetic computers*, Haken, H. (ed.), Springer, Berlin, 2–15

Recognition of faces and city maps:

Fuchs, A., Haken, H. (1988): Pattern recognition and associative memory as dynamical processes in a synergetic system I + II, Erratum, Biol. Cybern. **60**, 17–22; 107–109, 476

Bestehorn, M., Haken, H. (1991): Associative memory of a dynamical system: The example of the convection instability, Z. Physik B **82**, 305–308

Invariance with respect to translation, rotation and scaling see for example

Fuchs, A., Haken, H. (1988), loc. ci.

For the treatment of deformations see

Daffertshofer, A., Haken, H. (1994): A new approach to recognition of deformed patterns, Pattern Recognition, **27**, No. 12, 1697–1705, Pergamon

Facial expression:

Vanger, P., Hönlinger, R., Haken, H. (1995): Automated coding of facial expressions of emotions with the synergetic computer (to be published)
Vanger, P., Hönlinger, R., Haken, H. (1995): Anwendung der Synergetik bei der Erkennung von Emotionen im Gesichtsausdruck. In: Schiepek, G., Tschacher, W. (eds.), *Synergetik in Psychologie und Psychiatrie*, Springer, Berlin
Vanger, P., Hönlinger, R., Haken, H. (1994): Applications of the synergetic computer in decoding complex facial patterns, Proceedings of the First International Conf. on Applied Synergetic and Synergetic Engineering, June 21-23, Erlangen, 111–117

16.2 The Role of Attention Parameters. Ambiguous Figures

Haken, H. (1977): *Synergetics, An Introduction*, Springer, Berlin

In this first edition (and the following ones) I established a connection between bistability and hysteresis of human perception with the behavior of order parameters of synergetic (self-organizing) systems. Section 16.2 is based on this connection

For Gestalt psychology see

Köhler, W. (1920): *Die physischen Gestalten in Ruhe und im stationären Zustand*, Vieweg, Braunschweig
Köhler, W. (1969): *The task of Gestalt psychology*, Princeton

16.3 Influence of a Bias

There is by now a considerable literature on ambiguous figures. For a recent book with corresponding articles and many references see

Kruse, P., Stadler, M. (eds.) (1995): *Ambiguity in mind and nature, multistable cognitive phenomena*, Springer, Berlin
with contributions by: Roth, G., Stadler, M., Kruse,P., Kanizsa, G., Kelso, J.A.S., Luccio, R., van Leeuwen, C., Zimmer, A.C., Caglioti, G. and others

Our presentation essentially follows:

Ditzinger, T., Haken, H. (1989): Oscillations in the perception of ambiguous patterns, Biol. Cybern. **61**, 279–287
Ditzinger, T., Haken, H. (1990): The impact of fluctuations on the recognition of ambiguous patterns, Biol. Cybern. **63**, 453–456

For further work along these lines see

Ditzinger, T., Haken, H. (1995): A synergetic model of multistability in perception. In: Kruse, P., Stadler, M. loc. cit. 255–274

16.4 The Role of Fluctuations of Attention Parameters

Experiments:

Borsellino, A., deMarco, A., Allazetta, A., Rinesi, S., Bartollini, B. (1972): Reversal time distribution in the perception of visual ambiguous stimuli, Kybernetik **10** (3), 139–144

Borsellino, A., Carlini, R., Riani, M., Tuccio, M.T., deMarco, A., Penengo, P., Trabucco, A. (1982): Effects of visual angle on perspective reversal for ambiguous patterns, Perception **11**, 263–273

Theory:

Ditzinger, T., Haken, H. (1990), loc. cit.

For further experiments and alternative theories see articles in Kruse, P., Stadler, M. (eds.) (1995), in particular by Kelso, J.A.S., and Schöner, G., and also

Riani, M., Simonotto, L. (1994): Stochastic resonance in the perceptual interpretation of ambiguous figures: A neural network model. Phys. Rev. Lett. **72**, 3120–3123

16.5 Learning Patterns

This section closely follows

Haken, H., Haas, R., Banzhaf, W. (1989): A new learning algorithm for synergetic computers, Biol. Cybern. **62**, 107–111

see also Haken, H. (1991): *Synergetic computers and cognition*, Springer, Berlin

16.6 A Model for Stereo Vision

There is a huge literature on stereo vision so we can only mention a few basic papers or reviews:

Dhond, U.R., Aggarwal, J.K. (1989): Structure from stereo – a review, IEEE Trans Systems Man Cybern. **19**, 1489–1510

Overview of computational and biological models:

Barnard, S.T., Fischler, M.A. (1990): Computational stereo, Comput. Surveys **14**, 553–572

Comparison between cooperative and non-cooperative, hierarchical and non-hierarchical models:

Hentschel, H.G.E., Fine, A. (1989): Statistical mechanics of stereoscopic vision, Phys. Rev. A **40**, 3983–3997

Stochastic approaches:

Geman, S., Geman, D. (1984): Stochastic relaxation, Gibbs distributions and the Bayesian restoration of images, IEEE Trans. Pattern Anal. Machine Intell. **6**, 721–741

Yeuille, A., Geiger, D., Bülthoff, H. (1991): Stereo integration, mean field theory and psychophysics, Network **2**, 423–442

Deterministic approaches:

Terzoupulos, D., Witkin, A., Kass, M. (1987): Stereomatching as constrained optimization using scale continuation methods, Optical and Digital Pattern Recognition **SPIE 754**, 92–99

Random dot sterograms were invented and studied by

Julesz, B. (1991): *Foundations of cyclopean perception*, University of Chicago Press, Chicago

The *dynamical* model we formulate in Sect.16.6 is due to

Reimann, D., Haken, H. (1994): Stereo vision by self-organization, Biol. Cybern. **71**, 17–26

For particularly relevant references see

Marr, D., Hildreth, H. (1980): Theory of edge detection, Proc. R. Soc. Lond [Biol] **207**, 187–217

Marr, D., Poggio, T. (1976): Cooperative computation of stero disparity, Science **194**, 283–287

Marr, D., Poggio, T. (1979): A computational theory of human stero vision, Proc. R. Soc. Lond [Biol] **204**, 301–328

Further applications are given by:

Reimann, D., Ditzinger, T., Haken, H. (1995): Vergence eye movement control and multivalent perception of autostereograms, Biol. Cybern. **73**, 123–128

17. Decision Making as Pattern Recognition

This has become a highly active field as is witnessed, for instance, by the Journal of Behavioral Decision Making, Ed. Wright, G., Wiley, Chichester

An important early reference is

Lewin, K. (1936): *Principles of topological psychology*, Mac Graw Hill, New York

In this chapter I follow a talk I gave at the University of Regensburg, July 1994

For heuristics see

Wagenaar, W.A. (1993): Heuristics: Simple ways for dealing with complex problems. Talk given at the symposium: Natural Sciences and Human Thought, Villa Vigoni, Italy, 29.3.–2.4.1993

18. The Brain as a Computer or Can Computers Think?

Zuse, K. (1990): *Der Computer, mein Lebenswerk*, Nachdr. der 2. Aufl., Springer, Berlin

Neumann, J. von (1958): *The computer and the brain*, New Haven, Connecticut

Turing, A.M. (1936): On computable numbers, with an application to the Entscheidungsproblem, Proc. London Mathem. Soc. Series 2, **42**, 230–265

Turing, A.M. (1950): *Computing machinery and intelligence*, In Mind **59**

Wiener, N. (1948): *Cybernetics, or control and communication in the animal and the machine*, Cambridge, MA

Winograd, T. (1972): *Understanding natural language*, New York
Weizenbaum, J. (1966): Eliza – A computer program for the study of natural language communication between man and machine, Communications of the Association for Computing Machinery **9**, 36–45

Connectionism:

Rumelhart, D.E., Mc.Clelland, J.L., and the P.D.P. Research Group (1986): *Parallel distributed processing: Explorations in the microstructure of cognition*, Vols. 1 and 2, Cambridge University Press, Cambridge

Neurocomputers:

Wassermann, P.D. (1989): *Neural computing, theory and practice*, Van Nostrand Reinhold, New York
Mc.Culloch, W., Pitts, W. (1943): A logical calculus of the ideas immanent in nervous activity, Bulletin of Math. Biophysics **5**, 115–133
Rosenblatt, R. (1962): *Principles of neurodynamics*, Spartan Books, New York
Sejnowski, T.N., Rosenberg, C.R. (1987): Parallel networks that learn to pronounce English text, Complex Systems **1**, 145–168
Hopfield, J.J. (1982): Neural networks and physical systems with emergent collective computational abilities, Proc. Natl. Acad. Scis **79**, 2554–2558

19. Networks of Brains

Bartlett, F. (1932): *Remembering, A study in experimental and social psychology*, Cambridge Univ. Press, Cambridge
Haken, H., Portugali, J. (1994): Synergetics, inter-representation networks and cognitive maps, to be published in Geoforum

Here relations to earlier work by Vigotsky, Gibson, Rumelhart, Lakoff, Donald, Edelmann, Portugali and Haken are also established.

Portugali, J. (1994): Private communication; see also in Geoforum, to be published
Stadler, M., Kruse, P. (1990): The self-organization perspective in cognition research: Historical remarks and new experimental approaches. In: Haken, H., Stadler, M. (eds.), *Synergetics of cognition*, Springer, Berlin, 32–52

It might be interesting to establish relationships between our approach and the studies on cooperative robots, cf.

Steels, L. (ed.) (1994): *The biology and technology of intelligent autonomous agents*, Springer, Berlin
McFarland, D., Steels, L. (1995): *Cooperative robots*, MIT Press, Boston MA

20. Synergetics of the Brain: Where Do We Stand? Where Do We Go?

Crick, F.H.C., Koch, C. (1990): The problem of consciousness, Scientific American **267**, 152–159
Edelman, G.M. (1992): *Bright air, brilliant fire*, Basic Books, New York
Fromherz, P., Offenhäuser, A., Vetter, T., Weis, J. (1991): A neuron-silicon junction: A retzius-cell of the leech on an insulated-gate field-effect transistor, Science 252–253

Kornhuber, H., Deecke, L. (1965): Hirnpotentialänderung bei Willkürbewegungen und passiven Bewegungen des Menschen: Bereitschaftspotential und reafferente Potentiale, Pflügers Archiv **284**, 1–17

For a review see

Deecke, L. (1990): Electrophysiological correlates of movement initiation, Reviews of Neurology (Paris) **146**, 612–619

Delbrück, M.(1986): *Mind from matter? An essay on evolutionary epistemology*, Blackwell, Palo Alto, CA

Freeman, W.J. (1995): *Societies of brains*, Lawrence Erlbaum Ass., Hillsdale, NJ

Tass, P., Haken, H. (1995): Synchronized oscillations in the visual cortex – a synergetic model (to be published)

Appendix A

Time-delay coordinates:

Packard, N.H., Crutchfield, J.P., Farmer, J.D., Shaw, R.S. (1980): Geometry from a time series, Phys. Rev. Lett. **45**, 712–716

Takens, F. (1980): Detecting strange attractors in turbulence. In Dynamical Systems and Turbulence, Warwick 1980, Lecture Notes in Mathematics **898**, 366–381, Springer, Berlin

Fractal dimensions:

Eckmann, J.P., Ruelle, D. (1985): Ergodic theory of chaos and strange attractors, Rev. of Mod. Phys. **57** (3), 617–656

Grassberger, P., Procaccia, I. (1983): Measuring the strangeness of strange attractors, Physica D **9**, 189–208

Correlation dimension:

Grassberger, P., Procaccia, I. (1983): Characterization of strange attractors, Phys. Rev. Lett. **50**, 346–349

Grassberger, P., Procaccia, I. (1983): Estimation of the Kolmogorov entropy from a chaotic signal, Phys. Rev. A **28** (4), 2591–2593

Rapp plot:

Rapp, P.E., Zimmermann, I.D., Albano, A.M., de Guzman, G.C., Greenbaum, N.N. (1985): Dynamics of spontaneous neural activity in the simian motor cortex: The dimensions of chaotic neurons, Phys. Lett. **110A** (6), 335–338

Subject Index

About the Author

Hermann Haken, born in 1927, earned his Ph.D. in Mathematics at the University of Erlangen. In 1956 he became a Lecturer of Theoretical Physics at that university. Since 1960 he has been Professor of Theoretical Physics at the University of Stuttgart. He has been a guest scientist, consultant, or visiting professor at various institutions in the United States, Great Britain, France, Japan, and the USSR. He has made numerous contributions to group theory, solid-state physics, laser physics and nonlinear optics, statistical physics, plasma physics, bifurcation theory, chemical reaction models, and theories on morphogenesis. He is the author of the monograph *Laser Theory*, the texts *Synergetics. An Introduction*, and *Quantum Field Theory of Solids*, and of various other books, including text books written jointly with H.C. Wolf on *The Physics of Atoms and Quanta* as well as on *Molecular Physics and Elements of Quantum Chemistry*.

In 1976 he was awarded the Max Born Prize and Medal of the British Institute of Physics and the German Physical Society for his outstanding contributions to the theory of excited states in solids and to quantum optics, in particular, laser theory. He received the Albert A. Michelson Medal of the Franklin Institute, USA, in 1981 for his work on laser theory and his pioneering efforts in synergetics. In 1982 he received an honorary doctorate from the University of Essen, in 1987 from the University of Madrid, in 1992 from the Florida Atlantic University, and in 1994 from the University of Regensburg. Among his further awards are the Max-Planck-Medal of the German Physical Society, the Honda Prize (Tokyo), and the Lorenz-Oken-Medal of the Gesellschaft Deutscher Naturforscher und Ärzte.

He is a member of several academies including the *Leopoldina*, the Bavarian Academy of Sciences, the Academia Europaea, London, and the Academia Scientiarum at Artium Europaea, Salzburg.

Springer-Verlag
and the Environment

We at Springer-Verlag firmly believe that an international science publisher has a special obligation to the environment, and our corporate policies consistently reflect this conviction.

We also expect our business partners – paper mills, printers, packaging manufacturers, etc. – to commit themselves to using environmentally friendly materials and production processes.

The paper in this book is made from low- or no-chlorine pulp and is acid free, in conformance with international standards for paper permanency.